INSIDE THE SINGULARITY

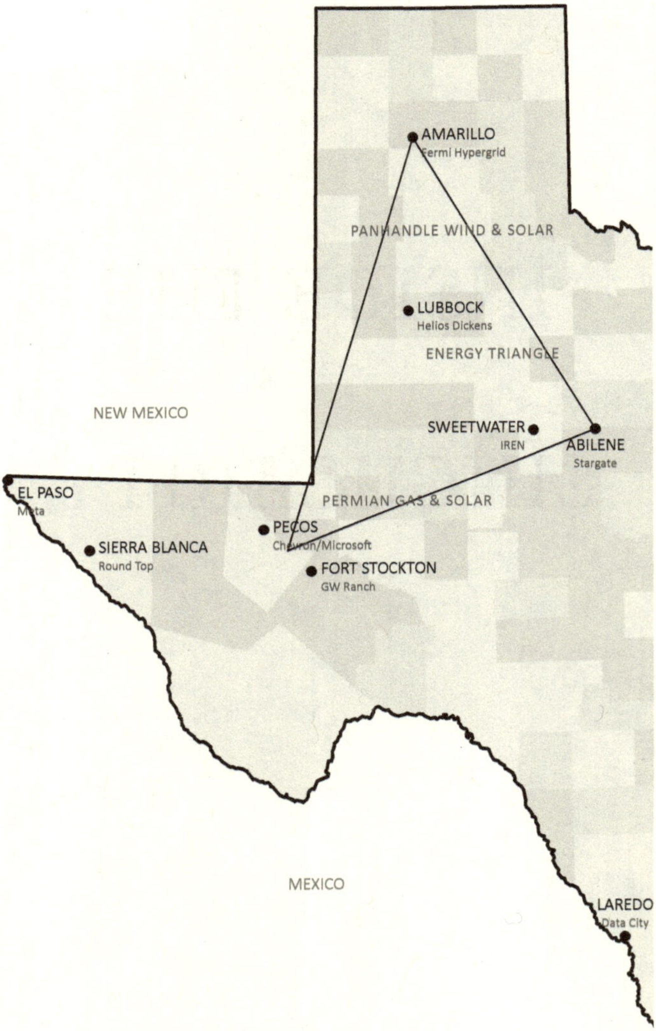

Inside the Singularity

WEST TEXAS AI-ENERGY GEOGRAPHY · 2026

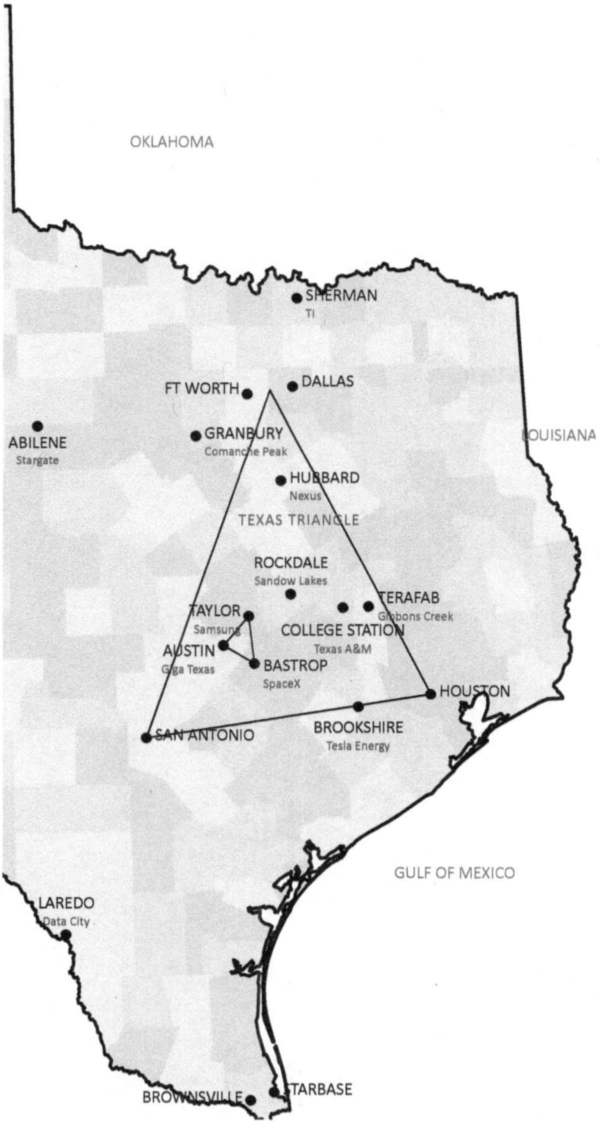

Inside the Singularity

EAST TEXAS AI-INDUSTRIAL GEOGRAPHY · 2026

INSIDE THE SINGULARITY

The AI-Industrial Revolution

From Austin to Alpha Centauri

DANIEL M. SISSON

Si P
SILICONPLANS
PRESS

Published in the United States by SiliconPlans Press
An imprint of SiliconPlans LLC, Austin, Texas.

ISBN 979-8-9966605-0-6 (hardcover)

ISBN 979-8-9966605-1-3 (paperback)

ISBN 979-8-9966605-2-0 (ebook)

Library of Congress Control Number: 2026916523

First edition: 2026

10 9 8 7 6 5 4 3 2 1

Printed in the United States of America.

For the new builders.

Stay Hungry. Stay Foolish.

— STEWART BRAND, WHOLE EARTH CATALOG, 1974

A country of geniuses in a data center.

— DARIO AMODEI, MACHINES OF LOVING GRACE, 2024

CONTENTS

Prologue

THE VIEW FROM OUTSIDE

IN THE SPRING OF 2020 I built a spreadsheet.

The columns were cities. The rows were filters: state income tax, property tax, land prices, regulatory posture, technology employment base, growth trajectory, crime rates, cultural attitudes toward businesses and building things. Four decades inside industrial control systems, electrical grid infrastructure, and enterprise cloud platforms had told me what the next industrial era would require. I was looking for the place that had it, before what was coming had arrived.

The spreadsheet kept pointing south. Texas in general, Austin in particular, the collar counties north of Austin most specifically. Williamson County, where the land was still ranchland, where the prices had not yet reflected what was taking shape, where the infrastructure was rising to accommodate growth whose source was not yet publicly legible. I booked a flight.

The first trip was reconnaissance. A realtor. A short list of communities. Metro filter applied. The filter drove me north of Austin, into the Georgetown–Liberty Hill–Leander nexus. The sky was the particular blue Texas produces in the fall — wide and high, carrying white clouds whose scale matches the landscape beneath them. Vast ranchland ran along both sides of the roads, much of it already developed for subdivision, some of it still untouched.

I settled on a community. Walked the lots. Got on a waiting list. A hundred-plus others had done the same. The developer would parcel and release the lots when ready. The builder representatives communicated by text — which streets were likely to open, which lots within those streets were most desirable, and which builders were jostling for which parcels. I zeroed in on a street. Waited.

At 9:04 on the morning of Sunday, January 3, 2021, a text arrived: the lots were officially released. The same text went to everyone on the list at once — locals who could drive over in twenty minutes, remote buyers spread across the country, speculators who had been waiting for this message. I responded almost immediately: were the lots still available? He said yes. First to respond, first in the queue. He would hold my position for 48 hours.

I booked the flight.

The next morning I drove out to the lots under the same wide Texas sky, the ranchland still stretching to the horizon beyond the staked boundaries of the new streets. As I drove up the unpaved road to the lots I was considering, a black-tailed jackrabbit broke from the scrub and bounded away in long arcing leaps — enormous ears, impossible speed, completely uninterested in the transaction about to occur on its territory. I walked the lots again. Made a selection. Cut the deposit check. Signed the paperwork.

That afternoon my realtor took me to The Brass Tap at The Domain — the mixed-use development in North Austin that anchors the metro's technology employment, where the surrounding office towers hold Amazon, Google, Meta, and IBM at one of the highest tech-industry densities in the state. Tesla has a retail store there, sandwiched between Gucci and Cartier, an Optimus gazing impassively out the window. We sat at the bar. I ordered an Electric Jellyfish Hazy IPA. The name fit. The Domain's parking lot outside held more Teslas per square foot than most any comparable lot in the country. The technology workers drinking around us were the workforce the buildout would require.

The realtor said the location was right. The January air said Texas. The Electric Jellyfish said celebrate.

Nine months later, Tesla announced its Austin headquarters relocation. Ten months later, Samsung announced its Taylor fab. The metro region and the corridor the spreadsheet had identified began announcing itself, one anchor node at a time.

The jackrabbit was long gone.

This book is what I have been watching form around that lot ever since — from a home office in Liberty Hill, 40 minutes from both Samsung Taylor to the east and Tesla Giga Texas to the south, under the wide Hill Country sky.

The argument is not that Texas got lucky when the new builders arrived. It is that Texas built the foundation over decades: the energy independence, the land economics, the regulatory posture, the university pipeline, the quasi-sovereign political frame. What arrived in 2021 and after is the clearest available proof. The stage was already built. This book is the account of what walked onto it, and why, and what it means — at scales that range from the Austin area to something considerably further out.

Austin is the center of the AI-industrial revolution. Not because a company moved there — because Texas spent 180 years, for reasons that had nothing to do with artificial intelligence, semiconductors, autonomous machines, or electric vehicles, accumulating the complete set of structural advantages the AI-industrial complex required. The builders did not make the place. The place was waiting for the builders. This book is the case for that claim, built from the ground, in the geography where it can be watched happening.

This is what I saw.

PART I

THE STAGE

One

THE QUASI-SOVEREIGN STATE

TEXAS IS THE ONLY STATE in the union that entered as an independent republic — a country.

That fact appears in every Texas history course and gets repeated at sporting events and on bumper stickers often enough that it has lost most of its punch. Restore it for a moment. The Republic of Texas existed from 1836 to 1846 as a functioning sovereign republic — issuing debt, conducting diplomacy, maintaining a navy, negotiating treaties. When annexation came, it came on negotiated terms: the right to subdivide into as many as five states, title to public lands that in every other territory passed to the federal government, and a political self-conception that has never entirely dissolved. Texas did not join the United States the way Ohio or Indiana did. It merged with it — and it has behaved ever since like an entity that remembers the difference.

That difference is not just sentiment. It is the deepest root of the argument: the structural advantages that makes Texas the host of the AI-industrial revolution were not built for that purpose, and they did not begin recently. They begin here, with a state that entered the union already understanding itself as something other than an ordinary state. The argument is underfoot, in the rotunda of the Texas State Capitol.

The Six Seals

Completed in 1888 at the head of Congress Avenue, the building is the largest state capitol in the United States by square footage, and stands fourteen feet taller than the Capitol in Washington. The height is not accidental. The architect was challenged to design a building that exceeded the national Capitol in scale. The exterior is clad entirely in Sunset Red granite quarried at Granite Mountain near Marble Falls, hauled to Austin on more than 15,700 railroad carloads. Texas granite facing the nation's marble. Built from the same geological base as the buildout itself.

The rotunda floor is where the argument lives. Inlaid in the stone are the seals of the six sovereign entities that have held authority over Texas: Spain, France, Mexico, the Republic of Texas, the Confederacy, and the United States. Six flags is a theme park cliche in Texas. In the Capitol rotunda it is a visible fact — the accumulated record of a jurisdiction governed by six distinct sovereignties, each of which believed its authority permanent, none of which was right. The United States seal is the sixth — installed alongside the other five in 1936. It is not visibly different from the other five.

I sat in the House gallery one session, watching from the second floor, as any citizen can. Below, on the rotunda floor, the six seals were gleaming. The quasi-sovereign state posture embodied in the stone halls here was not decided by committee. It was quarried from Granite Mountain for this rotunda, in a building designed to stand fourteen feet taller than Washington's.

This is not about Texas exceptionalism as mythology. It is about Texas exceptionalism as operating system — the specific structural features of the state that make it the logical host for the AI-industrial complex. There are five of them. Taken one at a time, each is a familiar fact about Texas. Taken together, they are the case.

Energy

The energy advantage is the first. Texas controls its own power grid. ERCOT — the Electric Reliability Council of Texas — operates in deliberate isolation from the Eastern and Western Interconnections that cover the rest of the continental United States. This was not an accident of geography or a failure of planning. It was a structural choice, made incrementally across the twentieth century and defended politically against repeated federal attempts to bring Texas transmission under interstate jurisdiction.

The consequence is that Texas sets its own electricity rates, operates its own ancillary services markets, and is not subject to Federal Energy Regulatory Commission (FERC) authority over its wholesale power markets. For industrial customers taking power at scale — semiconductor fabs, data centers, gigafactories — ERCOT rates are structurally the lowest available in the United States. The deregulated market design ERCOT operates produces wholesale prices large industrial customers can access directly, bypassing the utility markup that applies in regulated states. This is not a temporary incentive. It is a permanent structural advantage, written into the architecture of the grid itself.

Land

Land comes second. Texas retained its public lands at annexation. The federal government owns approximately 1.8 percent of Texas — the lowest federal land ownership percentage of any state with significant land area. In Nevada, the federal government owns 80 percent. In Utah, 63 percent. In California, 46 percent.

That difference is not cosmetic. Federal land ownership constrains industrial development, because federal permitting processes apply: NEPA review, Endangered Species Act consultation, Bureau of Land Management approval chains that operate on timelines measured in years. In Texas, a developer dealing with privately held ranch land faces state and local permitting only. Multiply such transactions by dozens of large master-planned communities in Williamson County alone, and by the dozens of data center campuses, fab sites, and gigafactory parcels transacting simultaneously across the Texas Triangle, and the structural advantage of private land dominance becomes quantifiable.

Financial Posture

The fiscal posture is the third. Texas has no state income tax. It funds state government primarily through sales tax and property tax — alongside two sovereign wealth instruments, the Permanent University Fund (PUF) and the Permanent School Fund, which together hold over $108 billion in assets as of 2025. The Permanent University Fund has an origin story worth telling, and a later chapter tells it.

The interlock is the point. UT Austin's research capacity — which underpins the semiconductor workforce pipeline feeding the Samsung fab in Taylor, Texas and the broader Austin tech cluster — is partially funded by PUF distributions from oil and gas royalties on state lands. The fiscal architecture that makes Texas competitive for corporate relocation is the same architecture that funds the university system producing the engineering talent those corporations need when they arrive.

Geography

The fourth feature is geography. Texas shares 1,254 miles of border with Mexico, the longest of any US state. The Port of Houston ranks first among US ports by foreign waterborne tonnage. These are not independent facts. Together they describe a state that sits at the intersection of North American continental logistics and Gulf Coast maritime access — direct overland connection to the manufacturing base of northern Mexico, and direct ocean access to global supply chains. For an industrial complex that requires rare-earth inputs from global sources, chemical feedstocks for semiconductor fabrication, and export capacity for finished goods, geographic position is structural rather than incidental.

Regulations

The fifth is regulatory posture. Texas has not confused permitting speed with regulatory abdication, though critics sometimes make that argument. The more precise description is that Texas has consistently resolved ambiguities in land use, environmental review, and industrial permitting in favor of development velocity.

The county-line regulatory arbitrage visible in the data center buildout — operators siting facilities in adjacent counties after primary counties imposed restrictions — is not a loophole being exploited. It is the market-discovery mechanism of a state that has deliberately maintained jurisdictional competition among counties rather than centralizing land use authority at the state level. The result is a permitting environment large industrial operators can navigate with confidence about timelines — a different kind of advantage than low taxes, but an equally significant one for capital allocation decisions.

The Accumulated Stage

None of these features was designed as an industrial recruitment strategy. The accumulated political decisions were made for reasons that had nothing to do with semiconductor fabs or AI training clusters. The energy grid isolation was about avoiding federal regulation. The land retention was about state sovereignty at annexation. The fiscal structure was about funding public education through resource revenues. The geographic position is simply where Texas is.

One feature remains, and it is the least visible. The structural advantages described above did not all emerge from cleared ground. The AI-industrial complex is not being built on entirely blank terrain. It is being built on the bones of prior industrial generations — inheriting their water rights, their power infrastructure, their permitting frameworks, their intake structures. This pattern — brownfield substrate inheritance — is one of the Texas buildout's least visible structural advantages and one of its most consequential. Greenfield operators face two-to-five-year water rights development timelines. Operators who inherit existing infrastructure compress those timelines substantially.

The AI-industrial complex did not have to build the stage. It found Texas's bones already set.

Two

THE ENERGY SUBSTRATE

THE AI-INDUSTRIAL COMPLEX RUNS on electricity. Not metaphorically — physically, at a scale that makes all prior industrial electricity demand look like a rounding error. A single hyperscale data center campus draws 500 MW to 1 GW of continuous power. A semiconductor fab at leading-edge process nodes draws 200 to 600 MW. Giga Texas draws between 350 to 500 MW. These are not peak figures or design maximums. They are the continuous baseline draw of single facilities, and they are coming online in the same geography at the same time.

The supply side is why Texas can answer that demand at all. Texas has more installed wind capacity than any other state. West Texas wind resources are among the deepest in the continental United States: consistent, high-capacity-factor generation across the Permian Basin and the Panhandle, producing electricity at levelized costs competitive with any generation source anywhere. The solar trajectory is just as aggressive; Texas now has more utility-scale solar capacity than any other state as well, on the same land availability and irradiance that make West Texas wind economics work. Permian Basin and Eagle Ford natural gas provide the baseload layer — cheap, geographically coincident with the renewable buildout, available at volumes no other US energy market matches at equivalent cost. Wind, solar, and gas together produce an electricity market where large industrial customers can buy power at rates that are structurally, not temporarily, the lowest in the United States.

The ERCOT Advantage

ERCOT's interconnection queue — the backlog of large-load projects formally requesting grid connection — has grown to a scale ERCOT's planning processes were never built to absorb, and the overwhelming majority of that requested load is data centers. ERCOT now projects peak demand reaching up to 218 GW by 2031: the steepest sustained load growth in the grid's history. The Texas Legislature's response, Senate Bill 6, restructured how large loads enter the queue.

ERCOT runs a nodal wholesale market: prices clear at 15-minute intervals based on real-time supply and demand at specific transmission nodes. Large industrial customers — fabs, data centers, gigafactories — can participate in that market directly, as their own load-serving entities, bypassing the utility retail markup that applies in regulated states.

The difference is concrete. A semiconductor fab in Arizona pays Tucson Electric Power's regulated retail rate. The semiconductor fab in Taylor contracts directly with operators, participates in the real-time market, and captures the spread between wholesale and retail that regulated utilities capture elsewhere. At the load scales involved, that spread runs into hundreds of millions of dollars per year in operating cost.

None of this was designed to attract data centers. The Texas Legislature restructured the electricity market in 1999 through Senate Bill 7, motivated by free-market ideology, incumbent utility opposition that was eventually overcome, and a genuine belief competition would lower residential rates. Residential rates did not fall as dramatically as proponents projected. Industrial rates fell substantially. The large-load economics of ERCOT became a magnet for energy-intensive industries the legislature was not specifically trying to attract. The AI-industrial complex is the beneficiary of a policy decision made for unrelated reasons.

Energy Sovereignty

At the frontier of the buildout, operators are moving behind the meter entirely — not remaining grid-connected customers but generating their own power on-site. Fermi America's nuclear-powered data center campus is the most extreme version: a purpose-built small modular reactor that will provide baseload power to a dedicated compute campus, with no grid connection required for primary supply. CloudBurst Evolve's vertically integrated developer-and-design-build model uses on-site natural gas generation as the primary source, with grid connection as backup.

The logic, from an engineering point of view, is straightforward. A grid-connected industrial customer is exposed to ERCOT's wholesale price volatility — the same market design that produces low average rates also produces extreme price spikes during supply stress, as the February 2021 Winter Storm Uri demonstrated at catastrophic scale. A 500-MW fab pays average ERCOT rates on normal days and potentially ruinous prices during grid stress. Behind-the-meter generation eliminates that exposure. It also eliminates transmission and distribution charges, which at industrial scale are material. And it provides the certainty multi-billion-dollar capital investments require: a fab cannot tolerate outages, and a data center service-level agreement cannot be written around ERCOT's dispatch stack.

The February 2021 event is the constraint argument made physical. Texas cannot import power from neighboring grids when its own generation fails. The winter storm knocked out roughly a third of Texas generating capacity simultaneously. Had Texas been interconnected with the Eastern or Western grid, neighboring utilities could have dispatched power across the interconnects. The structural advantage and the structural vulnerability are the same feature. Operators building behind the meter are not criticizing ERCOT's market design. They are building around its failure mode while continuing to benefit from its pricing for the loads they keep grid-connected.

THE ELECTRIC EXPRESSWAY

The grid's other problem is not price or reliability. It is geography. The generation is in the west. The load is in the east. Texas is now building the transmission backbone the buildout requires — moving Permian gas, West Texas wind, and West Texas solar eastward to the load centers where the data centers, fabs, and gigafactories are concentrating.

The $33 billion Strategic Transmission Expansion Plan — including three new 765-kilovolt import paths into the Permian Basin and associated upgrades, ordered by the Texas Legislature — also requires ERCOT to develop a reliability plan for the Permian Basin region, whose natural gas generation is the dispatchable foundation of the grid.

The 765-kV voltage class is not an incremental improvement on Texas's existing 345-kV system. It is a categorical upgrade — capable of moving three to four times the power per line, already in service across more than 2,400 miles of the Eastern Interconnection, but absent from ERCOT's infrastructure until now. Texas has been operating the most consequential industrial grid in the United States on a transmission system that predates the buildout by decades. The $33 billion program is the infrastructure response to a load growth trajectory the existing system cannot accommodate.

The three lines share an architecture — self-supporting steel lattice towers 15 to 18 stories tall, spaced every 1,000 feet — and a purpose: moving power from the Permian and from West Texas wind and solar east to the load centers. Together they form an electric expressway whose western origin is the Permian Basin and whose eastern terminus is the Bell County East substation, serving the same corridor where Samsung Taylor, Meta Temple, and the Williamson County data center cluster are all drawing from the grid at once.

But the buildout cannot wait for it. The data centers, fabs, and gigafactories are drawing power now — from a grid whose constraints they are managing through the behind-the-meter architectures, distributed generation, and demand response described previously. The behind-the-meter gas campus, the SMR-powered data center, the battery storage-integrated hub: these are not just operational preferences. They are load management strategies whose proliferation reflects an accurate assessment that the transmission infrastructure the buildout needs will not be fully built out until the decade's most critical period has already passed.

The Water Constraint

There is another constraint, and it does not appear in any power capacity announcement. Water — the variable that will determine the ultimate capacity ceiling of the Texas AI-industrial complex more decisively than land availability, labor supply, or transmission capacity.

Every layer of the buildout consumes it. Thermal power generation — natural gas, nuclear — requires cooling water. Semiconductor fabs require ultrapure water for wafer washing, at volumes that dwarf any prior industrial use in the regions where they are sited. Data centers require cooling water for thermal management, with water-usage-effectiveness ratios that translate to millions of gallons per day at hyperscale. The Colorado River basin, under LCRA jurisdiction, supplies the Hill Country corridor. The Trinity and Brazos systems supply DFW and Waco. Neither system was designed to absorb simultaneous demand from dozens of hyperscale facilities, plus the residential buildout at the Williamson County growth rate, plus agricultural baseline, plus municipal baseline.

The Brownfield Inheritance

The market's discovered solution to that constraint is brownfield inheritance. Gibbons Creek Reservoir's coal-plant water rights flow with the property to SpaceX Terafab. Sandow Lakes carries 70 years of Alcoa water rights into the Xebec campus. The Alcoa case is the pattern at its most complete. The aluminum smelter closed because US electricity prices made aluminum smelting uneconomical. The site is now becoming the largest solar cell manufacturing facility in the United States — powered by the same electrical infrastructure, drawing from the same Alcoa-era water rights, sitting on the same industrial land whose zoning, rail access, and utility connections were built for heavy industrial operations that no longer exist.

Operators who found pre-allocated water are building. Operators who require new freshwater allocations are navigating a permitting and priority system that was not designed for their load profile and does not move at their preferred pace. The energy substrate of the Texas story is therefore not simply about cheap wind, solar, gas, and a deregulated market, though those are real. It is about a state that accumulated — across decades of decisions made for unrelated reasons — an energy architecture that happens to be nearly optimal for the most energy-intensive industrial buildout in human history, and that is now discovering, in real time, which parts of that architecture were designed for a different demand profile than the one arriving.

The industrial platform outlasted the industry it was built for. The next industry found it ready.

Three

THE CRITICAL MATERIALS SUBSTRATE

ENERGY AND WATER ARE NOT the only inputs the AI-industrial complex draws from Texas geology. There is another category whose strategic significance is comparable to the energy substrate's and whose domestic supply chain vulnerability is, if anything, more acute. Critical minerals — lithium, rare-earth elements, graphite, specialty metals — flow through every battery, every semiconductor, every electric motor, every humanoid robot actuator, every wind turbine. They are the geological inputs whose absence cannot be substituted, engineered around, or imported from a friendly ally without strategic risk.

Two facts frame the chapter. China controls the global supply chain. Texas has more of the underlying material in the ground than anyone knew 20 years ago.

What follows is what is in the ground, what is being built to extract and process it, and why a domestic critical materials supply chain is the third leg of the substrate argument that makes Texas the logical geography for the AI-industrial buildout.

The China Dependency

The dependency is not marginal. China controls approximately 60 percent of global rare-earth mining, 85 percent of rare-earth processing, and 90 percent of rare-earth permanent magnet production. It controls the majority of global lithium processing despite not having the largest reserves. It produces approximately 65 percent of the world's silicon metal used in solar cells and semiconductors. It processes approximately 80 percent of the world's cobalt, mined primarily in the Democratic Republic of Congo.

Translate those percentages into the end products. Every motor in a Tesla produced at Giga Texas draws on neodymium and dysprosium whose processing chain runs through China. Every lithium-ion battery cell in a Megapack draws on lithium carbonate or lithium hydroxide refined through a similarly concentrated chain. Every semiconductor fab tool whose consumables include specialty gases and process chemicals draws on supply chains whose geographic concentration in China and a small number of allied nations represents a systemic risk.

China has demonstrated willingness to use that concentration as a strategic instrument. Export controls on rare-earths in 2010 triggered a global supply chain crisis and a temporary price spike of more than 1,000 percent for some elements. The export controls announced during the 2025 trade tensions — lithium battery components, gallium, germanium, rare-earth processing chemicals — are the more recent expression of the same instrument.

Texas's response is not a policy aspiration. It is a physical buildout already underway, running in sequence: from the geological resource, through processing and refining, to manufacturing and recycling. The supply chain is not complete. It is emerging, with a direction that is unambiguous and a momentum that is self-reinforcing.

The Smackover Formation

Start with the resource. The geological formation that funded the University of Texas system through petroleum production is now producing a second critical resource — from its brine.

The Smackover Formation is a deep sedimentary rock formation running from East Texas to the Florida Panhandle, deposited roughly 150 million years ago as a shallow evaporitic sea that concentrated its dissolved minerals into the carbonate and evaporite layers that now sit beneath thousands of feet across the Gulf Coastal Plain. Its petroleum-bearing zones have been producing oil and gas since the 1920s. Its brine — the formation water that comes up with petroleum production and has historically been treated as a disposal problem — contains dissolved lithium at concentrations ranging from 200 to 800 parts per million in the Texas and Arkansas portions. The most lithium-rich fluids ever found in North America were discovered beneath Smackover claims in late 2025.

The resource scale is significant. The Arkansas portion of the Smackover alone is estimated to contain enough lithium to meet global EV battery demand nine times over, according to US Geological Survey assessments. The East Texas portions extend the resource base further. The Smackover is not a marginal lithium deposit supplementing imports. It is a potential primary domestic lithium source whose development timeline — years rather than decades — could materially reduce US dependence on lithium imports from Australia, Chile, and Chinese-processed sources before 2030.

The enabling technology is Direct Lithium Extraction. Traditional brine recovery requires solar evaporation ponds, 18 to 24 months of residence time, an enormous land footprint, and geography constrained to the high-altitude South American salars where solar irradiance and low humidity accelerate evaporation. DLE pulls lithium directly from brine using selective ion exchange membranes or sorbent materials — cutting production from 18 months to days, and making economic recovery possible from brine concentrations and geographies evaporation-pond technology cannot serve. The

Smackover's deep formation brine is accessible through the same drilling and production infrastructure the oil and gas industry has built across East Texas over a century: brownfield inheritance at the geological resource layer.

EnergyX's Project Lonestar is the most advanced commercial DLE program in Texas. The demonstration plant near Texarkana processes Smackover brine, producing battery-grade lithium carbonate equivalent at 250 tons per year in the demonstration phase. EnergyX controls approximately 47,500 acres of Smackover Formation in Texas and Arkansas. The commercial phase targets 50,000 tons per year by 2030. A dedicated rail line at the Hooks facility moves processed lithium to downstream partners — battery cell manufacturers, cathode producers, the Tesla Lithium Refinery at Robstown — whose geography is concentrated along the Gulf Coast.

The Permian's Water

The Smackover is not the only Texas lithium resource. The Permian Basin, whose natural gas production is the dispatchable foundation of ERCOT's grid, produces approximately 8 million barrels of formation water per day alongside its oil and gas output. That produced water contains dissolved minerals — including lithium at concentrations typically ranging from 15 to 20 parts per million. Much lower than the Smackover's 200 to 800 ppm, but at volumes that compensate for the concentration differential. A second domestic lithium feedstock, advancing through Texas-based startups applying novel extraction technologies to an existing waste stream.

Element3, headquartered in Fort Worth, has opened a lithium carbonate production plant in the Permian Basin using oilfield produced water — the first facility of its kind in the United States in half a century. The plant's feedstock is the produced water Permian operators have historically disposed of through injection wells at significant cost. Element3's technology converts a disposal liability into a lithium carbonate revenue stream whose economics improve as lithium prices recover and as processing efficiency scales.

Maverick Metals, a San Antonio-based startup, is pursuing the same concept at smaller scale across multiple Permian operators — co-locating lithium extraction equipment at producing well sites and recovering critical minerals from the produced water stream before disposal.

And the produced water opportunity is not limited to lithium. Permian formation water contains dissolved concentrations of strontium, bromine, iodine, and other elements whose recovery economics are improving as extraction technology advances and the regulatory environment for produced water management evolves. Texas is developing a critical minerals industry from existing oil and gas infrastructure whose capital costs are already amortized — the same brownfield pattern, at the produced water layer.

Round Top Mountain

Rare-earth elements present a different supply chain challenge than lithium. Lithium is geologically abundant; its constraint is processing concentration, not geological scarcity. Rare-earth elements are distributed in low concentrations across many rock types, with economically mineable concentrations occurring in a relatively small number of deposit types globally. China's dominance reflects its control of the Bayan Obo deposit in Inner Mongolia — the world's largest rare-earth deposit — combined with decades of investment in the processing infrastructure that converts rare-earth ore concentrates into separated oxides, and then into the metals and alloys permanent magnet manufacturing requires.

Texas has one of the most significant domestic heavy rare-earth deposits identified in the United States: Round Top Mountain, about 85 miles southeast of El Paso in the Trans-Pecos region of far West Texas. The rhyolite deposit hosts virtually all of the rare-earth elements, including the heavy rare-earths (HREEs) — dysprosium, terbium, yttrium, holmium — whose supply chain density in China is more acute than the light rare-earths, because China's ionic clay deposits in the south are the world's primary source. Round Top also contains lithium, uranium, thorium, beryllium, gallium, hafnium, and zirconium: all minerals the Department of the Interior designates critical to US national security and the economy.

Round Top's strategic significance is structural rather than immediate: it is the domestic heavy rare-earth resource whose development eliminates the dependency on Chinese deposits for the dysprosium and terbium that every neodymium-iron-boron (NdFeB) permanent magnet requires to increase strength.

PROCESSING, REFINING, RECYCLING

A geological resource is not yet a critical material in the form the supply chain requires. The processing and refining stage — converting lithium brine to battery-grade lithium hydroxide; converting rare-earth concentrates to separated oxides, to metals, to alloys, to sintered magnets — is where China's dominance is most complete, and where the domestic supply chain's gaps are most consequential. Texas has the most complete processing and refining layer of any US state for the critical materials the buildout requires.

Tesla's Lithium Refinery in Robstown, adjacent to Corpus Christi, is the first large-scale domestic lithium hydroxide refinery in the United States. The facility currently processes lithium spodumene ore imported from Canada's only lithium mine, converting it to battery-grade lithium hydroxide monohydrate at a targeted capacity of approximately 25,000 tons per year in Phase 1. Battery-grade lithium hydroxide is the preferred cathode precursor for the cell chemistries whose energy density Tesla's vehicle and stationary storage products require.

MP Materials' Fort Worth Magnet Plant is among the first US facilities producing sintered NdFeB permanent magnets at commercial scale from virgin rare-earth feedstock. MP Materials' Mountain Pass mine in California produces rare-earth concentrates; the Fort Worth facility converts those concentrates through the full processing chain — separation, reduction, alloying, sintering, magnetizing — to produce the finished NdFeB magnets that EV motors, wind turbine generators, defense systems, and humanoid robot actuators require. The $2 billion Phase 1 total investment at

Fort Worth is the most significant single commitment to domestic rare-earth magnet manufacturing in US history.

Noveon Magnetics' manufacturing center in San Marcos occupies a complementary position. Where MP Materials addresses the virgin supply, Noveon addresses the circular economy dependency. It recovers neodymium, praseodymium, dysprosium, and terbium from end-of-life EV motors, wind turbine generators, hard drives, and defense systems, using its proprietary technology to reprocess them into new sintered NdFeB magnets without requiring virgin rare-earth mining inputs.

Metallium, operating a facility in Houston's industrial complex, applies flash heating technology to recover critical minerals from discarded consumer electronics — smartphones, laptops, hard drives, the full range of devices whose rare-earth content is currently lost to landfill or exported for Chinese recovery.

Together, the Texas manufacturers constitute the most complete domestic rare-earth magnet sovereignty architecture in the United States. Neither alone closes the supply chain gap. Together they address both the extraction dependency and the recycling dependency that define China's manufacturing dominance.

Another material may be joining the list. A San Antonio-based startup, Graphite Energy, is extracting graphite from methane gas — a pilot project whose commercial viability would address another supply chain gap currently dominated by China, which processes approximately 80 percent of the world's battery-grade graphite. The methane-derived process produces synthetic graphite whose purity and consistency exceed natural graphite's specifications for battery anode applications — the same quality argument that drove the semiconductor industry from natural to synthetic diamond abrasives. If the San Antonio pilot demonstrates commercial economics, Texas adds graphite to lithium and rare-earths: three of the critical material classes required.

Recycling closes the chain by addressing a dependency whose scale is difficult to appreciate. The rare-earth elements extracted from end-of-life equipment by American households and businesses are currently shipped to China for recovery. The United States exports its critical minerals waste to the country whose processing dominance it is attempting to reduce. The domestic recycling layer closes a supply chain loop whose closure is not just economically significant but geopolitically necessary.

The Geological Recursion

The brownfield inheritance pattern — water rights, power infrastructure, permitting frameworks inherited from prior-generation industrial facilities — operates in the critical materials context at the geological layer itself. The connection is most precisely expressed at a single Texas site: Gibbons Creek, Grimes County.

Gibbons Creek Reservoir was built by the Texas Municipal Power Agency as cooling water for the Gibbons Creek Steam Electric Station — a 470-MW coal-fired power plant that operated from 1982 until its retirement in 2020. The coal plant's closure left the reservoir and its water rights available as an industrial asset whose scale — 2,400 acres of water storage capacity with established withdrawal rights — is difficult to replicate at greenfield sites. The massive Terafab production facility being developed by SpaceX at the site inherits those water rights.

Now the recursion. The coal plant burned coal for 38 years, and coal combustion concentrates rare-earth elements from the coal mineral matrix into the ash residue, at concentrations roughly four times higher than in the original coal. The Bureau of Economic Geology at UT Austin has measured REE concentrations from the Gibbons Creek coal mine at up to 2,800 parts per million. The mine, adjacent to the power plant whose coal it supplied, is now closed and reclaimed.

The recursion is analytically precise. A coal mine extracted coal. A power plant burned the coal, concentrating rare-earth elements into its ash. The plant closed. A semiconductor fab is being built on the reclaimed site, inheriting the water rights the power plant required. The semiconductor manufacturing the fab will conduct —

chemical mechanical planarization, etch chemistries, deposition processes — requires rare-earth-based process chemicals and specialty materials whose supply chain runs through the same Chinese processing dominance documented earlier.

The substrate the coal plant produced as a waste product is the input the semiconductor fab that replaced it requires.

The Early Innings

Texas is not building a critical materials industry as a policy aspiration. It is building it as a physical response to a geological endowment whose strategic value the requirements have made newly legible. The Smackover brine was always there. The Round Top rhyolite was always there. The Gibbons Creek coal was always there. The oil industry's formation water was always there. The rare-earth elements in end-of-life consumer electronics were always there. What changed is the demand that makes their extraction and recovery economically necessary and strategically urgent.

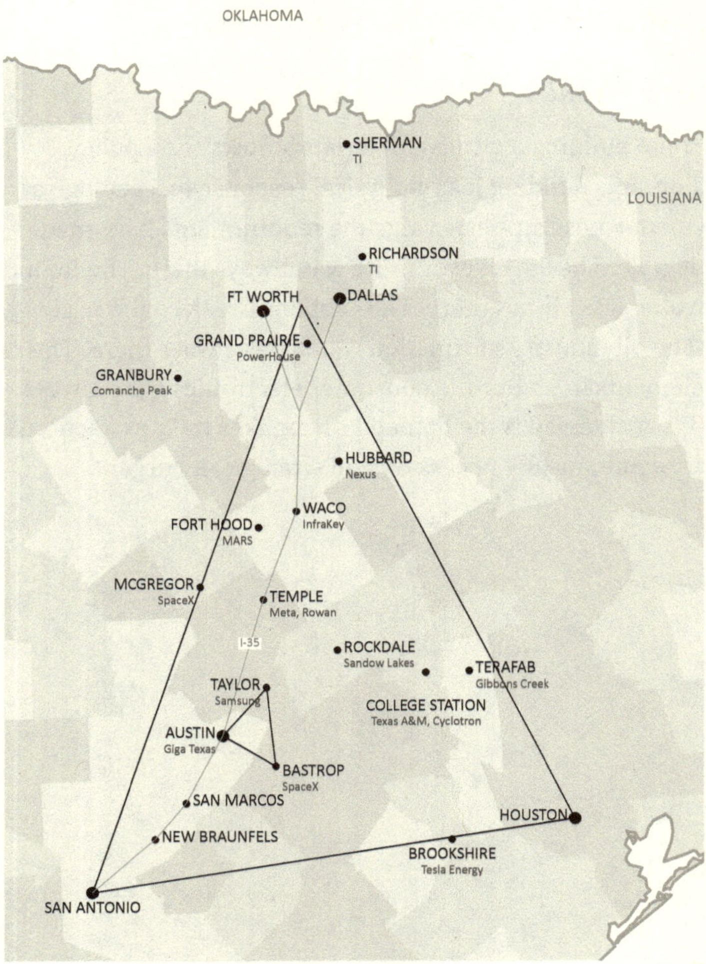

Inside the Singularity

THE TEXAS TRIANGLE · 2026

Four

THE TEXAS TRIANGLE

THE TEXAS TRIANGLE IS NOT A metropolitan area. It is a geographic description — the triangular territory defined by three of the largest cities in Texas: Dallas–Fort Worth at the northern apex, Houston at the southeastern, San Antonio at the southwestern. Inside the triangle nearest San Antonio sits Austin. The combined population exceeds 22 million. Measured as a sovereign economy, the Triangle's output would rank among the largest in the world.

The AI-industrial complex requires a complete set of inputs — semiconductor fabrication, AI compute infrastructure, energy generation at scale, advanced manufacturing and refining, launch capability, a defense industrial base, deep-water port access, petrochemical feedstock infrastructure, and a university research pipeline. The Texas Triangle contains all of them, in a single contiguous geography, with the pieces close enough to share labor markets, supply chains, and infrastructure.

DALLAS–FORT WORTH

Dallas–Fort Worth supplies three pieces of the Triangle: the logistics spine that moves the buildout's physical inputs, the capital intermediation layer that finances it, and the analog semiconductor depth that completes the silicon profile.

The logistics contribution is infrastructural. DFW International Airport is in the top ten busiest cargo airports in the United States, handling the air freight volume advanced manufacturing supply chains require. The I-35 corridor running south toward Austin and San Antonio is the overland spine of the Triangle's interior, carrying NAFTA freight, regional distribution, and the industrial supply traffic the buildout generates. The BNSF and Union Pacific rail infrastructure converging on the metroplex provides the bulk freight capacity air and truck cannot economically handle at scale. No other Triangle vertex offers this combination, and without it the interior's leading-edge fabs and gigafactories would carry a supply-chain exposure site selection cannot tolerate.

The silicon contribution is the one most easily missed. Texas Instruments anchors the metro's semiconductor identity: TI's Richardson headquarters, its 300mm analog fab, and the Sherman 300mm fab under construction with CHIPS Act support together make TI the largest US-based analog semiconductor integrated device manufacturer (IDM). Analog chips — power management, signal processing, sensing — are not the leading-edge logic that dominates semiconductor coverage, but they sit in every electrified system, every automotive platform, every industrial control application, and every data center power architecture. The Triangle therefore holds both halves of the silicon stack: leading-edge digital logic at Samsung Taylor (and someday Terafab), analog depth at TI.

Finally, DFW supplies capital intermediation. The financial infrastructure concentrated in Dallas — major bank headquarters, private equity platforms, real estate investment trusts — is the layer that translates institutional appetite for exposure into the project finance structures that individual fab expansions, data center campuses, and logistics developments require.

Houston and the Gulf Coast

Houston supplies the pieces the AI-industrial complex cannot manufacture for itself and cannot relocate: petrochemical feedstock infrastructure, the deep-water port, and — at the southern tip of the

Gulf Coast spine — the launch capability that carries the system's outermost loop into orbit.

Houston is the institutional head of a Gulf Coast industrial corridor that extends south from the Ship Channel through Robstown and Corpus Christi to Boca Chica near Brownsville. This corridor is not a vertex of the Triangle; it is the Triangle's extension along the coast, integrated with Houston's infrastructure, capital, and workforce. The Port of Houston ranks first among US ports by foreign waterborne tonnage — a position held for decades on the petrochemical trade through the Ship Channel, and one now acquiring new strategic weight: the inputs for semiconductor fabrication, battery manufacturing, and AI hardware enter the US supply chain through Gulf Coast maritime infrastructure. Specialty gases, chemical precursors, rare-earth compounds, and the industrial materials advanced manufacturing requires arrive at the Port of Houston before they arrive anywhere else in the Triangle.

The Ship Channel petrochemical complex is the most concentrated industrial chemistry infrastructure in the Western Hemisphere. Decades of chemical processing investment, pipeline infrastructure, permitting frameworks, and workforce development have produced a capability that cannot be replicated on greenfield land on any relevant timeline. The relevance is concrete: semiconductor fabrication requires dozens of specialty chemicals in ultrapure grades, battery manufacturing requires lithium compounds processed to exacting specification, and the industrial gases fabs consume — nitrogen, argon, hydrogen fluoride, ammonia — are produced and distributed through infrastructure the Ship Channel has accumulated over generations. This is the single piece of the stack least visible in AI coverage and least possible to build from scratch.

The southwestern coastal terminus is SpaceX's Starbase complex at Boca Chica, where Starship is built and flown. Starbase is institutionally tied to Houston through its NASA mission relationships, operationally tied to Austin-Bastrop through SpaceX's engineering presence and the Terafab silicon supply for orbital compute, and physically the launch pad for future orbital satellite missions, Starship production, and the orbital compute infrastructure still ahead.

San Antonio

San Antonio supplies the defense industrial base, the cybersecurity concentration that grows from it, and the demographic engine that feeds the Triangle's labor market — three pieces that are about people and institutions rather than physical plant.

Joint Base San Antonio (JBSA) — Randolph, Lackland, and Fort Sam Houston — is the largest military training wing in the US Air Force, and its combined military-and-civilian workforce makes it one of the city's largest employers by a wide margin. The cybersecurity industry that has taken shape around the military's cyber operations presence has made San Antonio one of the top three cybersecurity employment markets in the country, with a talent base and procurement relationship that extends up the I-35 corridor into Austin's commercial technology ecosystem.

The human capital contribution is demographic. San Antonio is the sixth-largest city in the United States, with a younger median age and faster population growth than most comparable cities. The workforce pipeline it feeds into the Triangle — across construction trades, manufacturing, logistics, and technology — is substantial and growing. The megaregion connecting San Antonio to Austin functions as a labor market integration mechanism, making the workforce of both cities available to employers at either end — an advantage neither city would hold alone.

Two projects near San Marcos show the corridor filling in. CloudBurst Evolve is a vertically integrated behind-the-meter AI campus powered by on-site natural gas rather than grid connection. Tract's Caldwell Technology Park is a multi-gigawatt master-planned data center foundation, with land, power, water, and fiber pre-developed to compress individual operator timelines.

New Braunfels, between San Marcos and San Antonio, has absorbed residential growth from both anchor metros and is also attracting data center development. These are not town pass-through points. They are the San Antonio-Austin megaregion filling in.

AUSTIN

Austin's contribution is different in kind from the vertices. It does not sit at a precise vertex at all, and its role is not to supply one more piece of the stack but to concentrate all of them.
Austin receives the logistics infrastructure from DFW, the energy and maritime outputs from Houston, the defense and human capital from San Antonio, and elsewhere in the Triangle — and concentrates them in a geography that also holds the university research pipeline, the technology company cluster, the semiconductor design ecosystem, and the operational center of the Musk enterprises. It is the node where the Triangle's compositional inputs converge, and converge densely enough to interact.

The gravitational claim rests on evidence, not civic boosterism. UT Austin's Cockrell School of Engineering and the Texas Institute for Electronics provide a research and workforce pipeline no other Triangle city matches for semiconductor and AI applications. The semiconductor company clustering alone — 20 notable semiconductor ecosystem companies in Austin proper with over a hundred smaller startups, plus the broader Silicon Hills cluster from San Marcos through The Domain — creates the supplier and talent density large operators require when they make site selection decisions. The capital inflows into Austin metro real estate, the EV adoption density visible on any arterial, and the growth velocity of the Williamson County corridor immediately to the north are all expressions of the same function: Austin pulls capital, talent, and industrial activity toward it and concentrates them in ways that compound the Triangle's overall capability.

That institutional confidence has a vertical expression downtown. The Waterline — a 74-story, 1,025-foot supertall at 98 Red River Street adjacent to the Rainey Street Historic District, funded in part by one of Canada's largest pension funds — topped out in August

2025 as the tallest building in Texas and the tallest in the Southern United States. Construction began in 2022, and the building is opening on schedule in late 2026, a delivery record comparable US megaprojects do not routinely achieve. A city whose construction industry can deliver a supertall on time while simultaneously building out the largest semiconductor fab, expand the largest gigafactory campus, and stand up out dozens of AI-scale data centers is a city whose industrial capacity has crossed a threshold institutional capital recognizes before planning authorities formally name it.

The Triangle as System

The Triangle's competitive position is not about any single asset — and the clearest way to see that is to subtract the system and look at each asset alone.

Samsung Taylor, without the Triangle's energy infrastructure, is a fab in search of affordable power. Giga Texas, without the Triangle's logistics network, is a gigafactory carrying supply-chain exposure. The DFW logistics hub, without the semiconductor and AI compute cluster to its south, is a freight node serving conventional manufacturing. The Houston port, without the industrial demand of the interior, is a petrochemical terminal. Every marquee asset, isolated from the rest of the stack, reverts to something ordinary. None of them is individually unique to Texas. What is unique to Texas is that all of them are present at once, contiguous, growing, and interacting.

In the Triangle, the pieces compound. The fab lowers the site selection risk for the next data center. The data center thickens the demand signal that justifies the next energy facility. The energy facility improves the economics for the next gigafactory. Each new facility draws in the suppliers, the talent, and the supporting infrastructure that make the following facility's site selection decision easier to resolve in Texas's favor.

The Triangle is not the sum of its parts. It is the product of their interaction.

Five

THE INSTITUTIONAL FOUNDATIONS

THE STANDARD NARRATIVE OF THE Texas AI-industrial complex, when it acknowledges Austin's institutional fabric at all, treats it as a pleasant consequence of the buildout rather than a precondition for it. The universities expanded because the technology companies arrived. The capital concentrated because the founders relocated. The startup ecosystem matured because the AI era made Austin fashionable. The institutional layer, in this telling, is downstream of what is happening.

The sequence is backward. Austin's institutional foundations — the university coordination structure, the federal research anchors, the capital concentration, the commercialization machinery, and the founder pipeline that feeds all of them — were assembled across six decades before Tesla and the others announced their headquarters relocations. The institutional base did not form in response to the buildout. The AI-industrial buildout landed in Austin because the base was already there. And unlike energy, or the quasi-sovereign political frame, whose origins reach back to statehood, the Austin institutional foundation has something the other structural features do not. It has a named origin and a specific date.

The date is 1966. The name is George Kozmetsky.

The Kozmetsky Foundation

George Kozmetsky co-founded Teledyne in 1960, building one of the defining conglomerates of the postwar American technology economy. In 1966 he left the company he had helped build and became dean of the University of Texas business school — a position he held until 1982.

What Kozmetsky put in motion in 1966 is still load-bearing six decades later. He founded the IC^2 Institute in 1977 to actively shape the relationship between university research, capital, and commercial formation. He led the recruitment of the Microelectronics and Computer Technology Corporation — the MCC research consortium — to Austin in 1983, when Austin was not the obvious choice. He did the same in 1987 for Sematech, the semiconductor manufacturing research consortium. He founded the Austin Technology Incubator in 1989, building the university-spinout commercialization infrastructure that pure private accelerators are not configured to provide. The Capital Network, and what became the Austin Technology Council, emerged from the same deanship and the decades that followed.

The Permanent University Fund

Kozmetsky's work explains the coordination structure — how UT Austin learned to operate as the connective tissue between research, capital, and commercial formation. It does not explain the financial base that made the university capable of building research capacity at that scale in the first place. That groundwork is older, and its origin reaches back before the institution it funds.

In 1839, the Congress of the Republic of Texas set aside fifty leagues of land — roughly 221,000 acres — to fund higher education in a republic that did not yet have a university to spend it on. The 1876 Texas Constitution expanded the endowment, setting aside land in West Texas to support what would become the University of Texas and Texas A&M systems. That land, today encompassing approximately 2.1 million acres, funds what became the Permanent University Fund.

The oil and gas revenue from those West Texas lands have transformed the Permanent University Fund into one of the largest university endowments in the United States — a position it still holds.

The energy substrate and the institutional foundation are not parallel structural features Texas happened to accumulate independently. At their origin they are one structure. Oil revenue from university lands is what capitalized the universities. Nobody who set aside West Texas land in 1876, and nobody who drilled a particular well in 1923, intended to fund a twenty-first-century semiconductor research consortium or a leadership-class supercomputing facility. The land was set aside and the well was drilled for reasons that had nothing to do with the buildout. But the Permanent University Fund is the reason both UT Austin and Texas A&M entered the modern era with the endowment base required to build research capacity at the scale the AI-industrial complex now draws on.

Now to the origin story promised earlier.

The Italian Nuns

Santa Rita of Cascia was a fifteenth-century Italian nun whose biography includes a stigmatic wound from a crown of thorns, a miraculous bee visitation in infancy, and a series of answered prayers so improbable that the Church eventually designated her the Patron Saint of the Impossible. Nearing the end of her life, she asked a cousin to bring a rose from her childhood home. Her cousin was surprised to find a single rose blooming in the cold January garden.

Santa Rita's name reached West Texas by a different route. In 1919, a group of University of Texas regents and petroleum geologists leased land in the Permian Basin, drilled into the Ellenburger limestone, and came up with nothing for three years.

Among the early investors were two Italian nuns. Before the unproductive site was abandoned, they prayed to Santa Rita for help. Her blessed rose petals traveled to the site and were tossed from the the derrick. On May 28, 1923, the newly named Santa Rita No. 1 well came in — the first major oil strike in the Permian Basin, producing from a formation that would become one of the most productive petroleum reservoirs in human history. Texas tea indeed.

Santa Rita No. 1 was drilled on University of Texas land. Under the Texas Constitution, oil and gas revenues from university lands flow into the Permanent University Fund.

The TIE NGMM facility at UT Austin — the $840 million semiconductor prototyping center whose process engineers will staff the Texas semiconductor ecosystem for the next generation — is funded in part by the compounded legacy of a Permian Basin oil strike that landed because someone scattered rose petals from a derrick and asked the Patron Saint of the Impossible for help.

University Structure

UT Austin is not a research university with industry partnerships in the conventional sense. It is the institution where federal programs, state programs, commercial industry, and defense modernization all run concurrently — across the same campus, the same faculty appointments, the same engineering enrollment pipeline. CHIPS Act and DARPA programs, the Texas Semiconductor Innovation Fund, the sustained relationships with Samsung and Tesla and Texas Instruments, and the United States Army's transformation command all touch the same handful of buildings and draw on the same talent pool. The campus couples directly to Giga Texas, to Samsung Taylor, to SpaceX Bastrop, to the surrounding startup base, and to the broader Texas Triangle research base — inside a single integrated labor market. Four programs make the coordination concrete.

The first is semiconductor research. The Texas Institute for Electronics houses the DARPA Next Generation Microelectronics Manufacturing (NGMM) facility for advanced semiconductor processes and three-dimensional heterogeneous integration packaging. Its consortium spans dozens of defense and commercial electronics companies and academic institutions.

The second is supercomputing. The Texas Advanced Computing Center (TACC) operates the university's production supercomputing systems. It is one of the most powerful academic computing facilities in the world, running systems that have repeatedly held top rankings on the TOP500 list. TACC trains the computational scientists and research computing engineers whose skills transfer directly into the AI training infrastructure at Tesla, Stargate, and others.

The third is engineering itself. The Cockrell School of Engineering runs seven academic departments, roughly 8,500 students, just under 300 tenure-track faculty, and an annual research enterprise in the hundreds of millions of dollars. The lithium-ion battery lineage, the named engineering departments, and the engineering pipeline that feeds Tesla, Samsung Taylor, the Austin chip design houses, the regional aerospace base, and the Army's software and modernization programs all operate out of Cockrell.

The fourth is defense. The United States Army Transformation and Training Command (T2COM), activated on the UT campus as the merger of Army Futures Command and the Army's training command, is the largest command in the United States Army. Applied Research Laboratories at UT, established in 1945 and one of fifteen Department of Defense University Affiliated Research Centers in the country, operates as the parallel defense research anchor under a multi-year US Navy contract.

Capital and Commercialization Layer

The Austin Technology Incubator, operating under the IC² Institute, has provided deep-tech and university-spinout commercialization infrastructure continuously since 1989. Across its history it has graduated hundreds of member companies, contributed billions of dollars in economic impact to Central Texas, and anchored a sustained record of public offerings and acquisitions.

The McCombs entrepreneurship complex, through its multiple centers and the graduate-level Master of Science in Technology Commercialization program, supplies the business-and-operator pipeline and the technology commercialization specialist pipeline that complement Cockrell's technical-founder pipeline.

Capital Factory, co-founded in 2009, anchors the downtown Austin startup density with co-working space, mentorship, investor connectivity, and a distinctive defense-technology accelerator. It has been among the most active early-stage investors in Texas for more than a decade, with operating presence now extending across the major Texas metros. The Austin venture capital community, which grew from a handful of regional firms in the early 2010s to a substantial late-stage capital base by the mid-2020s, concentrates capital across the seed-through-growth-stage phases through a mix of regional firms, satellite offices of coastal funds, and an increasing rate of local fund formation.

CONVERGENCE AT STARTUP SCALE

Austin has over two decades of results behind its institutions. The ecosystem has produced more than 25 unicorns — billion-dollar private valuations — plus dozens of additional notable exits, with cumulative ecosystem value on the order of $90 billion in recent reporting. That places Austin in the second tier of US unicorn-producing metros: behind the Bay Area and New York, comparable to Boston and Los Angeles.

Austin offers recycled operator expertise, recycled capital, and demonstrated precedent that lower the activation energy for the next company. This repeat-founder pattern — the prior generation's operators cycling back into new ventures — is the signature of a mature ecosystem rather than an emerging one.

Five companies make the point, and four of them are startup-scale expressions of the AI-industrial complex documented in this book. Firefly Aerospace, now public as of mid-2025, operates in the space and launch layer. Apptronik operates in the humanoid robotics layer. Base Power operates in the distributed energy storage layer. Saronic Technologies operates in the defense-autonomy layer. The fifth, ICON, operates on the residential buildout that the industrial complex is generating as a second-order consequence. The five together are not a representative sample of Austin's unicorn cohort. They are a deliberately selected one, and what they share is the point.

Firefly Aerospace, headquartered in Cedar Park, is the launch-and-lunar-lander company that landed the Blue Ghost Mission 1 spacecraft on Mare Crisium on March 2, 2025 — the first fully successful commercial soft landing on the Moon and the extension of Texas's orbital and cislunar layer beyond the SpaceX Starbase concentration on the coast. Apptronik graduated from the Austin Technology Incubator and was co-founded by a McCombs graduate; it is Austin's native entrant in the humanoid platform race that Tesla's Optimus, Figure, and 1X currently define at scale. Base Power, founded by Zach Dell — Michael Dell's son — builds residential battery storage and the distributed grid network behind it, a consumer-scale expression of the same behind-the-meter energy architecture the industrial cluster is deploying at gigawatt scale. The valuation trajectory is what stands out: a $200 million Series B in April 2025, a $1 billion Series C at a $4 billion valuation in October 2025, and by May 2026 a reported $12 billion Series D valuation — a threefold step-up in seven months, one of the fastest valuation ramps in the current Austin unicorn cohort. ICON builds large-scale 3D printing for housing, operating directly on the residential-buildout problem the Austin metro's AI-industrial migration is producing.

Saronic Technologies is a standout. Founded in 2022 by former SEAL Team Six commander Dino Mavrookas and co-founder Doug Lambert, Saronic designs and manufactures Autonomous Surface Vessels for the US Navy's hybrid fleet program: the 6-foot Spyglass, the 14-foot Cutlass, the 24-foot Corsair, and the 180-foot Marauder. In December 2025, the company received a $392 million Navy production contract for the Corsair. In March 2026, it closed a $1.75 billion Series D at a $9.25 billion valuation, more than doubling its Series C valuation from 13 months earlier and placing Saronic behind only Anduril in the current cohort of defense-technology unicorns. The investor lineup — Kleiner Perkins, Bessemer Venture Partners, Advent International, Franklin Templeton, Balyasny Asset

Management, Andreessen Horowitz, 8VC, Lightspeed, NightDragon, Point72 Ventures — is the defense-adjacent capital pool that has consolidated around Anduril, Palantir, Shield AI, and the emerging generation of defense-technology unicorns. This is not conventional venture capital deployed against a speculative platform. It is the capital consolidation that follows procurement validation from the customer.

The company's existing shipyard in Franklin, Louisiana is where the Marauder was built from initial design to first hull in nine months, which Mavrookas has publicly called the fastest ship built in the United States since World War II. The Franklin facility is a warm-up. Port Alpha is the main event.

Port Alpha is Saronic's next-generation shipyard, a $3.2 billion, four-phase, ten-year buildout designed to produce autonomous vessels at speed and scale. Located at the Port of Brownsville, Texas, it will span 4,387 acres. Pointe Isabel ISD, the school district that covers the port land Saronic will occupy, voted unanimously to approve a deal that gives the company $517 million in school tax abatements through 2034. Port Alpha will create up to 10,000 direct jobs across engineering, manufacturing, and operations.

The country's leading autonomous-vessel shipyard will be located 19 miles from the leading orbital launch complex at Starbase. An Austin-based defense and space startup now scales into a multi-billion-dollar production facility, with the corporate headquarters and engineering workforce concentrated in Austin and the industrial-scale manufacturing sited where land, power, and incentives make it possible.

Space launch, humanoid robotics, defense and autonomy, distributed energy storage, and 3D-print construction technology are not consumer-software categories. They are the deep-technology and dual-use domains expressing the AI-industrial convergence at startup scale. What makes that expression happen in Austin rather than elsewhere is the way the founder pipeline now couples directly to the AI-industrial cluster. Cockrell engineers and CS graduates supply the technical-founder inflow. McCombs graduates and the technology commercialization program supply the operator and commercialization specialists. And the cluster itself supplies the rest: Tesla engineers leave Giga Texas to start adjacent ventures, Samsung Taylor process engineers spin out semiconductor-startups, and officers transitioning out of the Army's Austin defense modernization commands move into defense-technology founder roles through Capital Factory's defense accelerator.

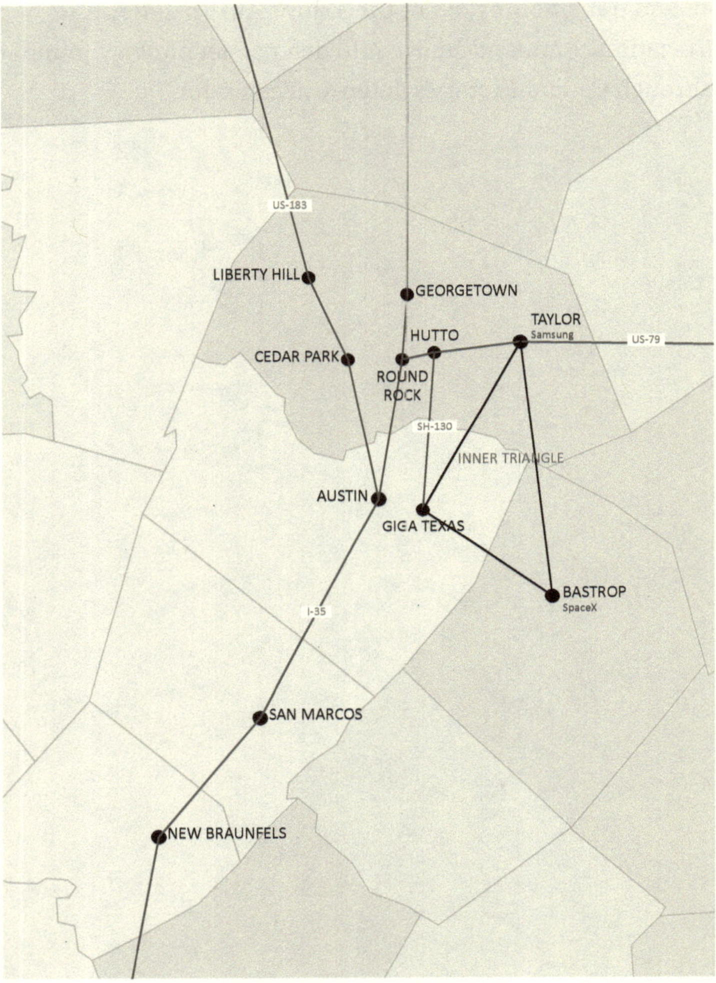

Inside the Singularity

AUSTIN METRO CORRIDORS · 2026

Six

THE METRO CORRIDORS

GEOGRAPHY IS NOT NEUTRAL. The physical distribution of industrial capital across a region follows infrastructure — power transmission, fiber, water, roads, rail — and infrastructure follows prior infrastructure, which followed terrain, which followed water. The AI-industrial concentration in the Texas Triangle did not distribute itself randomly across 268,000 square miles of available land. It concentrated along specific corridors, where prior development had already laid the base new development requires. Those corridors are now visible from above as threads of industrial activity connecting nodes of capital concentration. From the ground, driving them, they look like construction.

The Texas Triangle's four metropolitan economies together generate roughly 70 percent of the state's economic output. Inside the Triangle, the AI-industrial buildout is organized by transportation corridors that thread between the metros and within them. Four primary highway corridors define the architecture of the Austin-metro-anchored buildout, where the most capital-intensive concentration is taking shape. Dallas–Fort Worth and Houston have their own corridor architectures — the Alliance Corridor north of Fort Worth and the Energy Corridor and Ship Channel industrial belt threading through metro Houston — but the structural anchor of the

AI-industrial buildout is in Central Texas, where four corridors converge in a single county.

That convergence point is Williamson County, where I-35, US 183, SH 130, and US 79 liw within a span of roughly 30 miles. The county hosts Samsung Taylor near the northern end of SH 130, the data center cluster along US 79, and the technology employment concentration along US 183. The corridor cluster is what makes Williamson County the county anchor of the buildout — the subject of its own treatment later in this book. The corridors are how that convergence connects outward to the rest of Texas.

The I-35 Corridor

Interstate 35 is the north-south spine of the Texas interior, running from Laredo at the Mexican border through San Antonio, Austin, Waco, and into Dallas–Fort Worth. It has always been a logistics corridor — NAFTA traffic, agricultural transport, regional freight, and the supply chains feeding industrial manufacturers along its length. The buildout has added a second identity: AI-defense spine.

Fort Hood, the largest US Army installation by population, located near Killeen west of Waco, anchors the corridor's defense identity. Fort Hood's demand for autonomous systems, AI-enabled logistics, cybersecurity infrastructure, and advanced communications equipment creates a procurement relationship with the Austin technology cluster that operates at the scale of a major commercial customer. The base's modernization program — funded under recent defense appropriations and accelerated by the Golden Dome program's compute requirements — is the federal procurement engine pulling defense contractors into the Austin technology cluster's orbit. Interstate 14 — a short east-west spur connecting Temple, Killeen, and Fort Hood — extends Fort Hood's logistics reach into I-35 proper and reflects federal investment in the installation's regional connectivity.

The corridor's northern reach into Dallas–Fort Worth connects the Austin AI cluster to the DFW defense ecosystem: Lockheed Martin at Fort Worth, Bell Textron, Raytheon, and the dense defense contractor concentration around Naval Air Station Fort Worth Joint Reserve Base. I-35 is therefore not only a logistics route. It is the institutional spine connecting two of the largest defense procurement geographies in the United States. The AI-industrial buildout in Austin has, in a structural sense, been grafted onto a defense supply chain that I-35 has been carrying for 40 years.

The same geography also carries a commercial semiconductor capacity buildout of comparable scale. Texas Instruments' Richardson and Sherman campusus, extending 60 miles north of Dallas on US 75, is among the largest single-site additions of analog and embedded semiconductor capacity in the United States: a multi-fab, 300mm complex producing the power management, signal chain, and embedded processing parts that go into automotive, industrial, robotics, and defense systems. Sherman does not sit on I-35 itself, but it occupies the same North Texas industrial geography the corridor stitches together. The corridor's northern reach is therefore not only a defense procurement spine. It is also home to the largest American capacity in the analog silicon that every other tier of the buildout depends on: vehicles, robots, satellites, weapon systems.

The SH 130 Corridor

State Highway 130 connects the southern and northern anchors of the Austin metro buildout in a single uninterrupted run. Its northern terminus reaches Hutto, just west of the Samsung Taylor site. Its southern terminus connects to San Antonio. In between, it passes Giga Texas on Austin's eastern edge. That makes SH 130 the primary industrial artery linking Tesla's four-program campus to Samsung Taylor — currently the two most capital-intensive facilities in the Texas AI-industrial buildout — without entering Austin metropolitan traffic at any point.

The route carries heavy freight at a posted speed limit of 85 miles per hour, with none of the chronic congestion that makes I-35 operationally unreliable for time-sensitive logistics. That last fact is underappreciated. Supply-chain logistics between a gigafactory and its captive semiconductor foundry are not casual; components move on schedules production lines depend on. I-35 between Austin and Round Rock is congested, subject to incident delays, and structurally unsuited for the freight volume a mature Tesla–Samsung Taylor relationship will generate. SH 130 is not. It was built as a toll bypass for exactly this class of traffic, and its industrial-corridor identity has compounded accordingly.

The SH 130 Smart Corridor is an advanced technology pilot by the Texas Department of Transportation. The initiative uses AI-powered cameras to monitor traffic, spot road hazards, and guide autonomous and connected vehicles. The future of SH 130 is not as a bypass. It is as a primary industrial spine, with development pressure intensifying as Tesla and Samsung scale their facilities into full operation.

The US 79 Corridor

US 79 runs due east from Round Rock through Hutto and Taylor and on to Rockdale. It is the industrial spine connecting the semiconductor anchor at Samsung Taylor to the data center cluster along its corridor. This industrial real estate is now among the most valuable in the United States, with commercial demand and residential development rising together as the corridor absorbs the workforce serving its industrial buildup.

The data center concentration along US 79 is not limited to Round Rock-Hutto. More than 70 data center projects are being tracked between Temple and San Antonio, with the densest concentration in the Williamson County section of the corridor.

Behind the corridor sits the broader Austin semiconductor ecosystem — more than 20 major semiconductor companies representing the entire stack in Austin alone. That ecosystem supplies the talent base, the supplier ring of photomask, design IP, equipment service, and specialty chemicals, and the community college pipeline that makes the industrial node viable. Austin is one of the largest semiconductor clusters in the US, with a regional semiconductor workforce exceeding 60,000 people. The semiconductor concentration is what makes Samsung Taylor a viable anchor; without the ecosystem, the facility would be a stranded asset from its talent and supplier base.

THE US 183 CORRIDOR

US 183 is the buildout's internal arterial – connecting the technology employment anchor to the residential geography housing its workforce, and to the aerospace and defense cluster extending north from the Austin metro core.

The corridor starts from Austin proper and runs northward through Cedar Park and Leander, ending in Liberty Hill. Its southern anchor is The Domain, the mixed-use development in North Austin whose density of Apple, Google, Meta, IBM, and Amazon employment makes it the technology industry's primary employment concentration outside the downtown core. The corridor carries those employees northward each evening to the master-planned communities whose growth rate is the buildout's most visible residential expression. Parcels that were ranchland one quarter are framed by the next.

Cedar Park sits at the corridor's midpoint and is emerging as an aerospace defense anchor. Firefly Aerospace produces Alpha small-lift rockets and Blue Ghost lunar landers from its Cedar Park facility. The future Central Texas Spaceport and Defense Innovation Campus will anchor regional commercial space access infrastructure. The Cedar Park aerospace concentration is small relative to Starbase or to the Bastrop satellite complex, but it is the metropolitan complement to the coastal launch infrastructure: design, manufacturing, and small-payload operations within commuting distance of the Austin technology workforce.

The corridor is simultaneously a residential commuter spine and a technology industry arterial – the same road, at different hours, serving the buildout's workforce in both directions. That dual identity distinguishes US 183 from the other corridors documented in this chapter, whose character is primarily freight and supply chain. US 183 is where the workforce and the industrial buildout share the same pavement.

The DFW and Houston Corridor Systems

The Dallas–Fort Worth metroplex operates as a multi-corridor industrial geography rather than a single spine, with three distinct concentrations now relevant to the AI-industrial buildout. North of Fort Worth, Hillwood's Alliance development, originally an aviation and intermodal logistics platform serving aerospace and consumer goods distribution, has converted into one of the largest enterprise and colo data center concentrations in Texas as institutional capital flowed into the corridor through the late 2020s. The DFW defense cluster splits across two sub-geographies: Lockheed Martin Aeronautics and Bell Textron in west and central Fort Worth, anchored on Air Force Plant 4 and Naval Air Station Fort Worth Joint Reserve Base, and Raytheon's Intelligence and Space presence along the US 75 corridor in Richardson and McKinney.

The US 75 corridor itself is the densest analog semiconductor manufacturing geography in North America. Texas Instruments operates multiple 300mm analog wafer fabs along the Dallas–Richardson-Sherman axis: DMOS6 in Dallas, and the connected RFAB1 and RFAB2 fabs in Richardson, together holding more than 630,000 square feet of cleanroom and a combined production target above 100 million analog chips per day. Both RFAB fabs are QML-certified for space-grade production, putting space-qualified analog silicon in continuous output on US 75 for the past fifteen years. Sixty miles further north on the same corridor, the Sherman campus extends this concentration through a four-fab buildout under TI's $30 billion program, with SM1 in initial production and SM2 ramping. Sherman is not a new beachhead; it is the next stage of an analog manufacturing cooridor that has been compounding along US 75 since 2009.

Houston's industrial buildout follows two primary corridors. The Energy Corridor along I-10 in west Houston is anchored by the energy industry's engineering workforce concentration, with BP

America, Shell, ConocoPhillips, and the broader supermajor and oilfield services presence supplying the technical talent base. The Ship Channel industrial belt threads east from downtown Houston through Pasadena, Baytown, and Mont Belvieu, the densest petrochemical and refining concentration in the United States, now overlaying hyperscale data center pursuit of grid and behind-the-meter generation alongside the chemicals industry's existing power infrastructure. Both Houston corridors carry AI-industrial overlay rather than the AI-industrial concentration found in the Austin corridor system, but they are part of the same Texas Triangle geography and host the institutional capital flows documented elsewhere.

Seven

THE WILLIAMSON COUNTY BUILDOUT

WILLIAMSON COUNTY IS WHERE I LIVE. It is also the most fully emerged instance of a pattern that Bastrop County, Hays County, Bell County, and a lengthening list of regional counties are replicating — each on a compressed timeline, each drawing on the same foundation.

Georgetown grew from 67,000 people in 2020 to more than 115,000 by 2026. That is a 71 percent increase in six years — a more than nine percent annual growth rate, sustained across a period that included a global pandemic, a catastrophic winter storm, and a regional housing market that became one of the most expensive in the country before partially correcting. No comparable American city above 50,000 people grew faster over the same period. And the growth is not Georgetown specifically. It is the entire corridor from Cedar Park and Round Rock in the south, through Georgetown and Liberty Hill in the center, to Taylor and Hutto in the east — absorbing population at a rate that has put the county's planning apparatus, school districts, road network, and utility infrastructure into a state of permanent catch-up.

The residential, commercial, and industrial buildouts are happening at the same time, in the same county, drawing on the same infrastructure, competing for the same water, generating the same traffic. The result is a collision between two scales of capital deployment that regional planning has rarely had to manage at this pace. Understanding Williamson County means holding both scales in view at once. The residential phenomenon and the commercial-industrial phenomenon are not separate stories. They are two expressions of the same underlying force.

THE ROUND ROCK

I got lost in Round Rock one morning, on my way to a coffee shop to work on this manuscript, and the wrong turn is how I found the rock. The city is named for it: a round limestone rock in the bed of Brushy Creek that served, on the Chisholm Trail, as the marked low-water crossing for wagons and cattle. The rock is still there. So are the wheel ruts — actual grooves worn into the limestone of the creek bed by nineteenth-century wagon wheels.

I stood and looked at it for a while. Then I drove on to the coffee shop up the road and opened my laptop, which is a Dell, in the city that Michael Dell transformed, to write a piece about what replaced the cattle economy. I have used Dell computers for over 20 years now. The ruts in the creek bed are about 150 years old. Both facts sat there together.

THE CHISHOLM TRAIL

On Austin Avenue in Georgetown, there is a marker noting that the Chisholm Trail passed through this spot. My family stops to have their photo taken under it when we are in the historic district — the limestone commercial buildings along the square, the county courthouse at the center, the nineteenth-century stone facades that make Georgetown's downtown one of the best-preserved historic districts in the state. The marker is easy to miss. The Thai restaurant a few doors down is more immediately compelling.

It should not be easy to miss, because the trail it marks is the reason for everything around it. Between 1867 and the mid-1880s, more than five million Texas longhorns were driven up the Chisholm Trail — 1,000 miles from South Texas to the Kansas railheads, passing directly through what is now the county seat. The drovers and their herds followed the natural spine of the Texas interior, moving north along the same geography that I-35 now follows, that the supply chain now traverses, that the data center network assembling around Hutto now depends on. The cattle economy and the semiconductor economy are using the same geography for the same reason: it is the path of least resistance through the Texas interior, as it was in 1867 and as it is now.

Georgetown itself was founded in 1848, when a pioneer named George Washington Glasscock donated 173 acres for a county seat. Southwestern University, established in 1840 and the oldest continuously operating university in Texas, brought students and institutional permanence to the community. The limestone buildings that line the historic square were quarried from the same Hill Country geology whose water rights and aquifer systems the data center cluster around Hutto now depends on for cooling. The stone and the water come from the same ground.

The pattern repeats with energy. In 2017, Georgetown became one of the first US cities to reach 100 percent renewable power through its municipal utility – long-term wind and solar contracts sourcing power from West Texas wind farms and the 154 MW Buckthorn solar project, in place before most of the data center buildout that would need that power had even been announced. Georgetown built its renewable power base without knowing it was building the power base for an AI-industrial cluster that would form next to it. The trail, the stone, the water, the power: each was in place before the thing that would need it arrived.

So are the longhorns. A couple of them graze next to a middle school a couple miles from where I live. It is the same animal that was domesticated by Spanish settlers, went feral on the open plains of Texas, and was re-domesticated as an icon – the only large domesticated animal in American history to complete that particular arc. And now they stand in a field beside a suburban middle school in Liberty Hill while some of the fastest-growing cities in the United States form around them.

What forms around them is visible most plainly in the pickup lane. At the end of the school day that lane is, without intending to be, a fair census of the corridor. A Ford F-150 idles at the front — still the best-selling vehicle in Texas, still the default. A Cybertruck idles three cars back, a Mustang Mach-E behind it. Children come out to e-bikes leaned against the school wall. And just beyond the fence, the longhorns graze. Old Texas and New Texas are not in conflict. They are simply both there, at 3:15 in the afternoon — the longhorn as indifferent to the Cybertruck as it is to the F-150, as the rattlesnakes sunning on the newly poured sidewalks are indifferent, as the coyotes roll-calling in the night are indifferent, as the granite in the Capitol rotunda is indifferent.

Continuing on with longhorns, I live just a few miles up the road from Sunrise Ranch in Liberty Hill, home of the Bevo family of longhorns that have been the UT mascot since 1916. Bevo XV makes appearances at almost all home games at UT as well as many away games. Bevo is eleven years old. Since his birth, Liberty Hill increased in population by over an order of magnitude from ~1,400 in 2015 to ~15,000 in 2026. I drive by on the way to town and sometimes see him and his half-brother Two-Spot grazing near the fenceline. You can't miss their signature burnt orange and white hides.

Some geographies accumulate history faster than others. The trail that brought five million longhorns north now routes semiconductor supply chains through the same corridor; the county seat now hosts a good chunk of the data center buildout coalescing across the Texas Hill Country. Williamson County did not adapt to the AI age. It was already, for reasons that had nothing to do with it, standing exactly where the AI age would need it to be.

The Residential Phenomenon

The master-planned community is the dominant residential form of the buildout here — and the word "subdivision" does not capture it. These are not discrete neighborhoods platted by a single developer and built out over a few years. At the scale being deployed, a master-planned community is more accurately a privately developed municipality: thousands of acres, multiple homebuilders working at once under a single developer's master plan, amenity infrastructure that rivals an incorporated city, HOA governance structures that function as quasi-governmental entities for daily decisions, and buildout timelines measured in decades rather than years.

I live in Santa Rita Ranch — yes, named in honor of that Italian nun. It is a 6,500-home community with a 10,000-plus-home buildout plan, and the top-selling master-planned community in Greater Austin for six consecutive years per real-estate tracking. It is currently a seven-village development at already half the population of Liberty Hill, its nominal host city. At that scale it behaves less like a subdivision than like a Census-Designated Place: a self-sustaining residential geography with its own rapidly expanding retail, school, and amenity infrastructure, built to serve a population that will exceed a number of incorporated Texas cities at full buildout. Its developer, Ed Horne, has won Developer of the Year four times. Multiple national homebuilders operate on the property simultaneously; Toll Brothers opened its first 55-plus active adult community in Texas here. From here, Samsung Taylor is 45 minutes northeast. Giga Texas is 45 minutes south. It sits equidistant from the buildout's two most significant anchor nodes.

Santa Rita also sits on one of the densest and most advanced fiber deployments in Central Texas. The anchor is Ubiquity's Edge Data Center, a modular facility built inside the master-planned community. The site operates as open-access infrastructure, meaning multiple ISPs lease capacity through the same physical plant, and linking directly to the major metro data center hubs in Austin, Houston, and Dallas. The result is residential service with symmetrical speeds reaching 8 Gbps and a carrier-class transport posture that handles regional enterprise traffic on the same fabric.

The point is not the speed itself but the topology. The fiber serving Santa Rita Ranch homes is not a residential last-mile bolted onto a metro backbone. It is regional carrier infrastructure with a residential overlay. The workforce of the Williamson County buildout lives on a network architected the way the industrial buildout is architected: enterprise transport at the front door.

A few miles north along SH 29, Cimarron Hills occupies a different position in the residential hierarchy. Guard-gated, built around a Jack Nicklaus golf course, with custom homes reaching $3 million, it is among the most exclusive master-planned communities in the county. Its presence a few miles from the family-oriented developments illustrates the full wealth spectrum the area is absorbing at once: entry-level new construction, move-up tech-worker family housing, active adult communities, and high-net-worth custom residential — all emerging in the same county, in the same decade.

Wolf Ranch in Georgetown occupies 1,100 acres and represents a $1.1 billion development — roughly 2,600 single-family homes plus multifamily plus commercial plus retail infrastructure. One hundred of the homes are 3D-printed by ICON, the Austin unicorn. Hillwood, a Perot company, acquired the land from the Wolf family in 2014, after 65 years of continuous family stewardship; Jay Wolf's family had ranched it since the late 1940s. The transition from multigenerational ranch to master-planned community, accomplished in a single transaction, is the prototype of a pattern repeating across the county — families who have held ranch land for generations encountering offer prices no ranching operation can match, and development pressure no zoning framework was designed to contain.

Sun City Texas in Georgetown is in a category of its own. At 5,400 acres, with an approved buildout near 10,550 homes and a population around 18,000 residents, it is the largest age-restricted master-planned community in Texas. As of 2021 it housed roughly 15,700 residents with an average net worth above one million dollars per household. That is more than $15 billion in concentrated household wealth, in a single age-restricted community, in a county that did not appear on most national economic maps two decades ago. The wealth concentration is not incidental. It is the residential expression of the same capital flows building Samsung Taylor and the Giga Texas expansion: engineers, executives, investors, and retirees who accumulated wealth in the technology economy and are deploying it into residential infrastructure in the same corridor where that technology economy is now building its physical presence.

Williamson County hosts 625 homeowners associations — a figure that reflects not just the volume of master-planned development but the quasi-governmental density private residential development produces when it operates at this scale.

The Domain

The Domain sits just south of the Williamson County line, in northern Travis County — but its functional identity is inseparable from the corridor it anchors from the south. Developed initially as a retail destination around Neiman Marcus and Macy's, it has become Austin's second downtown: a high-density mixed-use employment center whose concentration of technology firms, Class A office space, and live-work-play infrastructure serves the northern technology workforce as a primary node rather than a suburban amenity.

US 183, is the thread that connects The Domain north through Cedar Park to Liberty Hill and the broader residential corridor. It is the commuter spine for the workforce employed at Domain-area technology companies — the same road, with a different functional identity depending on the hour.

The Domain's role here is to provide the southern anchor for the residential buildout's employment geography. Residents of Santa Rita Ranch, Rancho Sienna, Bar W Ranch, Morningstar, and Wolf Ranch who work in technology — and a substantial fraction do — are within 30 minutes of a major employment node without entering the congested core of Austin proper. That accessibility is part of what makes the residential proposition coherent. The corridor offers proximity to technology employment at The Domain and in the Austin core, proximity to the industrial buildout at Samsung Taylor and along the US 79 spine, and residential land economics that remain more favorable than Travis County despite years of rapid appreciation. The four overlays — residential, commercial technology, and industrial — are close enough to share a labor market and far enough apart to avoid the land-use conflicts proximity would otherwise generate.

THE BOULEVARD

There is a road that makes the argument without requiring any argument at all. Ronald Reagan Boulevard runs roughly north-south, parallel to the toll lanes of US 183, and it is the route I drive several times a week. It is the leisurely one — non-toll, semi-pastoral in places, still carrying stretches of open Hill Country between the construction. It is also, for that reason, the best vantage point on the buildout I have found. You see more from the slower road.

Driven south to north, the road is a cross-section of the corridor's two halves — and it is, in the most literal sense, a single road. The southern end is Parmer Lane, in Travis County: the Domain-anchored employment cluster where the work is — the Samsung Parmer Lane plant, the Apple campus, the H-P and eBay complex. This is the southern gravity well, the employment around which the corridor's whole residential logic is oriented.

To the north, in Cedar Park, the same road takes the name Ronald Reagan Boulevard and runs into Williamson County where the workers live: all the master-planned communities, and before the road arcs east toward Sun City Texas, the city within a city. It carries one name for the tech companies and another for the houses — a single ribbon of pavement physically connecting where the corridor's human capital sleeps to where it works.

What makes the drive worth describing is not the endpoints but the line between them — where everything is happening at once. The communities at the northern well are still expanding, framing crews and foundation pours visible from the road. The commercial buildout is exploding around them: at the Leander–Georgetown–Liberty Hill nexus, intersection after intersection is turning over all four quadrants into residential, commercial, and retail that did not exist a year ago. And Ronald Reagan Boulevard itself is being widened — the road under expansion at the same time as the communities it serves. There is no neutral ground from which to watch the buildout, because the ground itself is under construction.

Ronald Reagan Boulevard runs unusually electric. On any given day, it resembles a curated, traveling electric auto show. Teslas are ubiquitous across the Austin metro, but the corridor carries the full current range of EV production: Cybertruck, Rivian, IONIQ, Mustang Mach-E, ID.4, Polestar, BMW i4, the occasional Lucid and Mercedes, all passing on a single drive. The corridor concentrates young technology workers with the income and the disposition to adopt early, and the adoption rate runs years ahead of the national average. Park at the Leander H-E-B off the boulevard on a Saturday morning and the lot tells the same story. A Porsche Taycan Turbo GT in the McDonald's drive-thru across the lot — chef's kiss.

Austin's EV adoption rate is among the highest of any US metro. It reflects two things at once: the income levels of the technology workforce, and the professional orientation of a population that works in the industries producing such vehicles. The labor market and the product market are the same people.

The Diamondback

On a morning in the summer of 2024, walking the new sidewalks near my house with my phone pointed at the lush crepe myrtles against the sky — the Hill Country sky is worth photographing, the impossibly white clouds against the bright blue — I nearly stepped on a Western diamondback rattlesnake. It was stretched out full length alongside the sidewalk, perhaps four feet from head to rattle, completely motionless, entirely unimpressed by my approach. I had been looking up. The snake had been looking at me.

A Western diamondback on a master-planned community sidewalk is not a remarkable event in Liberty Hill, Texas. The community is built on what was ranchland eighteen months before the first families moved in. The sidewalks are new. The snake's territory is older than the plat. It had not read the development agreement.

I gingerly stepped around it. It did not move. I took a few photos, then kept walking. I always look down now when out walking. Now I get why the construction workers all wear cowboy boots in the summer.

The buildout is rising on top of a landscape that was something else entirely until very recently — ranch land, cedar scrub, caliche roads, jackrabbit and rattlesnake habitat — and that landscape does not vacate on the developer's schedule. The cement pumper truck's snake-like arm was visible and noisy two streets away. The snake was silent just inches away. Both things were true at the same time, in the same geography, on the same morning.

The Industrial Buildout

The industrial side of this transformation is the extensive subject of later chapters. What belongs here is the recognition that all of these are happening inside the same county, in the same labor market and the same water and power infrastructure, simultaneous with the residential phenomenon documented.

Williamson County is where the AI-industrial nodes converge densely in the Texas Triangle. Samsung Taylor anchors the semiconductor cluster from the northeast — the largest foreign direct investment in Texas history. Its equipment and process materials supplier ring extends to Round Rock and beyond. The design and engineering talent draws from the Giga Texas well in Austin.

The data center cluster along US 79 fills in between Taylor and Round Rock with the institutional-capital deployments — Prologis Skybox PowerCampus the most visible, with secondary operators like Colovore, Iron Mountain and Project Comal assembling around Hutto. Round Rock holds the older DNA — Dell since the early 1990s, anchoring the I-35 / Parmer Lane tech-employer corridor that predated the AI buildout by decades.

What is distinctive about Williamson County is not any single industrial node but the convergence — a county that hosts semiconductor fabrication, hyperscaler data centers, an aerospace company that landed on the moon, a master-planned community with its own micro edge data center, and a mature technology-employer corridor inside the same market, all expanding simultaneously, all served by the same geography.

The Local Water Constraint

The county's most consequential binding constraint is not power. It is water.

Water consumption sorts cleanly across the three primary facility types. A standard semiconductor fab consumes between 5 and 10 million gallons per day; a large multi-fab campus draws up to 14 million at full build. Most of that is not cooling. It is ultrapure water — the industrial solvent that rinses residue from silicon wafers between each of hundreds of process steps, water so clean it will leach salts and minerals from your skin on contact. Producing 1,000 gallons of ultrapure water requires 1,400 to 1,600 gallons of municipal input: a 40 to 60 percent loss ratio before the water ever touches a wafer. An advanced-node fab's ultrapure demand is irreducible. The physics of semiconductor fabrication require the rinse cycle, and the rinse cycle requires water at a purity level no recycling system can maintain at the input volumes a production fab consumes.

AI data centers are the second most water-intensive facility type, consuming between one and 19 million gallons per day depending on cooling architecture — comparable to a city of 30,000 to 50,000 people, but reducible through closed-loop and air-cooled designs at an efficiency cost. Gigafactories are the least water-intensive of the three. Giga Texas in Travis County consumes about 1.5 million gallons per day — becoming one of Austin Water's largest customers in the process.

Samsung Taylor's water sourcing decision reads as water-constraint management rather than infrastructure preference. At designed multi-fab capacity, Samsung Taylor would require 5 to 14 million gallons per day at full buildout. The municipal systems serving Georgetown, Round Rock, and the adacent communities draw from Lake Georgetown, Lake Stillhouse Hollow, and Lake Belton through the Brazos River Authority (BRA) system. Drawing Samsung Taylor's full fab water requirement from that system — already under pressure from the residential buildout documented — would create the same municipal supply conflict Austin Water is managing with Giga Texas. The Carrizo-Wilcox Aquifer Samsung Taylor draws from is a confined aquifer; its isolation from surface contamination and its depth give it a consistent supply, and its draw does not affect the BRA municipal systems serving the residential communities.

Farther east, Terafab's production facility at Gibbons Creek in Grimes County will inherit the former coal plant's cooling water infrastructure — the 2,400-acre reservoir whose water rights were sized for a 470 MW power plant's cooling demand. A power plant's cooling water requirement and a semiconductor fab's water requirement differ in character. Power plants withdraw and return most of their water. Fabs consume a larger fraction, through evaporation and the ultrapure production loss ratio. But the reservoir's scale provides a water infrastructure foundation whose municipal-supply alternative would take years of permitting and construction to establish. The brownfield water rights inheritance is not a convenience. It is the enabling condition for locating a high-demand ultra-massive semiconductor fab in Grimes County without straining the municipal systems of the nearest population centers.

The Central Texas data centers represent a cooling water demand whose full expression, added to the semiconductor fab demand and the gigafactory demand assembling at the same time, is being imposed on aquifer systems, river authority allocations, and municipal utilities across geographies experiencing moderate to extreme drought in 14 Central Texas counties. The Hays County aquifer is at historic lows. The Edwards Aquifer Authority's permit system is operating under stress conditions whose demand is compounding faster than the authority's management framework anticipated. The Hill County moratorium, the San Marcos City Council rejection, the Hays County moratorium proposal — these are the political expressions of a physical constraint the buildout's water demand is making acute in real time.

The Permanent Catch-Up

The roads share a similar problem. Williamson County's 2019 $412 million Road Bond has funded multiple simultaneous expansions across the county. These are not sequential projects. They are parallel construction programs responding to traffic growth that is outrunning the bond-funded projects meant to absorb it. The road network is being rebuilt while it is being used, at volumes it was not designed for, by a population still growing.

What makes this analytically distinctive is not that it is growing fast. Sun Belt metros grow fast. What is distinctive is the simultaneity. The county is not being built up. It is being transformed, from a semi-rural exurban fringe of Austin into something that does not yet have a name.

I live inside this transformation. The 360-degree view from the community's higher ground shows foundations being poured, framing going up, rooftops appearing, roads being widened, new roads being assembled, gas stations popping up, tech parks being built — in every compass direction at once. It has looked this way for three years. It will look this way for several more. The formation event is not complete. It is underway, and Williamson County, Texas, is one of the places where it is most visibly concentrated.

PART II

THE INDUSTRIAL CONVERGENCE

Eight

THE SINGULARITY CENTER

GIGA TEXAS IS THE CENTER OF THE singularity. Not metaphorically — physically. The address: 1 Tesla Road, Austin, Texas 78725.

There is a useful way to misunderstand Giga Texas. It is what you see from the elevated section of SH 130, where the facility becomes visible to the south — a vast low structure extending across the Austin prairie southeast of the city — and call it a car factory. The scale confirms the reading: the main assembly building alone covers approximately 10 million square feet, making it one of the largest buildings by floor area in the world. Cars go in as materials and come out as finished vehicles. That is what a car factory does, and Giga Texas does it at volume — Model Y, Cybertruck, and Cybercab today, and Roadster 2 soon to follow. The auto analyst's model is not wrong about what the building contains. It is wrong about what the building is.

Giga Texas is a four-program campus. The programs share land, power infrastructure, labor market, supplier networks, and the logistics corridor that connects the campus to Samsung Taylor to the north and to the broader Triangle supply chain in every direction. Each program would be a significant industrial facility on its own. Together they amount to something with no precise precedent in American industrial history: a single campus simultaneously producing consumer and industrial hardware at scale, training artificial general intelligence, and soon manufacturing humanoid robots and fabricating semiconductors. The convergence of those four activities in one place is not accidental. It is the physical expression of the vertical integration strategy that makes Tesla analytically distinct from every other company it gets compared to.

The vehicle program is the campus's most visible activity and its largest near-term revenue contributor. Cybertruck production began in late 2023, making Giga Texas the sole global production site for a vehicle that has no direct competitor in its category. Model Y runs alongside it, the best-selling vehicle in the world again in 2025, electric notwithstanding. Cybercab, Tesla's autonomous-native ride-hailing vehicle, is beginning to scale. Roadster 2 will be added to the lineup in late 2026.

The Car Factory

The consumer vehicle program's manufacturing innovations are themselves an extension of the campus's vertical integration thesis. Front and rear gigacastings replace 70-plus stamped and welded body parts each, reducing assembly complexity and labor. The on-site 4680 cell line supplies the structural battery pack that integrates the cells into the load-bearing chassis architecture rather than treating the pack as a removable subsystem. Motor and power electronics manufacturing runs on-site rather than through external suppliers. Together these innovations make the consumer vehicle operation one of the most labor-efficient in the world.

Cybercab takes that manufacturing thesis further and changes the vehicle class while doing so. It is the first vehicle Tesla has designed with autonomy as the architectural starting point rather than as a software layer over a human-driver vehicle. The two-seat single-volume cabin is the geometry that follows from removing the steering wheel, pedals, dashboard, and instrument cluster that human-driver vehicles require. Unboxed manufacturing, the parallel-subassembly approach that converges at final integration rather than running through sequential body-paint-trim stations, is co-designed with this autonomous-native architecture. Removing human controls removes most of the sequencing dependencies that force conventional automotive assembly lines. The result is a vehicle whose per-unit production cost is materially lower than retail Model Y at comparable volume, because the design removes consumer-vehicle cost entirely. Production economics resemble a commercial fleet vehicle more than a passenger car.

Cybercab runs the same AI inference silicon as the consumer lineup but with a Robotaxi-tuned variant of Full Self-Driving software. The chip families documented in Appendix D run both: AI4 today and AI5 starting in 2027. The campus is the production site, the operational hub for Tesla's Robotaxi network, and the first customer simultaneously. Training data from vehicles operating in commercial service flows back to Cortex on the same campus that built them.

The Robotaxi service is already operating, but it is operating today as a bridge. Tesla launched its commercial ride-hailing service in Austin using Model Y vehicles equipped with Full Self-Driving rather than Cybercabs, because Model Y carries the regulatory categorization of a conventional vehicle and Cybercab does not. A vehicle with no steering wheel and no pedals must clear a different regulatory and permitting threshold in every jurisdiction it operates in. Cybercab production is just beginning to scale into that clearance process. The Model Y Robotaxi fleet is the substrate that lets Tesla deploy, accumulate operational data, and refine the autonomy stack while Cybercab works through the scale-out production and the certification path the autonomous-native architecture requires. The substitution is just beginning.

The revenue model is where Cybercab structurally diverges from the consumer vehicle program and aligns with Optimus. Cybercab generates revenue per mile through the Robotaxi network, not per unit through sale. The unit-economics question becomes what revenue per vehicle-year rather than margin per vehicle sold. Cybercab is therefore a hybrid program in the campus's capital-flows architecture: a vehicle by manufacturing classification, a service-revenue program by economic function.

The vehicle program's analytical role on the campus, however, is not the manufacturing innovations or the volume or even Cybercab's emerging scale-out. It is the cash flow. Tesla's consumer vehicle business generates approximately $70 billion dollars in annual revenue, the substantial majority of the company's top line. That revenue is what finances the Cortex AI training cluster's gigawatt-scale buildout, the initial Terafab semiconductor investment, the Optimus humanoid development at production scale, and the broader Texas-integrated supplier ecosystem. The other three campus programs are pre-revenue, captive, or operating at margins that do not yet justify their capital intensity. So is Cybercab, for now. The consumer vehicle program is what makes the rest structurally possible. The car factory is not just one of four programs sharing the campus. It is the substrate that helps pays for the convergence.

The AI Training Cluster

The second program is compute. Cortex is Tesla's AI training infrastructure: a hyperscaler-class compute cluster operating inside the Giga Texas campus, currently in the H100-dominated generation and transitioning to D3 silicon once the captive-fabrication program at Terafab reaches production scale. Cortex 1 is fully operational. Cortex 2, partially operational, will make the combined cluster among the largest physical-AI training installations in the world by compute capacity. The campus's power infrastructure — transmission connections, substation capacity, 30 MW in rooftop solar generation — was not sized for a car factory. It was sized for a car factory plus a hyperscale AI training cluster, a combination that produces a load profile more similar to a large data center campus than to any conventional automotive manufacturing facility.

What distinguishes Cortex structurally from the other hyperscale AI data center clusters being assembled across the AI-industrial buildout is the closed captive loop. Microsoft operates Azure's training infrastructure for its enterprise and cloud customers.

Google operates its training clusters for its Google products and Google Cloud customers. OpenAI trains models for OpenAI's API and product business. Meta operates its clusters for Llama and the platform's recommendation systems. Tesla is the only operator whose hyperscale AI training cluster trains AI for its own physical products — vehicles, Cybercab, Optimus — using its own physical products as the training data source. The Tesla fleet has accumulated more than 11 billion FSD miles, growing with every delivery. That real-world driving data captured worldwide flows back to Cortex on the same campus that produced the vehicles generating it. The training output is captive Tesla-vehicle FSD inference capability for their vehicles, for Semi, for Cybercab, for Optimus, and the model improvements that propagate into the next production cycle. The car factory and the AI training cluster are not adjacent businesses sharing a zip code. They are two phases of the same feedback loop, operating on the same physical campus, in a configuration no competitor with separate manufacturing and training geographies can replicate.

The relationship to Colossus, the massive training clusters in Memphis, is worth marking. Following the SpaceX–xAI merger, the Colossus cluster sits inside the SpaceXAI division of SpaceX's broader portfolio. Cortex remains Tesla-anchored, trained on Tesla fleet data, serving Tesla-side products including Optimus and Cybercab. Colossus remains SpaceX-anchored, trained on broader-internet-scale data, serving Grok, Anthropic, Google, and Reflection AI inference.

The Optimus Factory

The third program is humanoid robotics. As of mid-2026 the Optimus factory at Giga Texas is in full steel buildout. When it completes, it will be the first production-scale manufacturing facility for a general-purpose humanoid robot — operating inside the same campus that produces the vehicles the robot will eventually help build, maintained by the AI training infrastructure that improves the robot's capabilities in real time. Optimus Gen 3 production is scheduled to begin in late 2026 with an initial run of roughly 5 to 10 thousand units first in Fremont, scaling to approximately a quarter million in 2027 per Tesla's disclosed targets, with the longer-horizon trajectory aiming at millions of units annually by the end of the decade and eventually billions over the program's lifetime.

The strategic significance of building Optimus at Giga Texas is the test environment the factory itself provides. Boston Dynamics builds robots that work in Hyundai facilities in pilot-stage development today. Figure builds robots for the home to help speed iterative improvement. Tesla is building Optimus for large-scale industrial factories that need it — high-volume, 24/7 manufacturing environments whose demands for reliability, repeatability, and cost-effectiveness are the actual standard the robot must meet. Optimus units are not evaluated against controlled-environment benchmarks but against the broad operating requirements of a production line whose downtime is measured in vehicle output. The Giga Texas factory is the production site, the test bed, and the first customer simultaneously. The robot that can work reliably in this environment is a robot that can work in any comparable manufacturing environment globally. The training data generated by Optimus units on the production floor is training data with immediate economic value, flowing back to Cortex on the same campus that produced both the robot, the cars, and the environment it operates in.

Optimus runs on the same silicon roadmap that runs Tesla's vehicles. The chip families documented elsewhere in this book — AI5 in initial Optimus Gen 3 deployment, AI6 designed specifically for the power envelope of a continuously-operated Gen 4 Optimus at production scale — are co-designed with Tesla's vehicle program and fabricated through the same captive-and-merchant supply chain the vehicles draw from. The motors, actuators, power electronics, and structural components share the supplier ecosystem the vehicle program anchors. The result is an automotive-supply-chain cost structure for a product class that competitors building outside the automotive ecosystem must source at materially higher per-unit cost. Tesla's Optimus cost target — on the order of $30,000 per unit at production scale, falling further as volume accumulates — is achievable on the Giga Texas manufacturing base in a way it is not yet for Boston Dynamics, Figure, or the broader humanoid-robotics field.

The economics of Optimus at scale are not yet visible in any Tesla financial disclosure, because the program is pre-revenue. The rough production economics, however, suggest a revenue stream that at full scale exceeds Tesla's current automotive revenue by a substantial margin. The global humanoid robot market is projected conservatively to be approximately $40 billion dollars by 2035 — a figure that almost certainly underestimates the addressable market for a manufacturer with Tesla's cost structure, manufacturing base, and AI training infrastructure. If Optimus reaches the production trajectory Tesla has stated, the humanoid program becomes the campus's largest revenue stream within a decade, financing the next captive-fabrication moves the way the vehicle program currently finances Cortex, Terafab Research fab, and Optimus itself.

The Terafab Research Fab

The campus's fourth program is a semiconductor fab. Terafab is the Tesla-SpaceX coordinated program to fabricate and rapidly iterate on the custom silicon the other three programs — the vehicles, the AI training stack, the humanoid robots — and SpaceX's orbital compute data centers will increasingly depend on. As of early 2026 it exists, like the Optimus factory, in site preparation rather than operational. But its presence on the campus is the detail that completes the four-program thesis: Giga Texas is not a manufacturer that will buy its chips on the merchant market. It is a manufacturer building toward fabricating them.

Terafab is not a single facility. It is a two-node program with distinct operators and distinct mandates, distributed across Texas. The first node is the Tesla-operated research and pilot facility within the Giga Texas campus footprint — the environment where chip designs and fabrication processes are designed, iterated on, and developed before production volumes are committed. The second is a SpaceX-operated production facility in Grimes County, on the Gibbons Creek brownfield coal-plant site whose reservoir water rights are sized for production-scale fabrication, with D3 — the radiation-tolerant chip for orbital compute — as its anchor program. Terafab at Giga Texas is under construction and Terafab Gibbons Creek is still in permitting as of mid-2026.

What Terafab fabricates is the AI5, AI6, and D3 chip families: the silicon roadmap that runs Tesla's vehicles, Cybercab and Optimus, and SpaceX's orbital systems respectively. Why some fabrication is captive — why the chips could not simply be contracted to a merchant foundry — is the argument of a later chapter. Here it is enough to register the fact: the fourth program closes the campus's vertical integration, putting chip fabrication on the same ground as the products the chips run in.

The Supplier Ring

A four-program campus aggregates suppliers at a density a single-program facility doesn't. Vehicle production requires a tier-one automotive supply chain — body stampings, powertrains, battery modules, glass, seating, electronics — that has been building around Giga Texas since the campus opened. The AI training infrastructure requires server hardware, networking equipment, cooling systems, and power management electronics: a second, entirely different supply chain. The Optimus factory will require joint actuators, sensors, precision mechanical components, and the specialized materials humanoid robot manufacturing demands — a third supply chain, with minimal overlap with the first two. Terafab requires semiconductor equipment, specialty chemicals, ultrapure water systems, and the metrology and process control infrastructure fab operations depend on — a fourth, one that overlaps with Samsung Taylor's requirements and therefore benefits from the supplier concentration Samsung Taylor is attracting.

The overlap between these four supply chains is partial and specific rather than comprehensive. But where overlap exists — batteries, motors, power electronics, thermal management systems, precision manufacturing equipment, advanced materials, engineering services — the four-program campus creates demand density that justifies supplier investment in local presence. A supplier that serves Giga Texas's vehicle production program, and can also serve the Optimus factory's mechanical component requirements and Terafab's equipment maintenance needs, has a business case for establishing Texas operations that no single program would provide. The supplier ring that forms around a four-program campus is not four times the supplier ring of a single program. It is larger than that — because the overlapping demands justify supplier commitments that isolated demands would not.

Nine

THE CAPTIVE FAB

IN THE SUMMER OF 2021, SAMSUNG announced it would build a semiconductor fabrication facility in Taylor, Texas. The announcement was covered as a foreign direct investment story — South Korea's largest company committing capital to a new American manufacturing facility, the largest single foreign direct investment in Texas history, a win for the industrial recruitment diplomacy that would feed into the CHIPS and Science Act of 2022. That framing was accurate as far as it went. It did not go far enough.

Samsung's decision to site in Taylor was not a simple capital allocation, and it was not driven by Tesla. Samsung chose Taylor for reasons that had nothing to do with any single anchor customer: robust local government incentives, available infrastructure, and proximity to Samsung's existing Austin semiconductor operation on Parmer Lane — a facility Samsung has operated since 1997, when it selected Austin for its first US manufacturing plant and began producing 64MB DRAM chips there in 1998. Samsung's Austin presence is not recent. It is nearly three decades old. The institutional familiarity with the Central Texas operating environment — labor market, utility relationships, regulatory framework, supplier ecosystem — accumulated across those decades informed the Taylor siting decision as directly as any financial incentive package.

The Tesla relationship came later, and on a different timeline than the facility's conception. In July 2025, four years after the Taylor announcement, Tesla and Samsung Semiconductor announced a major long-term agreement to produce Tesla's AI5 and AI6 chips at Taylor Fab 1. AI5 will serve the current Tesla vehicle lineup, Cybercab, and Optimus. AI6 will serve next-gen Cybercab, Optimus, and Cortex. The deal is valued at $16.5 billion for an initial term and is expected to run through 2033, with production targeted to begin in the second half of 2027. It was announced against a backdrop of Samsung having proceeded with the Taylor buildout while still seeking high-volume clients whose commitments would match the factory's advanced-node capabilities to sustained demand. Tesla's entry as an anchor customer resolved a customer problem that had generated coverage questioning the facility's near-term utilization.

The sequence matters analytically. Samsung built the fab for strategic reasons of its own: proximity to existing Austin operations, the local incentive environment, the pivotal Oncor power infrastructure upgrade after Winter Storm Uri, and confidence in the Texas Triangle's long-term AI-industrial trajectory. Tesla then entered a long-term supply agreement that transforms Samsung Taylor from a fab seeking anchor customers into a fab with a $16.5 billion starting domestic customer committed through 2033. The result is a captive relationship — but one that emerged from two independent strategic decisions converging, rather than from a coordinated plan.

Samsung Taylor Campus

Fab 1's capital history is the campus's most instructive single thread. Samsung's original 2021 announcement committed approximately $16.5 billion to a 4nm fab. Midway through construction, Samsung pivoted the process target from 4nm to first-generation 2nm — a response to TSMC's announcement that the Arizona Fab 21 expansion would run a 2nm process. The pivot ballooned the Fab 1 capital commitment to a range commonly cited at $37 billion to $44 billion.

Fab 1's product mandate is the Tesla AI5 and AI6 chip families. The fab's primary process focus is leading-edge high-performance computing, AI accelerators, and automotive-grade silicon — the product mix Tesla's captive contract requires. Fab 1 is the only fab in the planned ten-fab Taylor cluster that will operate as a Tesla-only facility under the current agreement. The remaining nine, when and as they come online, will serve other anchor customers and the broader merchant market.

Fab 2 is the merchant foundry node. Its process target is second-generation 2nm — SF2P in Samsung's nomenclature — a refinement of Fab 1's first-generation process. Its customer profile is global tech clients whose chip designs require leading-edge capability without the captive integration Tesla's relationship with Fab 1 provides; Google, AMD, and ByteDance have all appeared in coverage of Fab 2's commercial discussions. Both fabs are approximately 2.7 million square feet at full buildout. The two-fab starting architecture, and the eight-fab expansion option behind it, is itself analytically significant. Samsung is not building a captive Tesla fab. It is building a leading-edge campus whose Fab 1 happens to be operating as a Tesla-captive node, while Fab 2 and the subsequent buildout phases are positioned to capture the broader merchant demand Samsung's overall semiconductor strategy requires.

The campus's Fab 1 power draw at full operational capacity already places it among the largest single industrial electricity consumers in the ERCOT grid system, and the full ten-fab buildout would multiply that demand by an order of magnitude. The Oncor Electric power guarantee that accompanied the Taylor siting decision was sized for the original 4-nanometer Fab 1 specification.

The city of Hutto lies just 10 miles west. It is undergoing a transformation, not only from the Samsung Taylor semiconductor orbit, but also from the data center buildout coming from Round Rock along the US 79 corridor. The drive out to Hutto is still largely pastoral — open country, then the Co-Op, a number of big-box and other retail anchors, a few good places to eat.

I pulled into the parking lot of Casa Jalisco for lunch one afternoon and stopped, because across the pavement at the Co-Op stood something that had not been there very long: a gleaming four-story building, cobalt blue glass, with ASML in tall white letters across the top. The most strategically constrained piece of equipment in the global semiconductor industry, the company whose extreme ultraviolet lithography machines every leading-edge fab on Earth depends on, had put up a building across from a taqueria in Hutto, Texas.

I went in and had lunch. The building was still there when I came out. That is the pace of it — you go for enchiladas and the supply chain of the AI age has moved in across the parking lot.

The CHIPS Act Overlay

The federal government's $4.7 billion CHIPS Act allocation to Samsung Taylor is the largest single CHIPS Act award to a foreign-operated facility. Its scale reflects a strategic calculation that goes beyond supporting one company's expansion: the US cannot build a domestic leading-edge semiconductor supply chain without foreign-operated fabs, because no US-headquartered company currently has the process capability at 4-nanometer and below that Samsung and TSMC possess. The CHIPS Act awards to foreign operators are not concessions to globalization. They are acknowledgments that domestic semiconductor sovereignty, at leading-edge process nodes, requires foreign operational expertise deployed on US soil under US regulatory frameworks.

The conditions attached to the award — restrictions on expanding leading-edge capacity in China, requirements for domestic workforce development, commitments to technology transfer that build US process engineering capability over time — are the federal government's mechanism for converting a foreign direct investment into a national industrial asset. Samsung operates the fab. The federal conditions shape what Samsung can do with the capability the fab represents.

The Texas Semiconductor Innovation Fund's $250 million allocation to Samsung Taylor compounds the federal investment with state capital. The TSIF allocations to surrounding suppliers — Tekscend Photomask, SoulBrain, Linde, ENF, Nissin Ion — are the state's mechanism for building the cluster around the anchor, rather than leaving cluster formation to market dynamics alone.

The Dual-Source Architecture

Tesla's AI5 and AI6 chips are not exclusively sourced from Samsung Taylor Fab 1. The dual-source architecture — Fab 1 providing the domestic captive fabrication node, TSMC's Arizona facility providing a second supply node — is a supply chain risk management decision whose logic is straightforward. A chip family as central to Tesla's AI training and inference stack cannot be single-sourced from any facility, however capable and however strategically aligned. Fab outages, process yield excursions, and natural disasters are all risks that dual-sourcing mitigates.

The architecture reflects a broader strategic principle in Tesla's semiconductor approach: no single external dependency for any critical component. Samsung Taylor Fab 1 provides the domestic node — supply chain independence from Taiwan, CHIPS Act compliance, geographic proximity to Giga Texas, and the process customization access a long-term committed relationship enables. TSMC Arizona provides the process technology depth and yield maturity TSMC's manufacturing excellence has accumulated over decades. The two sources are complementary rather than redundant. Each provides something the other does not, and together they provide supply security neither could provide alone. D3 sits outside this dual-source architecture entirely; its radiation-tolerance requirements and orbital deployment profile make Terafab the only viable production path.

The Foreign Operator Question

What does it mean strategically when the most advanced domestic semiconductor manufacturing capability is operated by a company whose headquarters, shareholders, and ultimate strategic interests are in South Korea?

Samsung's nearly three decades of Austin operational presence creates genuine institutional alignment with the Central Texas operating environment that a purely transactional foreign operator would not have. The CHIPS Act conditions constrain Samsung's ability to use its Taylor investment as a platform for expanding Chinese market access. The Tesla long-term supply agreement creates an alignment of interests between Samsung Taylor's operational success and a domestic operator's strategic requirements through 2033.

What it does not create is US ownership of the process technology, or US control of the facility's strategic direction if Samsung's corporate interests diverge from US national interests. The captive fab is a bridge. It provides the domestic leading-edge fabrication capability US operators need now, while the domestic process engineering base that would make the bridge unnecessary is being built.

The Captivity Argument

Everything described so far is a commercial and strategic fact: a purchase obligation, a fab whose Fab 1 mandate is a single customer's chip families, a dual-source architecture pairing that domestic node with TSMC Arizona, and a federal-and-state capital overlay that converts a foreign-operated facility into a national industrial asset. At that level, the Samsung Taylor relationship is captive in the ordinary sense — an anchor customer with a long-term committed volume large enough to define the facility's near-term identity.

But there is a second, deeper sense in which a fab relationship can be captive, and it is not visible in the contract value or the customer. It is visible in whether the customer participates in developing the fabrication process itself — rather than designing a chip within process rules the foundry has already fixed. That process-level question — what becomes possible when chip architecture and fabrication process are co-designed rather than transacted across a fixed interface — is the strategic core of Tesla's entire semiconductor program. It is the analytical thread that ties Samsung Taylor to Terafab, to an Intel partnership, and to the three new chip families.

Ten

THE CO-DESIGN THESIS

START WITH THE MODEL THE ARGUMENT departs from. The semiconductor industry has a standard way of relating chip design to chip fabrication. A chip designer — Apple, Qualcomm, Nvidia, any of the hundreds of fabless design houses that make up the modern semiconductor ecosystem — develops a chip architecture and takes it to a foundry. The foundry provides a process design kit (PDK): a defined set of design rules, device models, electrical parameters, and physical constraints describing what its manufacturing process can and cannot do. The designer works within those constraints. The foundry fabricates the result. The process belongs to the foundry. The chip design belongs to the designer. The relationship is a transaction between two independent parties, each optimizing for its own interests within a defined interface.

This model has produced extraordinary results. TSMC's process technology roadmap, executed in service of hundreds of fabless customers at once, has driven semiconductor capability forward at a pace no vertically integrated device manufacturer (IDM) has matched in two decades. The fabless-foundry model works. It is the dominant model for a reason.

It also has a ceiling. The ceiling becomes visible when a chip designer's requirements cannot be satisfied within any foundry's standard process design kit. The application demands a specific combination of performance, power, and physical characteristics — and meeting it requires process parameters the foundry has not optimized for its general customer base. The foundry will not optimize them for a single customer whose volume, however significant, does not justify the process engineering investment. At that ceiling, the fabless-foundry transaction produces a chip as good as the foundry's standard process allows. It does not produce the chip the application actually requires.

Tesla is building above that ceiling. The chip-and-fab co-design thesis is the framework for understanding why — and what it means for the competitive position of the products that co-designed chips make possible.

The Four Chip Families

Tesla's silicon roadmap, as of mid-2026, comprises four chip families — distinct application profiles, distinct process requirements, distinct sourcing architectures. Each sits at a different point on the spectrum from standard foundry relationship to fully captive co-designed fabrication. Taken in sequence, they are the clearest way to see the thesis.

AI4 is the legacy baseline — the chip family running every production Tesla on the road today, fabricated in Korea on a standard Samsung 7nm process. AI4 is the merchant-foundry endpoint of Tesla's silicon strategy: a chip designed to the limits of what an external foundry's standard process design kit can provide, and produced at a foreign fab where the only Tesla-specific accommodation is contract volume. Musk has stated explicitly that AI4 is sufficient for unsupervised Full Self-Driving. That sufficiency is what finances everything that comes after.

AI5 is the next chip in the roadmap — the silicon that will be running in all Tesla's vehicles in 2027, handling the Full Self-Driving compute stack, as well as Optimus Gen 3. However, AI5 will be specifically insufficient for the next generation of applications Tesla is building toward. Its power envelope, thermal profile, and compute density will be inadequate for the constrained operating environments of future Cybercab and Optimus. AI5 is the foundation — the chip whose revenue finances the next two generations. The roadmap runs AI5 to AI6 to D3 in ascending capability, ascending process sophistication, and ascending dependence on captive fabrication rather than merchant foundry sourcing.

AI6 is the low-power chip designed for next-generation Cybercab and Optimus — the inference silicon that has to deliver sufficient autonomy capability within the required power budget. The power envelope is not a preference. It is a hard constraint. A Cybercab Gen 2 cannot carry the thermal management infrastructure a high-performance inference chip requires, and an Optimus Gen 4 unit's battery cannot sustain the draw AI5-class inference would demand. AI6 needs a process variant optimized for its specific power-performance tradeoff — a variant no external foundry's standard PDK provides, because the fabless-foundry model optimizes process parameters for the broadest possible customer base rather than for any single customer's constrained application. AI6 is Terafab's primary rationale at the research and pilot level: the chip whose requirements drove the decision to develop native captive fabrication rather than accept the constraints of merchant foundry process design kits.

D3 is the default radiation-tolerant chip for SpaceX orbital compute — the Starmind AI1 orbital satellites and their successors operating in low Earth orbit. Its process requirements are not a matter of performance optimization. They are a matter of physical survival in an environment that destroys standard semiconductor devices on timescales measured in months. The process modifications D3 requires demand changes to the foundry's baseline that affect yield, design rules, and process engineering investment.

That is the crux. No merchant foundry will develop a radiation-hardened process variant for SpaceX. The volume does not justify it, the customer base for radiation-tolerant LEO compute silicon is narrow, and the process engineering investment is substantial. The Terafab production facility being permitted now at Gibbons Creek, anchored on Intel's 14A node, exists because D3 requires a fabrication environment where the process can be modified to meet the chip's requirements rather than a chip compromised to fit the foundry's standard process. AI5 and AI6 will also be produced at Terafab as additional capacity, but their existence at Samsung Taylor and TSMC Arizona means they are not what makes Terafab necessary. D3 is what makes Terafab necessary. That is the chip-and-fab co-design thesis at its most unambiguous: the application requirement defines the process, not the reverse.

What gives these four chip families their weight is not the technical sophistication of any single design but the scale of their cumulative deployment. Tens of millions of vehicles. Potentially billions of Optimus units and successor humanoid form factors as production scales toward stated targets. Millions of chips in terrestrial AI training and inference data centers across Cortex, Colossus, and other clusters. Millions more in Starmind orbital compute satellites as SpaceX deploys its compute constellation. And an indeterminate population of edge inference devices and infrastructure gateways such as Tesla Megapod that will inherit the silicon as volume economics mature.

This is the largest concentration of custom inference compute ever attempted under a single corporate program in history. The full specifications for each chip family's process nodes, foundries, tape-out dates, performance metrics, and the engineering basis for radiation hardening are documented in Appendix D.

The Intel Partnership

Tesla is a new entrant in semiconductor fabrication, and new entrants face a specific problem capital alone cannot solve. The semiconductor equipment supply chain is constrained in ways purchasing power does not simply unlock.

ASML's extreme ultraviolet lithography systems are the most visible binding constraint in any leading-edge fab buildout. EUV tools are produced at roughly 50 to 60 units per year globally, each priced above $200 million, allocated years in advance to customers with established relationships and committed volume. But EUV is not the only long-lead, high-complexity equipment category where queue position and vendor relationship determine a new entrant's timeline. The full wafer fabrication equipment stack at leading-edge nodes includes deposition systems, etch tools, ion implantation equipment, chemical mechanical planarization systems, metrology and inspection platforms, and advanced packaging infrastructure. Each category has its own tier-one vendors, its own allocation dynamics, and its own preference for customers with established process development relationships over new entrants whose requirements and volume trajectories are unproven.

This is the constraint the Intel partnership can address — across the full equipment stack, not just at the EUV layer. Intel's established relationships with leading semiconductor equipment vendors, built across decades of leading-edge captive fabrication, would give Terafab access to equipment allocation and vendor engagement a standalone new entrant could not negotiate on any acceptable timeline. The queue position is not transferred; Intel retains its own allocation. What the partnership provides is the credibility and relationship infrastructure that lets Terafab enter the queue as a serious operator rather than an unknown entity. In a supply chain where relationships and track record determine access as consequentially as capital does, Intel's sponsorship is a material accelerant.

Intel also brings build-to-process expertise accumulated across decades of captive fabrication — the knowledge of how to take a process specification from development to high-yield production that no fabless chip designer has and no new fab entrant can acquire quickly. The partnership is not primarily a technology transfer arrangement. It is a supply chain, queue position, and process-execution arrangement that compresses Terafab's path from site preparation to operational fab by years relative to what Tesla or SpaceX could achieve independently.

The Trusted Foundry Program

The Department of Defense has a semiconductor sourcing problem. The Trusted Foundry Program – the DoD's framework for certifying semiconductor fabrication facilities as approved sources for defense and intelligence community silicon – was designed to ensure that chips used in weapons systems, secure communications infrastructure, and intelligence platforms are fabricated in facilities where supply chain integrity, process security, and personnel access can be verified and controlled. The program works within its current scope. Its limitation is that the leading-edge process capability the DoD increasingly requires for its most demanding applications – AI inference at the tactical edge, radiation-tolerant orbital compute, advanced signal processing for next-generation weapons systems – is concentrated in facilities that are either foreign-operated or outside the Trusted Foundry framework entirely.

Samsung Taylor is South Korean-operated. TSMC Arizona, however domestically sited, is Taiwanese-operated under a corporate structure whose ultimate strategic direction is determined in Hsinchu. The US defense and intelligence community needs leading-edge process capability for Golden Dome missile defense compute, for SpaceX Starshield communications, and for the AI-enabled weapons and communications systems that define the next generation of US military capability. That capability is currently accessible only through foreign-operated facilities.

Terafab's integrated footprint addresses this in a way no existing facility in the US semiconductor landscape will. The design-to-fabrication-to-advanced-packaging-to-certification pipeline – chip architecture development, process co-design, wafer fabrication, packaging, and security certification – under a single secured building footprint, operated by a domestically controlled entity, eliminates the supply chain handoffs where foreign operator exposure currently exists. A chip designed at the Giga Texas Terafab research facility, fabricated at the Gibbons Creek production facility, packaged and tested in a co-located secure environment, and certified through a DoD-recognized security framework without ever leaving a controlled domestic custody chain is a categorically different national security asset than the same chip fabricated at Samsung Taylor or TSMC Arizona.

The strategic significance is specific to the most demanding applications. D3's radiation-tolerant process requirements — designed for SpaceX orbital compute serving both commercial and national security satellite missions — require exactly the integrated secure fabrication environment a Trusted Foundry certification would provide.

The certification path is not trivial. It requires facility security infrastructure, personnel vetting frameworks, process documentation standards, and audit regimes that add cost and operational complexity to a fabrication environment already operating at the frontier of process difficulty. The timeline for achieving that certification, running in parallel with the fab's process development and equipment installation, is measured in years rather than months. But the strategic value of the endpoint justifies the investment. A US-operated, DoD-certified, leading-edge fabrication facility capable of producing radiation-tolerant AI silicon for orbital compute and hypersonic defense applications, with design-to-packaging integration under one secured roof, is an asset the United States currently does not have — and the combined Tesla-SpaceX-Intel-DoD ecosystem is building toward it.

That asset, when it exists, will be the most strategically significant semiconductor facility in the United States.

The Missile Defense Architecture

A hypersonic glide vehicle traveling at Mach 20 covers roughly four miles every second. The discrimination window — the interval between identifying a target with confidence and generating an intercept solution — is measured in seconds, not minutes. The inference workload that discrimination and fire control represent at that speed needs chips whose latency, throughput, and reliability under electromagnetic and radiation stress meet specifications the commercial market has never been asked to satisfy. Foreign-operated fabs cannot be trusted to produce them under the security frameworks a weapons system supply chain requires.

Golden Dome's compute requirement is therefore a direct driver of the Trusted Foundry program, whose physical expression will be Terafab. From outside, the line between a missile defense architecture and a semiconductor fab in Grimes County is not obvious. From inside — with the chip roadmap, the radiation certification framework, the Texas A&M Cyclotron Institute, and the defense procurement relationships are all in view.

The Two-Fab Architecture

The Trusted Foundry framework just described will not be a single facility. It is a two-node architecture distributed across the Texas Triangle — a research and pilot node at Giga Texas, and a volume production node at Gibbons Creek. The two are operationally integrated but geographically distinct, connected by the same corridor infrastructure, the same workforce pipeline, and the same program management structure that makes the distributed architecture function as a single Trusted Foundry rather than two independent facilities.

The first node is Terafab Giga Texas — the Tesla-operated research and pilot fabrication facility within the Giga Texas campus in Austin. This is where the chip-and-fab co-design thesis is physically executed: chip architectures developed jointly with the process parameters, rather than against a fixed PDK from a merchant foundry. The Intel partnership's 18A and Intel 14A process access operates here. Pilot production runs validate process modifications before they transfer to volume production. The Cyclotron Institute's radiation qualification program at Texas A&M validates D3's performance against the LEO and SSO radiation environment. The research node is where the program's intellectual content — the process co-design, the radiation tolerance specifications, the chip architecture — originates.

The second node is Terafab Gibbons Creek — the SpaceX-operated full-scale production facility in Grimes County, whose capital commitment places it among the largest fab investments in US history. This is the volume node at commercial scale, from front-end wafer fabrication through back-end assembly, under one roof. The security certification framework — the verified supply chain integrity, the cleared workforce, the physical security architecture — is centered here.

The geographic distribution between Travis County and Grimes County is not a weakness. It is a security feature. Separating research from volume production across two physically distinct sites reduces the single-facility vulnerability a concentrated program would present to adversarial disruption. The corridor infrastructure connecting the two nodes — the broader Texas Triangle workforce market, the supplier ring assembling around both anchors — operates as the physical base that lets the distributed architecture function as a single integrated program.

THE COMPETITIVE MOAT

The chip-and-fab co-design thesis is not primarily about cost reduction, though cost reduction is a consequence. It is about capability access — the ability to build chips whose characteristics are not available from any merchant foundry at any price, because the process parameters those characteristics require do not exist in any standard PDK.

Apple demonstrated the strategic value of this capability with A-series silicon. When Apple internalized chip design with the A4 in 2010 and developed its co-design relationship with TSMC, it was not primarily seeking to reduce the margin it paid Qualcomm — though that was a consequence. It was seeking chips whose performance and power characteristics would differentiate iPhone in ways a chip designed within Qualcomm's standard architecture could not match. The A-series delivered that differentiation consistently across multiple generations, and the competitive gap between iPhone's silicon and Android flagship silicon — measured in CPU performance, GPU performance, neural engine throughput, image signal processing — widened for years before competitors began closing it.

The moat chip-and-fab co-design creates is not a cost moat. It is a capability moat. Competitors building autonomy systems, robotics platforms, and orbital compute infrastructure from merchant foundry silicon are working within process constraints Tesla's captive fabrication is not subject to. The products that emerge from unconstrained chip-and-process co-design are not incrementally better than products built within merchant foundry constraints. They are qualitatively different — capable of things the constrained products cannot do, at power levels the constrained products cannot achieve, in environments the constrained products cannot survive.

The Silicon Chain

The broader observation is that the silicon supply chain for the Texas AI-industrial buildout is not a single layer but four, each with its own process technology, customer base, and production geography.

Leading-edge logic — the digital compute silicon at the center of every AI training cluster, every autonomous vehicle, every humanoid robot, every orbital satellite — is produced at Samsung Taylor and later on at the Terafab production facility at Gibbons Creek. The integration is captive on the Terafab side, with Tesla and SpaceX as silicon design partners holding operating control through Intel 14A licensing, and merchant on the Samsung Taylor side, with Tesla as the largest customer at the node. Either way, all production sits in Texas.

Analog and mixed-signal silicon — the power management, signal chain, data conversion, interface, and clocks-and-timing layer that surrounds every digital compute chip — is produced at the Texas Instruments Richardson and Sherman 300mm fab complexes, the world's largest concentration of analog and embedded processing semiconductor manufacturing. The supply chain that produces this silicon for space and defense applications runs through Texas at Texas Instruments as well. The integration is merchant rather than captive, but the production geography is the same: US-soil fabrication of the chips the Texas operators consume.

High-bandwidth memory — the stacked DRAM that feeds every AI accelerator training or inferencing at scale, and without which the compute silicon is bandwidth-starved and effectively idle — is a three-producer market: SK Hynix, Samsung, and Micron. Each HBM assembly is a vertical stack of DRAM dies bonded to a logic base die through hundreds of thousands of through-silicon vias, then integrated with the compute silicon through advanced 2.5D packaging — TSMC's CoWoS predominantly, with Intel's EMIB-T plus Foveros Direct positioned as the emerging alternative in the Intel Foundry ecosystem. The production geography for the DRAM stacks themselves sits offshore of Texas: SK Hynix and Samsung in South Korea, Micron primarily in Taiwan with advanced packaging still tied to partner facilities in the TSMC ecosystem. Micron's announced US buildout adds leading-edge DRAM fabs in Boise, Idaho and an HBM assembly facility at Manassas, Virginia — the first end-to-end domestic HBM capacity — but the tonnage lands in Idaho and Virginia, not Texas, and first US HBM output is not expected before the back end of the decade.

The constraint on HBM is compound yield across a stacked assembly. Each additional die layer in the stack multiplies yield loss and thermal load; 12-Hi HBM3E and HBM4 are the current commercial ceiling, with 16-Hi HBM4E under development for late-2026 supply and 20-Hi to 24-Hi stacks not projected until the HBM7 and HBM8 generations of the mid-2030s. Advanced packaging capacity is the compounding bottleneck. TSMC's CoWoS lines have been sold out for years and are the rate-limiting step across the entire AI accelerator industry. HBM allocation across all three producers is sold out through 2026, with capacity contracts extending through 2027, and the memory suppliers are prioritizing AI datacenter orders to the point of cutting consumer memory allocations by 30 to 40 percent. Intel's EMIB-T plus Foveros Direct is the only credible non-TSMC path at production scale, and it is the packaging technology natively associated with the Intel 14A process node that Terafab is licensing.

That last fact opens the analytically important question. Intel's own advanced packaging roadmap has publicly demonstrated a reference configuration of 16 compute tiles paired with 24 HBM sites in a single package, with 18A-PT as the base-tile layer and 14A/14A-E as the compute-tile layer above it. The base die, the compute silicon, and the packaging integration all fall inside the process technology envelope Terafab is licensing. What Terafab cannot produce, absent a DRAM fab it has not announced, is the stacked DRAM itself. Terafab can fabricate the HBM base logic die — an emerging foundry opportunity as memory vendors migrate their HBM4 and HBM5 base dies from in-house DRAM lines to leading-edge foundry nodes — and the compute silicon that sits above the HBM sites, and the advanced package that integrates them. The DRAM stacks themselves would still have to arrive from South Korea, or from Micron's Idaho-to-Virginia pipeline once it comes online at the end of the decade.

HBM is therefore a partial-closure story rather than the full-closure story of layers one and two. The DRAM stack is the gap. Whether a Texas operator with HBM demand at the volumes Tesla, SpaceX, and the compute-scale hyperscalers building in the Austin corridor are underwriting will move that piece of the supply chain onshore is the same integration question that applies on the power semiconductor side, with the same answer implicit in the volumes: at demand levels that exceed the capacity of the global merchant supply chain, captive fabrication is what closes the gap.

Power semiconductors — the wide-bandgap silicon carbide (SiC) and gallium nitride (GaN) devices that convert energy across every electrified system in the buildout, from EV inverters to grid-scale batteries to satellite power management to humanoid joint actuators — are only partially produced in Texas. The production geography of SiC sits at Wolfspeed in upstate New York, STMicroelectronics in Italy and France, Infineon in Germany, onsemi in Czechia and Arizona, and at the major Japanese producers including ROHM,

Mitsubishi Electric, and Sumitomo Electric. Texas is among the world's largest consumers of wide-bandgap power semiconductors. Texas Instruments produces GaN devices only, which will be widely used in humanoid joint motor actuators.

The SiC supply chain is constrained at the physical layer in ways the digital logic supply chain is not. Silicon carbide boules grow two orders of magnitude slower at the chemistry layer. A silicon boule yields over 1,000 wafers. An SiC boule, after weeks of growth, yields 50 to 80. The industry is still transitioning from 150mm to 200mm SiC wafer diameter, a generation behind silicon's leading-edge geometry, and yield losses to micropipes, dislocations, and polytype switching are an order of magnitude higher than silicon's.

The constraint is not capital allocation or commercial willingness — Wolfspeed's financial difficulties scaling substrate capacity demonstrate that capital alone does not solve the problem. The constraint is the chemistry and physics of wide-bandgap crystal formation, and no amount of capacity expansion accelerates the growth rate of a single boule. The result is structural supply tightness that does not relax on commercial timelines. The industry now characterizes SiC as the bottleneck across the three fastest-growing semiconductor demand sectors — electric vehicles, AI data centers, and renewable energy — converging on the same constrained substrate base.

Electric vehicles, humanoid robots, battery-storage systems, solid-state transformers, and AI compute racks each have dozens to hundreds of SiC modules in them. Multiplied across the production targets Texas operators have announced — millions of vehicles annually, eventual billions of humanoid units, gigawatts of grid-scale storage, hundreds of gigawatts of AI compute — the cumulative demand exceeds the global SiC wafer base by an order of magnitude on a multi-year horizon.

The structural question that follows is whether the Texas buildout will integrate the two offshore layers — HBM and SiC — the way it has integrated the leading-edge and analog layers. The captive-fabrication logic that will produce Terafab on the digital compute side maps cleanly to the HBM DRAM dependency and to the wide-bandgap dependency on the power side, with the physics making the case more acute rather than less on both fronts: when supply is constrained at the process level, whether by boule growth chemistry or by stack-and-package yield, captive fabrication becomes the only path that scales with the consumption volumes the largest operators have committed to producing.

The completeness of the Texas silicon chain is the integration thesis at its most ambitious framing — and the captive-fabrication announcements on the HBM and SiC sides, whenever they come and from whichever operators, will close the chain.

Eleven

THE AI COMPUTE LAYER

THE DATA CENTER IS THE ARCHETYPAL building type of the AI-industrial complex, and the Texas Triangle is filling with them faster than the map can keep up.

The comparison to the factory is not rhetorical. A factory in the Second Industrial Revolution was not simply a building where things were made. It was the physical infrastructure through which electrical power, mechanical force, and human labor were organized into productive systems at a scale craft production could never reach. Its concentration of machines, materials, and workers under one roof was the organizational innovation that made industrial-scale output possible. And the geography of that density — Pittsburgh, Detroit, Gary, Youngstown — set the geography of industrial wealth for a century.

The data center does the same organizational work for this revolution. It is the physical infrastructure through which electrical power, silicon compute, and software are organized into AI capability at a scale distributed computing cannot reach. Its concentration of servers, networking, cooling, and power management under one roof, or across one campus, is what makes industrial-scale AI training and inference possible. And the geography of data center concentration is already setting the geography of AI-industrial wealth.

The numbers are still catching up to the ground. Texas is the second-largest data center market in the world, slightly behind Northern Virginia, on a trajectory that industry practitioners expect will make it the largest in the world well before the decade is out. As of late 2025 the state hosts 484 data centers — 297 operational, 109 under construction, 124 planned — per Baxtel data. Forecast capacity passes 70 GW by 2030.

Set that against the grid it draws on. ERCOT's total installed generation capacity, for the entire state of Texas, is roughly 140 GW. The data center sector alone is projected to need half of that — as new, incremental load — within five years. Fifteen years ago the sector did not exist in Texas at meaningful scale.

The AI Workload Difference

The same nameplate megawatt capacity means very different things across data center workload classes. Rack density and capacity factor are the two variables that matter; multiplied together, they span more than an order of magnitude from the bottom of the taxonomy to the top.

Enterprise colocation — 4 to 10 kW per rack, 25 to 45 percent capacity factor. Mixed enterprise tenants, disaster recovery, web hosting. Air-cooled. This is the Dallas concentration.

Hyperscale/cloud — 15 to 25 kW per rack, 50 to 70 percent capacity factor. AWS, Azure, and Google Cloud baseline workloads. Predominantly air-cooled. This is the current baseline.

AI inference — 30 to 80 kW per rack, 40 to 70 percent capacity factor. The production layer behind ChatGPT, Claude, and Gemini. Diurnal, latency-sensitive, regionally distributed. Hybrid air and liquid cooling.

AI training — 120 to 150 kW per rack today, climbing to 200 to 230 kW with Nvidia Vera Rubin NVL72 in late 2026, 600 kW with Rubin Ultra Kyber NVL576 in late 2027, and 1 MW or more with Feynman in 2028. Capacity factor of 90 to 95 percent, continuous over weeks to months. Liquid cooling required. Behind-the-meter natural gas increasingly the energy substrate.

A single Nvidia Blackwell AI training rack consumes roughly 15 to 30 times the continuous power of an enterprise colocation rack. Nvidia's Rubin Ultra in the Kyber NVL576 configuration, arriving in the second half of 2027, raises that figure another fivefold to 600 kW per rack — the equivalent continuous draw of roughly 400 American homes packed into a single 19-inch cabinet.

Feynman, the generation after Rubin and scheduled for 2028, takes the trajectory further into megawatt-class racks — 1 MW or more per cabinet, 700 American homes worth of continuous draw concentrated in a single rack position, with per-GPU packages at 4.4 to 6 kW forcing 800 VDC distribution to replace the conventional 54 V arrays the industry was designed around.

A nameplate megawatt of AI training capacity consumes 100 times the annual energy of a nameplate megawatt of enterprise colocation, and behaves on the grid like aluminum smelting rather than office buildings — flat, sustained, with no diurnal relief and no weather-driven curtailment. This is the load shape ERCOT was not designed for, and that the conventional grid interconnection process assumes does not exist.

THE SCALE OF THE BUILDOUT

Texas data center load currently runs around 35 GW as of mid-2026, accumulated over more than two decades of conventional colocation and early hyperscale buildout. The announced and permitted AI buildout is over ten times that figure — 438 GW in interconnection requests — with new commitments arriving at a regular cadence. The gap between what exists and what is planned is the first signal that something structurally different is underway. The roster of major commitments sorts cleanly by capacity tier.

At the operational end of the spectrum, Google's Midlothian campus in Ellis County is the longest-running Texas hyperscaler site, with four buildings at an estimated 600 to 800 MW, and the adjacent Red Oak campus brought its first building online in late 2025. Tesla's Cortex AI training cluster at Giga Texas runs at hundreds of megawatts today and is scaling toward gigawatt-plus capacity for

FSD, Optimus, and Cybercab development. Stargate Abilene has four of eight buildings operational at approximately 300 MW live, scaling toward 1.2 GW and combining with the adjacent Microsoft buildout for a 2.1 GW node by 2027. Together with the established DFW colocation operators — Compass, Aligned, Digital Realty, CyrusOne, QTS — these account for the bulk of Texas's current data center load.

The largest single commitments under development sit at 5 GW and above. Fermi America's Project Matador in Amarillo tops the list at 20 GW planned across four phases through 2036 — gas in 2027 and 2028, conventional nuclear by 2032, small modular reactors by 2036 — the largest single planned data center campus in the United States.

Pacifico Energy's GW Ranch in Pecos County holds a TCEQ air permit for 7.65 GW of gas-fired generation, the largest power-generation permit ever granted in the US. Nexus Hubbard in Hill County is permitted for 7.2 GW on a sovereign-grade campus anchored by Anthropic. Data City in Webb County plans 5 GW on the largest single-site footprint in Texas by acreage, transitioning from gas to hydrogen. Chevron's partnership with Microsoft and Engine No. 1 in Pecos County targets 2.5 GW initial expandable to 5 GW in the first energy-major-anchored AI campus in the dataset. Sailfish Investors' Comanche Circle in Hood County plans 5 GW across nine campuses adjacent to Vistra's 2.4 GW Comanche Peak Nuclear Plant.

In the 2 to 3 GW range, MSB Matrix in Hopkins County is permitted for 3 GW phased across Bloom fuel-cell and geothermal substrates. The Stargate program adds 1.4 GW at Vantage Frontier in Shackelford County. LandBridge and PowerBridge's Alpha Digital Campus in Reeves County plans 2 GW initial on Permian surface-rights acreage near the Waha hub. Provident–PowerHouse plans 2 GW in Grand Prairie adjacent to Google Midlothian, densifying the DFW Corridor. Amp Z's Project Lufkin in Angelina County plans 2.1 GW on the former Southland Paper Mill site, the first major data center in East Texas. Tract Capital Management's Caldwell Technology Park near San Marcos plans 2 to 4 GW of turnkey capacity. HC Capital Partners' Energy Ranch in Atascosa County plans 1 to 3 GW as the oil-and-gas-capital pivot into AI compute. The

Sandow Lakes Ranch in Milam County has assembled 3.9 GW of Oncor switchgear capacity supporting Xebec's 1.2 GW rebuild on the Alcoa brownfield plus the SoftBank Stargate Milam node co-located with T1 Energy's solar-cell manufacturing plant.

At the 1 – 2 GW tier, CloudBurst's San Marcos campus runs 1.2 GW of behind-the-meter gas; InfraKey Capital's Waco campus runs another 1.2 GW; Meta's El Paso campus is 1 GW; Project Jupiter in Santa Teresa, New Mexico, the Borderplex cross-border anchor, is 1 GW initial scaling to multi-gigawatt over 30 years on a $165 billion OpenAI–Oracle commitment; Fort Bliss / Carlyle in El Paso is 100 MW initial scaling to 3 GW by 2029 on the first defense-anchored AI buildout in the dataset.

Adding the major commitments produces a roster of approximately 400 GW of planned and permitted AI data center capacity across Texas, against roughly 35 GW operational today. This is the scale problem the next section addresses: how the operators are sourcing power for a workload class that exceeds Texas's entire current data center installed base by an order of magnitude, on a timeline the grid cannot accommodate.

The Power Constraint Solution

Every AI data center in Texas is an answer to the same question: where does it get reliable power, at predictable cost, fast enough to matter? AI training is the most power-hungry computing workload ever run at commercial scale, and the grid was not built to absorb it on the timeline the AI development cycle demands. The facilities look like undifferentiated sprawl until you sort them by how they answer that question — and then a small number of clear approaches emerge, each with a different exemplar and a different risk.

Some facilities take grid power and compete on everything else. Tesla's Cortex cluster is the purest case of building compute for one's own use: its output is FSD capability, Cybercab autonomous driving and Optimus control, not cloud revenue, and it is sized and scheduled entirely to Tesla's roadmap. Meta's Temple complex is owned the same way but built for a different end — training Llama models and running recommendation systems at Facebook and Instagram scale, on Meta's capital and timeline. Its location: sitting on I-35 between Austin and Dallas, marks the corridor becoming one continuous AI infrastructure geography rather than two metros joined by a highway. Google's longstanding Midlothian campus completes the grid-anchored hyperscaler triple — four buildings, approximately a million square feet of operational capacity, and a 100 percent renewable energy commitment satisfied through power-purchase agreements with West Texas wind and solar developers rather than on-site generation. The model worked through the 2018-to-2024 hyperscale buildout when grid interconnect was tractable; it became a transitional model the moment AI training loads arrived.

Tract Capital Management's Caldwell Technology Park, east of San Marcos, takes grid power too but inverts the model. Tract builds no compute at all. It assembles large land positions, pre-develops power, water, and fiber, and sells turnkey sites to operators. Tract's 25-plus GW national portfolio is a bet that demand will outrun purpose-built campuses, and that the sustained value will sit with whoever sells time-to-deployment. Prologis runs a parallel version of the institutional real-estate thesis at the Skybox PowerCampus in Hutto — six to eight buildings, 600 MW in Phase 1, $10 to $20 billion in committed buildout. The same logistics-REIT capital that built the country's warehouse infrastructure is now deploying into hyperscale at a scale that conventional real-estate operators did not reach a decade ago.

A newer model sits between grid procurement and behind-the-meter independence. Google's $40 billion Texas commitment and its acquisition of Intersect Power for $4.75 billion together signal a structural shift: power-first co-location, in which the data center and the generation infrastructure are designed and built as a single integrated investment rather than separate procurements. At the Meitner Energy Center near Pampa, Google's $3 billion campus comes online alongside more than a gigawatt of wind, solar, and battery storage developed by Intersect Power, with on-site gas-fired generation for reliability firming. The same template is being deployed at Armstrong County and at Wilbarger County. The structural significance is that Google is now the energy developer. Vertical integration into the generation pipeline gives Google control of the entire generation-to-compute path without crossing the ERCOT interconnect queue while keeping the renewable-energy profile intact. It is grid-anchored procurement re-engineered to behave like behind-the-meter in deployment terms.

The more consequential facilities escape the grid entirely. Behind-the-meter generation removes wholesale price exposure, transmission and distribution charges, and — most decisively — the interconnection queue. What differs among these facilities is the power source, and each source carries its own risk.

Fermi America is the corporate entity behind Project Matador — the most ambitious single buildout in the dataset, a four-phase, 20 GW private energy and AI data center campus in the Texas Panhandle designed to operate independent of the public grid. The architecture is staged for execution. Project Matador comes first: 6 GW of natural gas generation expected operational in 2027, providing speed-to-power. Project Matador East follows: another 5 GW of gas. The deeper bet is the second half. Fermi Nuclear, 4.4 GW of conventional nuclear via four Westinghouse AP1000 reactors, is targeted for 2032. The economics on the nuclear legs are genuinely strong: AI training runs at sustained 95-percent-plus utilization for weeks at a time, and nuclear generation — capital-heavy, fuel-light, low marginal cost once built — matches that load almost perfectly. The risk is timeline. SMR technology is mature in physics but thin on commercial deployment at these scales, and NRC permitting runs in years. Matador is the right long-run answer at the largest scale anyone is proposing, with the gas phases providing the bridge to the nuclear destination.

Behind-the-meter natural gas is the dominant near-term strategy. CloudBurst buys grid independence faster and dirtier: on-site natural gas turbines deliver freedom from ERCOT price exposure and grid capacity limits on a commercial permitting timeline rather than a regulatory one. The 1.2 GW San Marcos campus is the first phase of the model. InfraKey Capital's $10 billion Waco campus runs the same model at scale — also 1.2 GW of behind-the-meter gas.

The largest single commitments sit in the Permian Basin, which has consolidated over the past eighteen months into the densest behind-the-meter concentration in Texas. Pacifico Energy's GW Ranch in Pecos County at 7.65 GW of gas-fired generation is larger than every operating Texas power plant. The distinctive move is location: GW Ranch sits inside the Permian Basin proper, 8,000 acres along SH 18 north of Fort Stockton, co-locating compute with the gas source rather than transporting gas to compute. Chevron's partnership with Microsoft and Engine No. 1, also in Pecos County, introduces a structural first: the first energy-major-anchored AI campus in the dataset. The $7 billion initial phase targets 2.5 GW expandable to 5 GW, with seven GE Vernova combined-cycle gas turbines on order and Chevron supplying gas directly from its Permian operations. This is vertical integration in a new direction — the oil-and-gas major as power foundry, the hyperscaler as captive customer, with no ERCOT crossing in either direction. Adjacent in Reeves County, LandBridge and PowerBridge are deploying the Alpha Digital Campus at 2 GW initial near the Waha hub, on a surface-rights lease model that turns LandBridge's existing oil-and-gas-tenant royalty business into a data center royalty business. Together with GW Ranch, the three projects make Pecos and Reeves Counties the largest single behind-the-meter AI concentration in Texas at roughly 12 to 15 GW of permitted or announced capacity, and the first place in the dataset where the AI buildout and the Permian Basin oil-and-gas substrate are integrated end-to-end.

The Texas Panhandle is the second emerging concentration and a different topology. Fermi's 20-GW centralized buildout in Carson County combines with Google's distributed footprint across Gray, Armstrong, and Roberts Counties for combined planned capacity of 22 to 25 GW — the largest of any emerging supernode in the dataset. Pantex, the only US final-assembly and disassembly site for nuclear warheads, sits adjacent to Fermi in Carson County and provides the structural rationale for the nuclear-AI siting. The Panhandle wind generation base supplies the renewable substrate for both Fermi's renewables component and Google's power-first Intersect Power co-location.

Nexus Hubbard in Hill County completes the multi-gigawatt-class behind-the-meter set: 7.2 GW of continuous generating capacity across 31 combined-cycle and simple-cycle turbines plus 24 reciprocating engines, 2,000 acres between Waco and DFW on the I-35 spine. Nexus's distinguishing positioning is sovereign-grade infrastructure backed by Anthropic's anchor lease. The cost across all of these — CloudBurst, InfraKey, GW Ranch, Nexus Hubbard, Chevron-Microsoft, Alpha Digital, and the Fermi gas phases — is the carbon profile, though the industry's revealed preference is that power certainty outweighs carbon when training infrastructure is at stake.

Sailfish Investors' Comanche Circle proposal sits at a different intersection. Five gigawatts across nine planned data center campuses on 2,600 acres in Hood County, adjacent to Vistra's 2.4 GW Comanche Peak Nuclear Plant. Comanche Circle is not building nuclear — it is leveraging existing nuclear. The first 600 MW comes online within 24 months via on-site power islands; subsequent phases deploy through 2028 via additional islands and possible grid integration with Comanche Peak's output. The structural pattern — data centers siting next to existing nuclear plants to capture clean-baseload economics without the construction risk and timeline of building new reactors — has parallels at other US nuclear sites and may become the dominant path to nuclear-served AI capacity, well in advance of new SMR commercial deployment.

Xebec Holdings' Sandow Lakes campus in Milam County neither builds nor buys its power so much as inherits it. The 33,000-acre master plan sits on brownfield water rights from Alcoa's 1954 aluminum operations and rebuilds 1.2 GW of generation on the footprint of the old Sandow Power Plant — land whose long industrial history makes its permitting and environmental frameworks far more navigable than greenfield siting at the same scale. T1 Energy's co-located 5-GW solar cell manufacturing plant adds the option of behind-the-meter clean generation that compute-only campuses lack. The brownfield-revival model has a second exemplar in East Texas: Amp Z's Project Lufkin in Angelina County rebuilds 2.1 GW of capacity on the former Southland Paper Mill site, leveraging brownfield grid interconnect, water rights, and rail access that greenfield sites must construct from zero.

Not every behind-the-meter campus runs on gas. MSB Matrix in Hopkins County is the technically distinctive exception in the Texas dataset: 3 GW phased across a Bloom Energy solid-oxide fuel-cell microgrid (1.5 GW Phase I), a geothermal microgrid (1.2 GW Phase II), and a hybrid third phase. The fuel-cell and geothermal architecture is the rare non-gas behind-the-meter substrate in the dataset, and its execution will determine whether the model can scale beyond the gas-dominant default.

Energy Abundance Development Corporation's Data City project in Webb County is the experimental edge of the behind-the-meter model. Five gigawatts on a 50,000-acre footprint at the southern terminus of I-35 near Laredo at the US-Mexico border — the largest single-site data center footprint in Texas by acreage — initially gas-powered but planning a transition to green hydrogen supplied by pipeline from the company's Hydrogen City project in Duval County, where salt-cavern storage at the Piedras Pintas Salt Dome scales toward six terawatt-hours of hydrogen energy capacity. Hydrogen City is not a captive supply: its pipeline network serves a 1-million-ton-per-annum green ammonia production facility at Corpus Christi, sustainable aviation fuel production at the Port of Brownsville, statewide hydrogen power-plant fuel-switching customers, and Data City as one anchor offtaker. The Data City and Hydrogen City pairing — physically separated by approximately 90 miles and integrated by pipeline rather than geography — is the only entry in the dataset where the energy and compute infrastructure are designed and built as a single integrated program by a single operator, with the clean-fuel transition pathway embedded from the start rather than retrofit later.

Beyond the individual exemplars, the consequential geography is the concentrations — places where multiple operators converge on the same geography for related but distinct reasons. The Abilene supernode has drawn three major operators at once: Stargate Abilene, the $500 billion OpenAI-SoftBank-Oracle program's first anchor, with $100 billion committed to Phase 1 alone; Vantage's $25 billion, 1.4 GW Frontier campus, purpose-built for frontier training with racks exceeding 250 kW; and Crusoe's Microsoft-dedicated campus at a combined 2.1 GW buildout. Combined cluster capacity exceeds 4 GW. The Abilene operators converged for one reason — energy economics: ERCOT access, proximity to West Texas wind and Permian Basin gas, and land costs.

The El Paso-Borderplex cluster is the second concentration node and the first cross-border one. Meta's grid-connected El Paso campus is one anchor; the second is Project Jupiter in Santa Teresa, New Mexico, a $165 billion OpenAI-Oracle frontier-training cluster scaling to multi-gigawatt over 30 years; the third is Fort Bliss / Carlyle Group, the first DoD-anchored AI buildout in the dataset, scaling to 3 GW by 2029. Three operators, three structural firsts — cross-border geography, DoD-anchored AI, frontier-training — and 5 GW of capacity within 30 miles of each other.

Stargate itself has expanded into a multi-site program — additional anchors in Shackelford County and Milam County, where the Freebird campus partners OpenAI with SB Energy — and across the state line at Project Jupiter. A single AI infrastructure program now anchors multiple Texas supernodes plus the cross-border Borderplex node, the largest single-program footprint in the appendix graph and the clearest demonstration that the buildout has moved past individual mega-campuses into multi-site program supernode execution at national scale.

What this resolves into is not a catalog but a finding. The heaviest capacity commitments in Texas are being built to bypass the grid rather than connect to it. When the largest industrial-scale compute is engineered to escape the public grid, the grid has stopped being the platform for this kind of load. It has become the fallback.

The Compute Layer as System

The most consequential interaction runs between the compute layer and the silicon layer. Tesla consumes chips that Samsung Taylor and Terafab will produce. Stargate consumes chips from TSMC Arizona and other leading-edge foundries. Add up the chip demand the Texas compute concentration sends into the silicon supply chain, across the full cluster, and the aggregate is large enough to move things that move slowly — fab utilization decisions, process development priorities, equipment procurement timelines, at foundries whose allocation choices reach every AI company on Earth.

That demand signal is the Texas Triangle's most underappreciated form of industrial leverage. The cluster is not only a consumer of AI compute capability. It is a shaper of the silicon supply chain that produces the hardware AI capability runs on. Once the cluster's aggregate demand is large enough — and the trajectory says it is approaching that point — its procurement decisions begin to influence which process nodes get prioritized, which chip architectures get optimized, which fabrication capacity gets built. The consumer of silicon becomes a co-determiner of silicon's direction.

The SpaceX Compute Pivot

The AI compute layer this chapter has documented — the training clusters, the inference infrastructure, the hyperscale campuses, the power architecture — has a corporate structural feature whose implications reach into Chapter 21 and are worth marking here. The largest single operator of AI compute infrastructure in the Texas–adjacent geography as of mid-2026 is not a hyperscaler in the conventional sense. It is SpaceX.

The path to that position runs through the February 2026 integration of xAI into SpaceX, from which the SpaceXAI division emerged. Colossus — the AI training cluster in Memphis that xAI built between 2024 and 2026, reaching a compute scale that placed it among the largest AI training installations in the world — became a SpaceX asset. Grok, the frontier model xAI had developed, became a

SpaceX product. And the commercial compute relationships that xAI had begun to develop with third-party AI operators became, after the integration, SpaceX revenue streams.

The scale of those revenue streams is worth stating plainly. SpaceX's S-1 filings and subsequent industry reporting disclose that Anthropic has leased the full compute capacity of Colossus 1 at $1.25 billion per month through 2029, and that Google has leased roughly half the capacity of Colossus 2 for $920 million per month across a 32-month term to provide bridge capacity for its Gemini Enterprise platform. Reflection AI has taken additional Colossus 2 capacity at $150 million per month. The aggregate is a compute infrastructure business generating approximately $27 billion of annualized inference lease revenue as of mid-2026 — a business whose scale, on the revenue metric alone, is almost twice that of Starlink's consumer broadband operation.

That revenue base changes what SpaceX is. Through 2024, SpaceX was a launch company with a satellite internet business emerging alongside it. By 2026, it is a launch company, a satellite internet company, and a compute infrastructure company operating at three vertically integrated scales at once: terrestrial compute (Colossus 1 and 2 in Memphis), orbital compute (the future Starmind constellation, whose FCC filing documents up to one million satellites), and launch (the Starship platform that gets the orbital compute layer to orbit and gets any expansion of the terrestrial footprint's physical infrastructure to its site). Compute is not a side business acquired through corporate transaction. It is a third vertical of the same vertically integrated infrastructure company, with launch as the connective tissue between the terrestrial and orbital layers.

The analytical implication for the Texas AI-industrial complex is straightforward. The compute infrastructure documented — Cortex at Giga Texas, the Williamson County US 79 compute corridor, the Abilene and Panhandle concentrations, the emerging Central Texas clusters — has a peer operator whose compute scale rivals the largest hyperscaler campuses and whose deployment trajectory will extend the same physical infrastructure into low Earth orbit. Chapter 21 documents the orbital extension. This chapter documents the terrestrial substrate on which the orbital extension will be commercially validated before it ever deploys. The customers for orbital compute are already the customers for terrestrial compute. The infrastructure operator that will sell them orbital inference already sells them terrestrial inference. The bridge between the two is Starship's cost trajectory. Everything else is already in place.

The Foundational Microgrid

Back to Earth. A microgrid is the physical implementation of energy autonomy: the architecture that lets a facility, a fab, or a fleet run without continuous dependence on the grid. The pattern is not new. Off-grid industrial sites and military installations have operated this way for decades. What is new is where it is turning up. The microgrid is becoming the default energy architecture for hyperscale AI compute, semiconductor fabs, gigafactories, seaports, EV fleet charging depots, and logistics hubs — and that shift is the buildout's most consequential change in how industrial power is organized in the United States.

The reason is blunt. The traditional utility grid cannot deliver power to new large sites fast enough, reliably enough, or at the power quality that sensitive industrial processes and high-density compute demand. So the microgrid is no longer a premium option for operators with unusual needs. It is increasingly the default for any new industrial load above a threshold the buildout's own pace keeps pushing downward.

Training clusters drawing 100 to 500 MW need redundant high-voltage feeds, on-site battery storage, and often full microgrid capability to hold their uptimes while grid interconnection queues stretch to five and ten years. Every Texas Triangle data center operator has hit that constraint and answered it with a microgrid architecture — behind-the-meter natural gas, nuclear co-location, brownfield power inheritance, or renewable-plus-storage.

What began as a pattern observed at individual facilities is now the architectural norm. The interconnection queue is the constraint that drove operators to the microgrid in the first place. The transformer, the chip, and the high-bandwidth memory (HBM) stack are the constraints that decide which microgrid deployments actually reach operational status on time.

The Three Chokepoints

The bottleneck has moved. The hardware constraints that used to gate data center expansion are no longer the binding ones. Those binding constraints now sit at three layers that ten years ago were not constraints at all: grid interconnection, large power transformers, and AI accelerator memory.

Grid interconnection is the first. Queues in PJM — the regional transmission organization serving the Mid-Atlantic and parts of the Midwest — now run roughly eight years for new large-load requests. ERCOT's queue is shorter, which is one of the structural reasons the Texas Triangle has captured a disproportionate share of the buildout. But ERCOT's queue is extending too, as aggregate load requests overwhelm transmission planning. This is not a permitting delay capital can compress. It is a transmission infrastructure development horizon whose physical minimum is set by right-of-way acquisition, environmental review, equipment procurement, and construction sequencing — none of which an operator can speed up by paying more.

Large power transformers for fabs, data centers, and gigafactories are the second. The grid-scale transformers that step utility voltage down to facility distribution levels now carry lead times of four to five years for some classes, as aggregate demand has overwhelmed a small set of qualified manufacturers, most of them foreign-operated. A site with cleared interconnection and signed power purchase agreements still cannot energize until the transformer arrives. It is the most operationally specific bottleneck in the buildout today — and the one whose partial relief, through Tesla's Megablock product line and the broader industrial response, will most directly decide which programs hit their announced timelines.

AI accelerator memory is the third, and the one furthest from the conventional infrastructure framing. High-bandwidth memory, at the densities and speeds frontier AI training demands, is sold out at the dominant supplier, SK Hynix — with Samsung's and Micron's HBM capacity fully allocated as well. The GPUs and TPUs operators are buying at unprecedented scale depend on HBM stacks whose production capacity is the binding constraint on accelerator output, no matter how much fabrication capacity Nvidia, AMD, or Google's TPU partners hold at TSMC.

On the first two chokepoints, Texas has a structural advantage — ERCOT's shorter interconnection queue, and the in-state response to the transformer constraint through Tesla's Brookshire grid-level Megablock deployment. Those are the specific reasons the region is taking a disproportionate share of the buildout. The third chokepoint, HBM, falls on every operator equally, wherever they are.

THE PUSHBACK IN CONTEXT

The engineering responses — brownfield rights inheritance, behind-the-meter generation, closed-loop water systems — are the industry's substantive answers to the resource-consumption critique. They are not marketing. They are capital deployments measured in billions of dollars and engineering programs measured in years. They represent the industry's actual position on whether data center buildouts can be sited responsibly: not in argument, but in construction.

These responses are also largely invisible to the public conversation in which the buildout is being contested. The data center pushback — the moratorium in Hill County, the data center ban in San Marcos, the review period in Hays County over water protection, and negative coverage in the local press — is occurring in a public information environment where the engineering responses have not been properly communicated, the historical analogs have not been drawn, and the long-arc rationale for the buildout has not been articulated by the operators executing it. The industry has solved more of the resource problem than its critics know, and has communicated less of what it has solved than the moment requires.

The resistance is rational on present evidence. The local political-economy concerns are legitimate: a hyperscale data center campus represents a step change in a county's electrical load, water demand, industrial-zone footprint, and noise environment. The concerns of residents who watch a 200 MW facility rise next to their subdivision are not the product of misinformation. They are the product of accurate observation of a physically large industrial facility appearing in a place that did not previously contain one. Whether the facility's resource consumption is offset by behind-the-meter generation, whether its water cycle is closed, whether its waste heat is captured for district use — these are technical details that may not have been part of the local conversation in most of the jurisdictions where it is happening.

The historical analog is the transcontinental railroad. The Pacific Railway Acts of 1862 and 1864 authorized a buildout whose resource consumption — steel for rails, timber for ties, land for right-of-way, labor at scales the American workforce had never been organized at — triggered immediate and substantively correct public resistance. The labor conditions were genuinely terrible. The land grants were corrupt. The resource concentration in the hands of the railroad operators was unprecedented. The Credit Mobilier scandal was real. The labor exploitation was real. The indigenous land seizures were real. The resistance was not the product of misinformation; it was the product of accurate observation of an industrial project whose costs were borne disproportionately by people who would not see its benefits for decades.

The railroad was built anyway. Its consequences — the national logistics system that enabled refrigerated shipping, mail-order retail, standard time zones, and the transcontinental newspaper industry — were not visible at the time of construction. The case for the buildout could not be made on the use cases of 1869, because those did not yet exist. They emerged in the four decades after Promontory Summit, as the network's existence made them possible. The railroad's defenders in the 1860s could not have articulated the world the railroad would create. The railroad was correct in its tendency anyway.

This is the structure of every major infrastructure transition in the industrial era. Steam, rail, electrification, telecommunications, the interstate highway system, the public internet — each required physical infrastructure built before its applications were understood, each triggered substantively rational resistance on present evidence, each was completed anyway, and each enabled a use case landscape that no defender of the original buildout could have specified in advance. The data center buildout is the same pattern. The engineering responses above address the local resource consumption critique. The use-case critique — that AI data centers are consuming gigawatts to produce annoying video slop, collecting massive public surveillance data, and widespread labor displacement — is the harder question, and one this book takes up in its closing pages. What is documented here is the substrate: what is being built, where, with what engineering, against what constraints. The applications layer is downstream of the substrate, and like every prior infrastructure transition, its full shape is not yet legible to anyone, including the people building it.

The moratoriums and bans will not stop the data center buildout. They will shape its geography however. Facilities sited in jurisdictions that have done the work of integrating community concerns into engineering specifications — water-cycle closure, noise abatement, behind-the-meter generation, decommissioning bonds — will proceed. Facilities sited in jurisdictions that have not done that work will be relocated to ones that have. The compounding pressure on operators is to build the communications discipline they currently lack: to explain, in language the affected publics can engage with, what is being built and why. The industry has not yet treated this work as engineering work. It is engineering work. The communications failure is itself a constraint on the buildout, as binding in its way as the water ceiling or the grid capacity, and the operators who treat it as a peripheral concern will find themselves outcompeted by operators who treat it as central.

Twelve

THE AUTONOMY DEPLOYMENTS

THERE WILL NEVER BE AN autonomous internal combustion vehicle. That is not a prediction about regulation or consumer taste. It is an observation about systems architecture. Autonomy requires continuous high-bandwidth compute — perception, prediction, planning, and control running at once, at millisecond latency — and that draws more power than an ICE vehicle's electrical system can sustain without a large battery. It requires over-the-air software updates whose frequency and data volume demand always-on connectivity that ICE vehicles were never designed to carry. It requires sensor arrays — cameras, LiDAR, radar, ultrasonic — whose power and thermal-management demands are native to electrified platforms and parasitic on ICE ones. And most importantly, it requires precise, low-latency actuator control, which electric motors deliver instantaneously, and hydraulic and mechanical drivetrains approximate poorly, with lag.

So autonomy and electrification are not two parallel transitions running at the same time. They are one transition with two visible faces. The electric drivetrain is the hardware the autonomy runs on. Every autonomous system in commercial development — Tesla FSD, Cybercab, Waymo, the robotruck programs, the port automation pilots, the yard robot fleets — runs on electric power. The electrification transition is therefore not only an energy transition. It is the infrastructure buildout that makes the autonomy transition physically possible.

That has a direct consequence for the Texas thesis. The same geography is building the largest concentration of EV manufacturing capacity in the United States — Giga Texas producing Model Y, Cybertruck, Cybercab, and soon Roadster 2, with the broader supply chain feeding battery cells, power electronics, and drivetrain components. It is also the geography where autonomous deployment is most advanced, most densely concentrated, and most aggressively supported by the regulatory environment that has made Texas the primary autonomy deployment state in the country. The manufacturing base and the deployment ground are emerging in the same place, at the same time.

THE FLEET-SCALE TRAINING LOOP

The Tesla vehicle fleet is backed by a data moat the autonomy industry has no analog for. Full Self-Driving has accumulated more than 11 billion miles of real-world driving data, adding approximately 29 million new miles per day across a fleet exceeding nine million vehicles globally. The HW4 fleet generates the highest-quality training data. HW3 vehicles continue on a supervised FSD branch whose long-term unsupervised capability is limited, but their miles still feed the broader training corpus that improves every subsequent model version.

The data infrastructure connecting the fleet to Cortex on the Giga Texas campus closes the loop at the geographic level. Every Tesla in operation generates training data for the next training cycle that runs on the same campus that produced the vehicle. FSD version updates deploy to the entire fleet simultaneously through over-the-air update, on a cadence measured in months rather than years. The consequence is a training and deployment loop whose iteration speed is fundamentally different from any competitor's engineering cycle.

The moat's analytical significance is not raw mileage. It is the ratio of training data to the operational design domain the training data is required to cover. Camera-only end-to-end neural network autonomy requires training data volumes that were physically unavailable until Tesla built the driving video data collection infrastructure into the fleet. No competitor has the equivalent data pipeline, and no competitor can build one on a timeline that matters. This is what makes the architectural bet on camera-only autonomy structurally credible for Tesla and structurally implausible for any other operator.

Tesla Cybercab Economics

The Cybercab is a purpose-built platform. No steering wheel, no pedals, two seats, designed from inception for autonomous operation rather than converted from a human-driven vehicle. It is the first vehicle in Tesla's lineup whose entire architecture assumes the absence of a driver. Inductive charging closes the autonomy loop at the fleet level: vehicles dock themselves at charging pads without human intervention. Autonomous cleaning between rides completes the human-labor elimination that ride-hailing economics at scale require.

The analytical case for Cybercab is not the per-ride fare. It is the unit economics of running an autonomous fleet at scale, and the specific structural asymmetries that determine whether unit economics converge at profitability.

Vehicle capital cost is the first asymmetry. A current-generation Waymo vehicle combines an approximately $75,000 Jaguar I-PACE with a fifth-generation Driver sensor and autonomy suite that Waymo has publicly discussed at roughly $100,000, producing an all-in vehicle cost of approximately $175,000. The sixth-generation Waymo Driver reduces sensor count and cost to approximately $20,000 per kit, and Waymo is transitioning that kit onto the purpose-built ZEEKR Ojai platform manufactured in China and the Hyundai IONIQ 5 that starts at approximately $45,000 in fleet configuration. The all-in cost for the sixth-generation Waymo vehicle therefore lands in the range of $65,000 to $95,000 depending on platform. Cybercab targets an all-in cost of under $30,000, sensors, compute, and FSD included, with no retrofit required because the vehicle is designed as an autonomous platform from inception. Amortized across 300,000 lifetime miles per vehicle, the per-mile capital contribution is approximately 10 cents for a Cybercab against roughly 28 cents for the new sixth-generation Waymo vehicle and roughly 58 cents for the current fifth-generation fleet. For consumer Teslas joining the network as owner-operator nodes once FSD unsupervised operation reaches the installed base, the incremental vehicle capital cost approaches zero.

Passenger optimization is the second asymmetry. Two seats match the roughly ninety percent of ride-hailing trips that carry one or two passengers. A two-passenger platform avoids the mass overhead of a five-passenger SUV design and the energy consumption penalty that mass carries across every mile driven. Waymo's platforms optimize for OEM commonality with consumer vehicle designs. The mass mismatch between platform and mission is a permanent efficiency drag on Waymo's per-mile energy cost.

Charging infrastructure is the third asymmetry. Tesla owns approximately 7,000 Supercharger stations globally with more than 65,000 stalls and roughly 7 GW of aggregate permitted grid capacity. The infrastructure is already built, already permitted, and already amortized against consumer vehicle sales. Adding autonomous charging pads to selected Supercharger sites is a marginal-cost operation whose economics compound globally rather than city by city. Waymo has no equivalent infrastructure. Every operating city requires either paying retail rates at third-party charging networks or building proprietary depot charging from scratch, at capital costs that must be recovered through per-ride pricing in that specific city.

Manufacturing integration is the fourth asymmetry. Cybercab is produced at Giga Texas on the Unboxed assembly line, targeting one vehicle every ten seconds and two million units annually at full ramp. The manufacturing is domestic, insulated from tariff exposure, and scalable through Tesla's existing manufacturing engineering capability. Waymo does not manufacture vehicles. Every fleet vehicle is purchased from an external OEM at negotiated cost, with the ZEEKR Ojai platform manufactured in China and subject to import duties that have ranged from 25 to over 100 percent depending on the political moment, and the Hyundai IONIQ 5 sourced partly from Hyundai's Georgia Metaplant and partly from Korea. Waymo's cost-reduction pace is capped at the pace of supplier negotiation cycles. Tesla's cost-reduction pace runs on internal manufacturing engineering cycles that are structurally faster.

Compounding these factors, Cybercab's projected structural cost per mile targets $0.30 at mature scale. Waymo's current per-mile structural cost, based on approximately $4 billion in annual operating losses against roughly 500,000 weekly paid rides in mid-2026, sits in the range of $6 to $12 per mile, running almost entirely on the fifth-generation Jaguar I-PACE fleet whose per-vehicle capital cost was approximately $175,000. The unit economics gap in mid-2026 between the two architectures is 20 to 40 times on cost per mile. The sixth-generation Driver kit on the ZEEKR Ojai and Hyundai IONIQ 5 platforms will compress Waymo's per-mile cost as those platforms replace the Jaguar fleet through 2027, with the defensible projected range at mature scale falling to approximately $4 to $8 per mile depending on how far the cost compression continues. Alphabet has absorbed $4 billion in Waymo operating losses annually for seven consecutive years, sustained by strategic optionality rather than by per-ride economics. The unit economics gap between the Waymo and Tesla platform architectures at future scale is 10 to 25 times on cost per mile. At any material scale, that differential is existential.

Charging and Compute Infrastructure

Tesla's Supercharger network architecture, the largest EV charging network in the world by station count, charger count, and geographic coverage, is primarily consumer infrastructure. The evolution toward true autonomous charging, wireless inductive charging pads, automated vehicle positioning, depot infrastructure designed for fleets rather than individual cars, is the transition full autonomy requires.

Autonomy at scale requires autonomous charging at scale. The relationship is direct and non-negotiable: an autonomous vehicle that cannot charge itself without human intervention is not fully autonomous. A robotaxi fleet that needs human drivers to position vehicles at chargers and clean them before an autonomous shift has not eliminated the labor cost autonomy was supposed to eliminate. A robotruck that needs a human to plug it in has not reached the operational efficiency that makes autonomous long-haul economics compelling.

The scale of that transition is set by the deployment it has to support. A scaled Cybercab fleet operating in Austin at real commercial scale, thousands of vehicles rather than dozens, needs charging depot infrastructure whose power draw, footprint, and automation are qualitatively different from the Supercharger stations serving human-driven Teslas today. A scaled robotaxi depot is not a bigger consumer Supercharger station. It is a specialized industrial facility with a microgrid whose design follows fleet-operations logic rather than individual-convenience logic: vehicles arriving and departing on schedules set by demand algorithms, charging at rates optimized for fleet utilization, managed by software that balances charging load against grid pricing and battery state across hundreds of vehicles at once.

The Supercharger network has a second emerging role that inverts the conventional relationship between compute and grid infrastructure. In June 2026, Tesla filed a USPTO trademark for MEGAPOD, described as a modular data center hardware system for artificial intelligence computing, comprising servers, AI processing hardware, networking equipment, power distribution units, and liquid cooling. The filing followed the announcement of the Digital Optimus initiative, which disclosed that Tesla holds approximately 7 GW of available power capacity across its Supercharger network. That capacity is already permitted, already interconnected, and not fully utilized at any given moment. Megapod is the product that turns stranded interconnection into distributed compute.

The architectural pattern Megapod instantiates was not merely predictable; it was documented before the filing. I derived and published the Fleet Energy Depot (FED) framework in late 2025 on ElectronsX, my industrial electrification reference site. The framework lays out a multi-layer stack extending from utility interconnection through grid-forming battery storage, microgrid control, onsite generation, and vehicle energy delivery to the edge compute and data layer. That layer anticipates Megapod directly: industrial edge gateways, local compute nodes, real-time telemetry processing, protocol translation, and latency-critical decision logic that must run locally because operations cannot depend on cloud uptime. Two adjacent architectures extend the framework. Fleet Energy Corridors (FECs) describe route-level networks of interconnected FEDs supporting long-haul freight, intercity mobility, and multi-metro robotaxi rebalancing. Energy Autonomy Yards (EAYs) describe the operations envelope where autonomous vehicles, robotic charging, and humanoid support workflows share yard space with fleet charging. Megapod is the specific product instantiation of the FED framework at Tesla's proprietary charging network, with the compute layer implemented as inference silicon rather than the more general edge compute node the framework describes. The AI4 chip that powers Full Self-Driving in the vehicle fleet is the same Tesla inference silicon that will sit inside the Megapod, extending Tesla's vertical integration from the vehicle to the Megapod compute node in a single silicon lineage.

Two use cases emerge from the Supercharger-sited compute layer, and both connect directly to the autonomy thesis. The first is edge processing for the vehicle fleet itself. A Cybercab or a consumer Tesla arriving at a Supercharger for a charge cycle can offload sensor data, receive over-the-air software updates, and coordinate with the fleet-management stack over local rather than backhaul connections. The compute is physically adjacent to the vehicle for the 20 or so minutes it takes to charge. Fleet telemetry and OTA distribution move from centralized cloud infrastructure to a distributed edge that follows the vehicles wherever the

Supercharger network reaches. The second use case is generalized AI inference. When a Supercharger site is not saturating its grid capacity with charging load, the Megapod consumes the excess for external compute workloads. Inference revenue is paid for by the same interconnection that already earns revenue from charging. The economics compound with peak-load arbitrage against the Megapack storage adjacent to the site: cheap overnight power charges the batteries, which feed the Megapod during peak-price hours when neither the vehicles nor the grid rewards drawing directly from the utility.

The Megapod completes what is emerging as a three-tier distributed compute topology built on a single silicon lineage. The mobile tier is the AI4 chip in every vehicle, running Full Self-Driving inference while the vehicle is in motion. The static second tier is the same silicon idle when the vehicle is parked, which for a private consumer vehicle is roughly twenty-two hours per day and for a robotaxi is whatever fraction of each day demand does not fill. Musk's Digital Optimus disclosure framed parked-vehicle idle compute as a fleet-scale pool that could execute AI workloads on behalf of the vehicle owner or aggregated across millions of vehicles on behalf of third-party customers. The station third tier is the physical Megapod at the Supercharger site, sized to consume the grid capacity that charging does not saturate and to provide predictable low-latency compute at fixed geographic points. All three tiers share the same chip, the same software runtime, and the same fleet-management orchestration layer. Mobile compute rides with the vehicle. Static compute pools across the parked fleet. Station compute runs at the charging touchpoint. Workloads route across the tiers according to latency requirement, power availability, and pooled idle capacity.

The homogeneity of the architecture is what makes it work, and the vertical integration required to build it is what makes it exclusive. Traditional distributed compute stacks combine heterogeneous silicon, heterogeneous ownership, and heterogeneous network topology, which imposes orchestration overhead that scales with the fragmentation. Tesla's three-tier topology combines silicon homogeneity through a single AI4 lineage, ownership homogeneity through Tesla operating all three tiers, and topological predictability because the station locations are known and fixed. One binary compiles for all three tiers. One security posture covers all three. One billing and metering layer accounts for compute consumed anywhere in the stack. Assembling the architecture requires simultaneous ownership of the inference silicon, a vehicle fleet in the millions, a proprietary charging network with several gigawatts of aggregate grid capacity, and a unified software runtime spanning vehicle and station. No competitor holds all four. The topology cannot be built from any other single company's starting position, and it cannot be assembled by consortium without reintroducing the fragmentation costs that homogeneous ownership avoids.

The Austin Competitive Landscape

Waymo, Tesla, and Zoox operate in Austin as of mid-2026. The three-operator concentration in one metropolitan geography is the structural signal that Austin has become the proving ground for the autonomous ride-hailing industry outside China. AVride adds a fourth operator with testing in Austin and pilot deployment in Dallas.

Waymo began paid commercial ride-hailing in Phoenix in October 2018 and operates a fleet of approximately 3,000 vehicles across ten US cities delivering more than 500,000 paid rides per week. Its Austin operation launched in 2024 and expanded through 2025. The fifth-generation Jaguar I-PACE fleet is being succeeded starting in 2027 by the Zeekr Ojai and the Hyundai IONIQ 5 running the sixth-generation Driver. Waymo's commercial lead reflects both genuine engineering execution over seven years and a capital subsidy no independent company could have sustained. The lead is real. The economics that produced it are not durable however without continued Alphabet backing.

Tesla launched paid ride-hailing in Austin in June 2025 using Model Y vehicles with supervised FSD, transitioning to unsupervised operation in December 2025. As of June 2026, approximately thirty unsupervised vehicles operate across the full Austin metro geofence, with Model Y capacity held roughly constant while Cybercab production ramps at Giga Texas. Tesla also has Robotaxis operating in Dallas, Houston, Miami, and the Bay Area. Cybercab entered volume production in April 2026 and is not yet operational for public ride-hailing as of June 2026. The current fleet-size gap against Waymo reflects the seven-year deployment gap between the two companies. Tesla's year-one operational scope in Austin, full metro coverage, unsupervised operation, no HD map dependency, exceeds Waymo's year-one operational scope in Phoenix in 2018 on the metrics that matter.

Zoox is the third operator, coming from farther behind but with Amazon's balance sheet behind it. The company launched paid commercial ride-hailing in Las Vegas and San Francisco in 2025. As of mid-2026, Zoox has accumulated approximately two million autonomous miles and more than 350,000 cumulative riders, with Austin and Miami in initial rollout through Explorer program access. The purpose-built bidirectional vehicle carries no front or back and can drive either direction. The Amazon parent's $2.4 trillion market capitalization provides a loss-absorbing capital cushion comparable in kind to what Alphabet provides Waymo, though the strategic priority Amazon assigns to Zoox is meaningfully less than the priority Alphabet assigns to Waymo. Zoox's Uber partnership targets a Los Angeles launch in mid-2027, extending the network into California through partnership rather than direct expansion.

AVride, the Austin unicorn, is the fourth operator in the Austin autonomy concentration, with an analytical significance that runs past its current operational scale. The company relocated its headquarters to Austin after separating from Yandex in 2022, doing autonomous vehicle testing on Austin streets alongside a smaller-scale autonomous delivery robot business. In 2024, AVRide announced a partnership with Uber to deploy autonomous ride-hailing service in Dallas, extending its Texas operational footprint from Austin testing to Dallas commercial operation through the largest ride-hailing platform in the world. AVride runs its Driver system on Hyundai IONIQ 5 vehicles, placing it in the same OEM-partnership vehicle category as Waymo. The company's commercial ride volume is smaller than Waymo's, Tesla's, or Zoox's current operations, but its Austin headquarters and Dallas deployment plan add a fourth operator to the Texas ride-hailing autonomy concentration that no other US metro can match.

The four-operator concentration in Austin and Dallas has analytical significance beyond the industry publicity value. Every non-Chinese autonomous ride-hailing operator that matters is running commercial or pre-commercial operations in the same city, at the same time, in the same regulatory environment. Whichever architecture wins the reliability, cost, and scaling contest will demonstrate that outcome in Austin first. The competitive contest that will determine the shape of the autonomous ride-hailing industry is being conducted within the Texas Triangle, on the same roads, in front of the same passengers, with the same weather, traffic, and infrastructure conditions applying to all four fleets simultaneously.

The outcome hinges on two variables the near-term timeline will settle. First, when Tesla's FSD v15 or a successor version crosses the threshold for scaled unsupervised operation. Second, whether Cybercab production reaches the two-million-per-year capacity target on the projected timeline. Both events are late-2026 to 2027 targets. Both are answerable by direct observation of Austin fleet performance and Giga Texas production data over the next eighteen months. When those two questions resolve, the competitive picture restructures on the first-principles unit economics basis this section has laid out, and the industry structure that emerges is not the industry structure the current Waymo to Tesla fleet-size gap suggests.

MEGAWATT CHARGING AND AUTONOMOUS FREIGHT

The autonomous deployment with the most demanding infrastructure requirements and the most immediately compelling economic case is not the robotaxi. It is the robotruck. Long-haul trucking moves roughly 70 percent of all US freight by value. Its economics are dominated by two things: driver cost, at 35 to 40 percent of operating cost per mile, and the hours-of-service rules that cap how long a human can drive before mandatory rest. An autonomous truck removes both at once. No driver cost, no hours-of-service limit, continuous operation bounded only by charge state and maintenance. At scale, the improvement over human-driven trucking is large enough to restructure the freight industry's entire cost base.

The constraint that gates robotruck deployment, and all electric long-haul trucking with it, is high-wattage charging. A Class 8 semi needs roughly one to 1.2 MWh to cover 500 miles, the range a meaningful long-haul mission requires. Refilling that pack in a commercially useful window means charging at MW-class power. The CharIN Megawatt Charging System standard targets one to 3.75 megawatts, enough to add 500 miles to a Class 8 truck in under 30 minutes.

Megawatt charging is not a larger version of passenger DC fast charging. It requires power delivery systems, cable management, thermal management, and grid connection infrastructure that are industrial-scale rather than consumer-scale. A truck stop equipped with megawatt charging for an autonomous fleet is not an amenity. It is a power substation with vehicle interfaces, a behind-the-meter industrial power installation whose grid connection is measured in megawatts rather than kilowatts, and whose battery storage is essential rather than optional.

Texas leads every state in the Megacharger buildout, with 19 planned sites at Pilot Travel Centers across the state, each hosting four to eight stalls at up to 1.2 megawatts, construction beginning in 2026. The charging network's Texas density follows the same geographic logic as every other layer of the autonomy stack:

production campus, regulatory environment, freight corridor, and charging infrastructure all concentrating in one geography at once.

The battery-storage requirement follows directly from grid-connection economics. A hub serving dozens of autonomous trucks at once, each drawing one to three megawatts in thirty-minute windows, arriving on algorithmically optimized schedules, presents a load profile whose peak would demand a grid connection sized for the peak rather than the average. Peak-sized connections are expensive to build and carry stranded capacity cost off-peak. On-site battery storage lets the hub hold a steady grid draw sized for average demand and use stored energy to serve the peaks. It is not an optional efficiency improvement. It is the architecture that makes megawatt charging economically viable at fleet scale.

The I-35 corridor is the primary robotruck deployment geography in Texas, for the same reason it is the primary logistics corridor: it is the spine connecting the Triangle's freight origins and destinations. Autonomous truck operations run on I-35, I-45, I-10, and I-20 under a state framework that permits commercial autonomous vehicle operation without a human driver present, conditions most other states do not yet allow. Fleet and charging infrastructure are co-dependent. The fleet cannot scale without the charging, and the charging investment cannot be justified without the fleet scale that generates its utilization.

Aurora Innovation is the current commercial leader in long-haul autonomous freight in the United States. The Pittsburgh-based company launched fully driverless commercial trucking service between Dallas and Houston in 2025, becoming the first company to operate commercial self-driving heavy-duty trucks on public roads in the United States. Six months later, Aurora expanded to a 600-mile Fort Worth-to-El Paso route, the fastest scaling to a second market in the history of the US self-driving industry. By late 2025, Aurora had surpassed 100,000 driverless miles with zero safety incidents attributed to the Aurora Driver system. As of mid-2026, the company operates driverless service across five Texas and Southwestern lanes: Dallas-Houston, Fort Worth-El Paso, El Paso-Phoenix, Fort Worth-

Phoenix, and Laredo-Dallas. Customer contracts include FedEx, Schneider, Hirschbach Motor Lines, Uber Freight, McLane, Russell Transport, and Detmar Logistics. The 2026 target is deployment of hundreds of trucks, with next-generation hardware integrated into the Volvo VNL truck platform. Continental's spun-out mobility unit AUMOVIO will produce Aurora's third-generation hardware at its expanded New Braunfels, Texas plant beginning in 2027, further concentrating the autonomous freight supply chain in Texas geography.

Tesla Semi is the manufacturing-scale competitor. The Long Range variant covers 500 miles on a single charge using 4680 battery chemistry, the threshold that makes long-haul electric trucking operationally viable across the majority of US freight routes without intermediate charging stops. The Semi entered mass production in 2026 at Giga Nevada at a target of 50,000 units annually. The Semi is not yet operating autonomously, but it is built with autonomy in mind. The same FSD inference compute stack that runs in Tesla's passenger vehicles runs in the Semi, which means the path from human-driven electric truck to autonomous electric truck is a software progression, not a hardware redesign. Every Semi mile a human drives today is a training mile for the autonomous system that will eventually drive the same route without one.

The long-term Semi architecture includes platooning. Multiple Semis operating in coordinated convoy formation reduce aerodynamic drag on the trailing vehicles by significant fractions, cut energy consumption per ton-mile across the platoon, and increase highway throughput per lane. Platooning is architecturally straightforward for a fleet under unified software control. It is operationally impossible for a fleet of independently owned and operated trucks with human drivers, because the coordination overhead exceeds the platooning benefit. When Semi transitions to unsupervised autonomous operation, platooning becomes a native capability rather than a coordination challenge, and the per-ton-mile energy economics improve further.

Aurora and Tesla represent two different architectural bets in the same competitive space. Aurora is a software-and-hardware autonomy provider that integrates its Driver system into vehicle platforms manufactured by external partners including PACCAR, Volvo, and Navistar. Its scaling path depends on partner manufacturing capacity and vehicle platform availability, and its unit economics depend on how much of the total vehicle-and-Driver cost stack Aurora can compress relative to human-driven alternatives. Tesla is a vertically integrated vehicle manufacturer whose Semi is designed as a hardware-software co-optimized platform, whose autonomy software rides on the same FSD stack running across a fleet exceeding nine million consumer vehicles, and whose manufacturing scale is capped only by internal engineering constraints rather than by external supplier negotiations. Aurora has the commercial deployment lead today. Tesla has the manufacturing integration and the cost curve. The competitive contest will be settled by whether Aurora can scale its partner-vehicle deployment fast enough to hold operational lead before Tesla completes the transition of Semi to unsupervised autonomous operation. The same architectural asymmetries that shape the Cybercab-Waymo competition in passenger ride-hailing apply here, with a different competitor and a different mission profile but the same underlying question about whether integrated manufacturing and shared software stack economics compound faster than partnership-based commercial deployment can scale.

AUTONOMOUS CARGO AND DELIVERY PLATFORMS

Between the two-passenger robotaxi and the Class 8 semi lies a middle segment that both operators of the two extremes are now targeting: purpose-built autonomous cargo and delivery vehicles. The economic case for autonomy in this segment is different from long-haul freight but similarly compelling. Last-mile delivery, medium-duty urban freight, and the growing e-commerce logistics volume that feeds both are dominated by labor costs comparable to long-haul freight, but the mission profile is short-range and high-frequency rather than long-range and continuous. The infrastructure requirements are also different. A vehicle covering 100 to 200 miles per day within a metropolitan area does not need megawatt charging. It needs consumer-scale DC fast charging at existing charging network densities, which means the same charging infrastructure serving passenger vehicles serves autonomous cargo platforms without incremental buildout.

Tesla announced at its Q4 2025 earnings call that the Cybertruck production line will transition to a fully autonomous cargo delivery variant. Musk framed the use case as localized cargo delivery within a city or across a few hundred miles, which places the vehicle between the Cybercab's ride-hailing mission and the Semi's long-haul freight mission. The autonomous Cybertruck variant removes steering wheel, pedals, and driver-focused interior amenities, potentially with a modified rear body optimized for cargo rather than passenger seating. The stainless-steel exoskeleton, dent-resistant construction, full-size cargo bed, steer-by-wire, four-wheel steering, and tight turning circle that were designed for consumer utility translate directly into commercial urban logistics advantages. The vehicle competes against Rivian EDV in the electric commercial delivery segment, but with the differentiating capability of driverless operation once FSD unsupervised operation clears the threshold that Cybercab and Semi both depend on.

The autonomous Cybertruck completes a three-platform autonomy architecture in which Tesla operates vehicles across three classes serving three mission profiles. Cybercab handles the two-passenger urban ride-hailing segment at sub-thirty-cent-per-mile economics. Autonomous Cybertruck handles the mid-duty urban and short-range regional cargo segment at consumer-scale charging infrastructure. Autonomous Semi handles the long-haul freight segment with megawatt charging and eventual platooning. Each vehicle is optimized for its mission profile, each shares the same underlying FSD software stack, and each amortizes the same training data collection infrastructure across billions of consumer vehicle miles.

The three-platform autonomy strategy is not three separate business bets. It is one architectural bet whose economics compound across three vehicle classes simultaneously, on shared silicon, shared software, shared training data infrastructure, and shared manufacturing capability. This is the compounding logic that the substrate-reading methodology this book is built around predicts. The vehicles look different because their missions are different. The economics work because everything underneath the vehicles is the same.

Port and Hub Autonomy

Autonomy deployment in controlled logistics environments predates its deployment on public roads by more than a decade. It runs inside the geofenced spaces of ports, intermodal rail yards, airports, distribution centers, and warehouses, where the regulatory contest and edge-case tail that gate every public-road deployment do not apply. The invisibility of the deployment class is not incidental. It is its competitive advantage. Controlled-environment autonomy operates ahead of public-road autonomy on the maturity curve for the same reason it stays out of the news: the environmental constraints that eliminate visibility also eliminate the operational risks that force public-road programs into extended validation cycles.

Warehouse and distribution-center automation is the most mature segment. Autonomous mobile robots, the wheeled platforms that move inventory between storage and picking stations, have been operating at commercial scale for more than a decade. Amazon operates AMR fleets across its Texas fulfillment centers in North Fort Worth, San Marcos, Houston, Round Rock, and San Antonio, moving inventory through facilities whose throughput would be uneconomical without the automation. Walmart, third-party logistics operators, and every major retailer with e-commerce fulfillment volume run parallel deployments across Texas facilities. The distinction between a warehouse AMR and a public-road autonomous vehicle is one of operating environment, not underlying technology. The perception, prediction, planning, and control stack an AMR runs is a constrained version of the same AI inference stack that runs on public roads, operating in an environment whose known geometry, absent pedestrians, and fixed obstacle inventory reduce the model's decision complexity by orders of magnitude.

Intermodal rail yard automation is the next segment up the operational complexity curve. Intermodal yards move containers between rail and truck at the highest volumes anywhere in the freight system, with the labor-intensive yard-truck movements between rail sidings, container storage, and truck loading zones being the primary automation target. BNSF's Alliance intermodal facility north of Fort Worth is among the highest-volume yards in the country, and autonomous yard truck pilots and rubber-tired gantry crane automation have been running in intermodal environments across the state through 2025 and 2026. The equipment ecosystem is more concentrated than in warehouse automation: Kalmar, Konecranes, and a handful of specialty operators supply the electric yard truck and gantry crane platforms that the autonomy software integrates with.

Port terminal automation extends the same operational logic into the marine cargo environment. The Port of Houston's Barbours Cut and Bayport container terminals are among the busiest in the Western Hemisphere by tonnage, and their operations combine ship-to-shore crane movements, terminal tractor movements between crane and yard, and stacking crane movements within the yard. Fully automated terminals at Rotterdam, Long Beach, and Singapore have been operating for more than fifteen years, providing the operational playbook that Texas ports are adopting incrementally. The Texas deployment follows the same regulatory-permissiveness pattern as every other autonomy layer in the state: the framework that permits Aurora's driverless trucks and Tesla's unsupervised Robotaxi permits accelerated automation at Texas ports and yards.

HUMANOID ROBOTICS

Every autonomy deployment shares a defining characteristic: the vehicle rolls on wheels through an environment shaped for wheeled movement. Public roads exist because we built them for cars. Port terminals exist because container ships and gantry cranes require them. Humanoid robotics operate on a fundamentally different premise. The form factor is engineered to operate in environments built for humans, using tools designed for human hands, navigating spaces laid out for human bodies. The infrastructure the wheeled deployments require becomes optional. The mission profile shifts from moving people or cargo between points to performing physical labor in a place. That is not incremental automation of an existing category. It is the first automation category whose total addressable market is any physical task currently performed by humans.

Humanoid robotics is the emerging category whose commercial-scale outcome will determine whether general-purpose physical AI displaces purpose-built automation across manufacturing, logistics, construction, healthcare, and services. The Tesla Optimus factory is the Texas anchor. Optimus units operate on the Giga Texas production floor performing manufacturing tasks that were previously performed by human workers, with Tesla targeting two million units annually at full production ramp and a per-unit cost target of $30,000. Optimus rides on the same AI4 and soon AI5 silicon that runs Full Self-Driving in Tesla vehicles, sharing the compute substrate and the training data infrastructure with every other Tesla autonomy program.

Hyundai-owned Boston Dynamics has begun industrial deployment of its electric Atlas humanoid at Hyundai and Kia manufacturing plants. Figure AI has deployed its Figure 02 humanoid at BMW's Spartanburg plant since 2024. Austin-based Apptronik deploys its Apollo humanoid at Mercedes-Benz manufacturing and GXO Logistics facilities. A new 90,000 sq ft data-collection and testing site dubbed Robot Park is planned near the Domain. The competitive landscape is early, with at least five credible commercial-scale operators, each with different form factor bets, different customer commitments, different capital positions, and different scaling paths. The commercial-scale outcomes are not yet settled.

The long-term thesis is that embodied physical AI will restructure larger markets than wheeled autonomy will restructure. The wheeled deployments automate the transportation of people and cargo. That is a large market. Humanoid and quadruped robotics automate physical labor. That is many markets, and their aggregate size exceeds the transportation market by orders of magnitude when measured against the total labor value in the industries the form factor addresses. Manufacturing labor, warehouse labor, construction labor, agricultural labor, inspection and maintenance labor, hospitality and healthcare labor, security and emergency response labor: each is currently performed by humans, each is bounded only by the reliability threshold the physical AI must clear, and each represents a market whose restructuring compounds with the underlying autonomy technology. The semiconductor supply chain implications reinforce the analytical significance for the Texas thesis specifically. Each humanoid contains approximately 1,100 to 1,500 semiconductors, dominated by analog and mixed-signal devices rather than digital compute chips. If the category reaches its projected commercial scale, the end-market demand for the analog silicon Texas Instruments and others concentrate around a customer base that did not exist in 2020. Both markets concentrate their production geography in Texas. Both markets are being built at once

Texas Regulatory Posture

Texas's autonomous vehicle regulatory framework is the least celebrated and most consequential enabler of every deployment described in this chapter. California — the state most associated with autonomous vehicle development — requires permits for every test mile, mandates disengagement reporting, and has extended regulatory oversight into commercial deployment in ways that have slowed Waymo's operations and constrained other operators' timelines. Texas took a structurally different path.

Texas permits autonomous commercial vehicle operation — including fully driverless operation, no human present — under a framework that places primary responsibility on the operator rather than requiring pre-approval of specific operational design domains. The framework does not eliminate accountability. It moves accountability from pre-deployment regulatory approval to post-deployment operator liability. For an operator confident in its system's safety performance, that is a faster path to commercial deployment.

The practical result is that Texas has become the primary commercial deployment state for autonomous programs that would face longer permitting timelines elsewhere. Austin is currently the only city in the United States where four driverless commercial robotaxi services from competing operators run at once.

The posture reflects the same operating system Texas expresses across every domain: resolve ambiguity in favor of deployment velocity, hold the operator accountable rather than gatekeeping before deployment, and let market discovery identify the configurations that work before regulation tries to prescribe them. Applied to autonomy, that posture has made Texas the geography where the transition is most visibly underway — not because the regulatory environment is reckless, but because it is built to let deployment happen and hold operators accountable for outcomes, rather than preventing deployment until outcomes can be guaranteed.

Thirteen

THE SPACE AND DEFENSE LAYER

THE FALCON 9 BOOSTER COMES BACK. It separates, flips, relights three engines against the fall, and settles onto a landing pad or a drone ship with a precision that still looks, after more than 600 successful recoveries, like a special effect. That image — the booster returning — is what the public sees when it thinks of SpaceX, and most observers file the company under the word the image suggests: launch.

The launch frame is not wrong. It is incomplete — incomplete the same way the automotive frame for Tesla is incomplete. It describes the most visible current activity while missing the system that activity is funding. Launch is the largest single line item in SpaceX's disclosed finances. But launch is the access mechanism, not the enterprise.

The enterprise is a vertically integrated orbital infrastructure company. Launch is how every other layer reaches orbit. Starlink is the connectivity layer — global broadband from a constellation that has passed 10,500 satellites as of mid- 2026 and is still expanding. The Starmind orbital compute layer — AI inference pushed to low Earth orbit rather than round-tripped to ground-based data centers, running on the D3 radiation-tolerant chip Terafab will build specifically for it — is the layer being developed. And Starship is the infrastructure that makes all of it economically sustainable at planetary scale: the fully reusable heavy-lift vehicle whose economics, if achieved as designed, cut the cost of reaching orbit by an order of magnitude against any prior system.

These layers do not interact with the Texas AI-industrial complex as a separate enterprise that happens to sit down south. They are the orbital extension of the same compounding system. Understanding the space and defense layer means holding both the orbital system and its terrestrial anchor at once.

Starbase Complex

Starbase at Boca Chica is the first physical anchor of SpaceX's next orbital infrastructure program. It sits on the Gulf Coast near Brownsville, roughly six hours south of Austin – outside the Texas Triangle's geometric boundaries, but inside its operational system. The labor market connection is direct. SpaceX's Bastrop engineering presence is substantial, drawing on the same talent pool as Tesla and the broader Austin technology cluster, with engineers moving between Bastrop offices and Starbase operations on schedules that make the six-hour distance manageable. The supply chain connection runs through the same corridor infrastructure that serves the rest of the Triangle.

A single Starship launch consumes approximately 4,600 metric tons of propellant – roughly 3,400 tons in the booster, 1,200 tons in the upper stage – at a 78/22 mass ratio of liquid oxygen to liquid methane. At one launch per month, that is 55,000 tons annually. Liquid oxygen is produced on-site through air separation units, paired with merchant industrial gas supply for surge capacity. Liquid methane is produced through cryogenic liquefaction of pipeline-sourced natural gas. The same Gulf Coast natural gas and industrial gas infrastructure originally built for petrochemical and LNG export now feeds the orbital gateway at the southern tip of the state – a draw on the regional energy substrate that scales as a function of launch cadence rather than as a function of Starbase's physical footprint.

The water deluge system at the orbital launch mount releases on the order of 350,000 gallons per launch in the first seconds of engine ignition, absorbing acoustic energy and protecting the pad from thermal damage. The deluge water draws from on-site storage fed by Cameron County supply, with much of it vaporized to steam during the launch event — the system is not closed-cycle in any meaningful operational sense. At cadence, the combined draw of propellant production water, deluge water, and broader industrial cooling at Starbase becomes the unresolved infrastructure question for Cameron County water capacity over the long horizon.

The flame deflector the deluge protects, and the launch table and mechazilla catch infrastructure whose maintenance and inspection cycles determine pad availability, are the ground support constraints that determine turnaround time as directly as the vehicle's own refurbishment requirements. High launch cadence requires not just a fast vehicle but a fast pad — and a fast pad in an arid coastal environment requires a water supply chain Starbase has to support from a standing start.

Starbase's scope has expanded continuously since SpaceX chose the site for Starship development. The expanded Gigabay facility now encompasses Starship production, static fire testing, and orbital test launch operations — a vertically integrated production and launch campus whose footprint has consumed most of the available land in the immediate Boca Chica geography. Its operational tempo has transformed the surrounding South Texas community on a smaller scale than, but parallel to, what Giga Texas, Bastrop, and Samsung Taylor are doing further north.

The Rio Grande Valley geography adds something the Triangle's interior nodes do not have: proximity to the Mexican manufacturing base. The maquiladora industry along the border has produced decades of precision manufacturing capability, cross-border logistics frameworks, and supplier infrastructure that aerospace and defense operators can reach from Boca Chica and cannot reach from further north. SpaceX's regional supplier relationships follow that logic — a high-volume rocket production facility sited within reach of the border's manufacturing ecosystem.

McGregor Facility

Starbase is the visible face of SpaceX's Texas operations. The less visible but equally load-bearing component sits 460 miles north: the McGregor Rocket Development and Test Facility, southwest of Waco — the propulsion manufacturing and qualification anchor at the northern end of the Austin–Boca Chica axis.

McGregor is where SpaceX's rocket engines are built and tested, and its history runs deeper than the company's recent Texas concentration. SpaceX acquired the site in 2003. That makes it one of the company's oldest operational facilities — and one of the earliest signals that Texas would become the geographic center of SpaceX manufacturing and test. The facility's test stands have fired every major SpaceX engine family across two decades, accumulating propulsion testing infrastructure and institutional knowledge no other single site in the US rocket industry holds at comparable scale and cadence.

The current operational significance is Raptor. The Raptor engine — the full-flow staged combustion methane-oxygen engine that powers both the Super Heavy booster and the Starship upper stage — is produced and tested at McGregor at the scale Starship's launch cadence requires. Raptor V3, the current production variant, is being manufactured at volume as of early 2026. The production rate reflects the engine count high-cadence Starship operations demand: Super Heavy carries 33 Raptor engines, the upper stage carries between 3 and 6 depending on configuration, and full reusability with rapid turnaround requires engine hardware robust enough to fly multiple times without refurbishment between launches. The production volume at McGregor is not an engineering program. It is an industrial manufacturing operation whose output rate is the binding constraint on Starship launch cadence.

Raptor V4 is in testing at McGregor now. The V4 program targets improvements in thrust-to-weight ratio, manufacturing simplicity, and reusability cycle life — the three variables whose compound improvement determines whether Starship's designed economics are achievable at the cadence the orbital infrastructure program needs. Engine testing at McGregor is the qualification gate every Raptor variant must pass before flight: the controlled environment for the full-duration firings, thermal cycling, and anomaly characterization flight certification requires, without the launch-complex constraints that make anomaly investigation at Starbase more operationally complex.

STARSHIP

Starship is the load-bearing technology for everything the orbital infrastructure thesis depends on. Without Starship reaching its designed economics the Moon Base timeline extends by decades, Mars colonization stays beyond any near-term model, and orbital infrastructure deployment costs stay high enough to cap the constellation densities orbital compute requires. With those economics achieved, the cost of reaching orbit falls by an order of magnitude, and every downstream application whose economics depend on launch cost becomes viable on a compressed timeline.

Those economics are not guaranteed. Starship's development has included dramatic failures whose footage has been covered exhaustively. What that coverage consistently underweights is the methodology producing them. SpaceX's iterative hardware testing philosophy treats early failures as information, not as program-threatening events. The Falcon 9 followed this pattern: early failures, rapid iteration, eventual operational reliability that made it the most-launched orbital rocket in history. Starship is following the same methodology, at larger scale and higher complexity.

The architecture's key innovation is not its size — though the size, 200 metric tons to low Earth orbit for V3 in fully reusable configuration, is unprecedented. The key innovation is full reusability of both the Super Heavy booster and the Starship upper stage. It turns a vehicle that needs Earth-launched propellant for every mission into a vehicle that can refuel one day where it lands —

the architectural requirement that makes a Moon Base and Mars colonization economically tractable rather than a one-way expenditure of Earth-sourced propellant.

Starlink Factory

The ground infrastructure that connects the Starlink constellation to its users is manufactured in Bastrop, 30 miles east of Austin on SH 71. The Bastrop factory produces Starlink user terminal kits — the ground-based phased-array antennas whose global distribution makes the constellation's connectivity reachable by the vehicles, autonomous systems, commercial customers, and individual subscribers the network serves. Starlink satellite manufacturing itself stays in Redmond, Washington. The Bastrop factory is the largest printed circuit board manufacturing facility in the United States: over one million square feet, annual production capacity above 4.7 million units. SpaceX board member and investor Steve Jurvetson, after touring it, called the line "the closest thing to a fully automated alien dreadnought that I have ever seen."

Starlink's 10,500-plus satellite constellation provides global broadband whose strategic significance runs well past its consumer internet revenue. For the Texas AI-industrial complex, Starlink is the connectivity foundation enabling four specific capabilities whose value compounds as the autonomy and robotics deployments scale.

The first is over-the-air update delivery to autonomous systems operating anywhere on Earth. A Tesla FSD update, an Optimus control update, a Cybercab refinement — each requires connectivity between the training infrastructure at Tesla Cortex and the deployed hardware in the field. In cities with dense cellular coverage, terrestrial networks suffice. Across oceans, in the skies, in the developing-world geographies where autonomous agricultural equipment and mining trucks may operate, terrestrial coverage is absent or unreliable. Starlink, coupled with a vehicle-mounted Starlink Mini dish, is the global connectivity layer that makes truly global autonomous deployment viable rather than geographically constrained.

The second is remote-operation fallback for edge cases — scenarios rare enough that dedicated teleoperator infrastructure is uneconomical, but consequential enough that when they occur they require a human. Starlink's low-latency LEO architecture, 20 to 40 milliseconds for the Gen 2 constellation, makes real-time remote operation feasible where latency would introduce unacceptable control lag.

The third is the data uplink from deployed systems back to the training infrastructure. The FSD miles Tesla's fleet has accumulated had to travel from vehicles to Cortex. As the autonomous fleet scales — millions of Cybercabs, millions of Optimus robots, thousands of autonomous trucks operating globally — the uplink bandwidth requirement scales with it. Starlink provides the global uplink capacity terrestrial cellular networks cannot give a globally distributed fleet 24/7.

The SpaceX FCC filing proposes a million-satellite Starmind constellation carrying in-orbit compute rather than connectivity. It is the next-stage architecture, and its approval and deployment would add a fourth capability: inference at the orbital layer, eliminating the ground-station round-trip for latency-sensitive applications. The compute capacity it would provide draws on a power source — continuous solar generation in SSO, without atmospheric loss or nighttime interruption — that is structurally superior to terrestrial generation for sustained high-utilization workloads.

GIGASAT COMPLEX

The most strategically significant industrial buildout at Bastrop is the Gigasat complex — an eleven-million-square-foot, five-building vertical integration of the orbital AI satellite, dubbed AI1, supply chain on a single 1,000-acre campus, announced by SpaceX in June 2026 a few days ahead of the company's IPO. Bastrop is being assembled as the production hub for the orbital compute constellation, and the supply chain that constellation requires cannot be sourced from the existing supplier base at any price or any timeline.

The first building under construction is the solar cell factory, whose permit filings carry the name Project Echo Solar. The footprint is 1.1 million square feet at initial buildout, with an annual production capacity of 10 GW of aerospace-grade solar cells across two production floors rated at 5 GW each. The process equipment — diffusion furnaces, PECVD systems, screen printers, firing furnaces — is the conventional terrestrial silicon solar cell manufacturing stack, adapted to help meet the radiation tolerance and reliability that orbital deployment demands.

The space-grade solar cell market is currently a small, high-margin oligopoly — Boeing's Spectrolab, Rocket Lab's SolAero, a handful of European and Japanese suppliers — whose combined capacity was sized for the modest volumes of prior satellite generations. SpaceX is already the largest commercial purchaser of space-qualified solar cells in the world. A million-satellite Starmind constellation requires solar panel volumes that supplier base cannot deliver. In-house manufacturing is not the preferred option. It is the only option at this scale.

The radiation tolerance requirement connects Project Echo to the broader orbital compute program at the level of physics. Total ionizing dose (TID) accumulation damages silicon solar cell efficiency through the same particle bombardment that damages silicon chips. Damage from high-energy protons degrades power output continuously across the satellite's operational life. The radiation tolerance specification for Project Echo's cells is therefore not a performance optimization. It is a survival requirement, its engineering as demanding as D3's. Both programs qualify their hardware against the orbital radiation environment at the Cyclotron Institute.

Project Echo is the first of Gigasat's five buildings. Beside it: a solar ingot and wafer plant providing the upstream silicon feedstock for the cell line; a user terminal, gateway, PCB, and silicon component manufacturing building consolidating ground-equipment production that was previously distributed across smaller facilities; the AI1 satellite development and testing facility; and the AI satellite production and assembly building, the largest of the five, where the satellites are assembled - power, compute, thermal, and the communications system.

The five buildings together formalize the supply chain on one campus: D3 silicon to solar to integrated satellite, with the ground equipment that completes the network in the same complex. SpaceX has stated production targets of 1 GW of orbital AI compute capacity per year by the end of 2027 — approximately six thousand AI1 satellites annually — scaling toward 10 GW per year by 2028 and aspirationally 100 GW per year by 2030. First-phase capital is in the tens of billions of dollars. The Gigasat complex alone will exceed the size of Giga Texas.

Bastrop Compound

From SH 71 east of Austin the Bastrop compound now reads as a cluster that did not exist three years ago — the original Starlink terminal factory, the Gigasat complex rising beside it, the Boring Company yard, X.com headquarters, and the Snailbrook row, all visible from the highway, all assembled in the last three years.

Adjacent to the Starlink factory is the existing advanced semiconductor research and chip packaging facility focused on Starlink hardware. Its focus is the silicon on the terrestrial ground-equipment side of the system — the chips inside the phased-array, routers, and terminals — rather than the orbital compute silicon whose fabrication runs through the Terafab program at Giga Texas and Gibbons Creek. The Bastrop semiconductor facility brought the chip-packaging stage of the Starlink ground-equipment supply chain under SpaceX's direct control: the same vertical integration logic that produced the in-house terminal kit factory and soon Gigasat itself.

The Boring Company sits nearby on 73 acres with roughly 200 employees, continuing tunneling technology development whose potential lunar and Mars application is documented later.

Across the street sits Snailbrook — Musk's planned self-sustaining company town on 3,500 acres, providing housing, dining, and amenities for employees of the Bastrop and Giga Texas operations. Currently operational with a restaurant, pub, salon, and recreational facilities, Snailbrook is the first company town established by a technology operator in Texas since the oil patch era. It is the sovereign-autonomy thesis at the corporate-community level: an operator building the residential and social infrastructure for its own workforce rather than depending on the surrounding municipality to provide it.

At full buildout the Bastrop compound — Gigasat at 11 million square feet, the existing Starlink terminal factory, the semiconductor research and packaging facility, the Boring Company operations, X.com, and Snailbrook's 3,500-acre town — exceeds the integrated industrial footprint of Giga Texas and Samsung Taylor combined.

Starshield Program

The intersection of SpaceX's orbital infrastructure with the US defense and intelligence community is not a recent development. SpaceX has been a significant defense contractor since the early Falcon 9 era, launching national security payloads under contracts with the Space Force, the NRO, and other defense and intelligence customers. What has changed as the company's capabilities expanded is the depth of the integration — the degree to which US national security space architecture now depends on SpaceX's specific capabilities rather than treating it as one launch provider among several.

Starlink's role in the Ukraine conflict demonstrated a national security utility the DoD had not fully anticipated and has since worked to formalize. The Starshield program, SpaceX's defense-oriented variant of Starlink with enhanced encryption, inter-satellite link capability, and hosted payload capacity for government sensors, is that formalization.

Golden Dome's architecture creates a more direct integration still between SpaceX's orbital infrastructure and the US missile defense system. Hypersonic threat intercept means detecting, characterizing, and maintaining track on fast-moving maneuvering targets through their full flight profile — and that requires sensor coverage whose geometry is most efficiently achieved from orbit rather than from ground-based or geostationary assets. Three components combine to meet it: SpaceX's Starshield constellation infrastructure, the launch cadence Starship's economics will enable for rapid satellite replenishment, and the D3 orbital compute chip Terafab is developing for radiation-tolerant orbital inference. Their integration creates a space-based missile defense sensor and compute layer that the prior generation of US missile defense architecture — built around ground-based radars and a few expensive exquisite satellites — cannot match in coverage, responsiveness, or cost per defended intercept.

Texas Instruments Anchor

The radiation-tolerant silicon discussion has emphasized D3 and the Terafab production facility at Gibbons Creek — the digital compute fabrication required for the AI workloads the orbital constellation will run. That coverage is correct and proportionate to D3's strategic role. It also obscures a parallel supply chain that has been operating for decades, and produces the silicon that makes every satellite — including the AI1 — actually function: the analog and mixed-signal radiation-tolerant chip ecosystem anchored at the Texas Instruments' Richardson-Sherman fab corridor north of Dallas.

Every satellite requires a layer of silicon around the digital compute that converts solar-panel current into the regulated voltages the compute needs; converts sensor signals into the digital data the compute can process; converts compute outputs into analog drive signals for actuators, transmitters, and control surfaces; conditions and filters analog signals across the bus; and provides the interfaces and clocks that link every subsystem together. Power management, data conversion, signal chain, interface, clocks and timing — these are product categories that constitute the silicon ecosystem around the compute. By device count, a typical satellite contains far more analog and mixed-signal ICs than digital compute chips. The compute is one chip. The supporting silicon ecosystem is dozens.

Texas Instruments is the dominant US producer of this silicon for space and defense applications. The Sherman 300mm fab complex — Sherman Manufacturing Site 1 and Site 2 operational, with Sites 3 and 4 in the master plan — is the world's largest analog and embedded processing semiconductor manufacturing concentration. TI's space-grade product portfolio spans three qualification tiers. The supply chain that produces these tiers has been operating at industrial scale, on commercial terms, for decades.

The applications inventory is essentially the inventory of US defense and aerospace itself: the F-35's mission computers, avionics, and power management buses; the Patriot, THAAD, and Aegis radar and guidance chains; every major commercial and military communications satellite from Iridium to Starlink to the Starmind generation; Mars rovers, lunar landers, and the ISS visiting fleet; the autonomy stacks emerging at Fort Hood and the broader Army manned-unmanned teaming program; the Golden Dome homeland missile defense architecture as it materializes. TI Sherman and its predecessor fabs supply the analog and mixed-signal silicon for nearly every major US spacecraft and defense platform of the past two decades. The dominance is structural rather than incidental: TI's combination of breadth across the analog and mixed-signal product portfolio, mature space-qualification infrastructure, and US-soil 300mm production volume makes the company the default supplier for any program with space-grade or military-grade reliability requirements.

The structural point matters. The orbital silicon supply chain is not single-layered. The digital compute layer — D3 and its successors — does require captive fabrication for reasons documented in the Terafab discussion: no merchant foundry will develop a radiation-hardened process variant for SpaceX's specific orbital compute requirements at the volumes a single program requires. The analog and mixed-signal layer faces no such constraint. The merchant foundry model produces space-qualified analog silicon profitably and at scale, because the customer base is broad — every spacecraft built by any operator anywhere depends on it — and TI's investment in Sherman has already amortized across that base. The result is a hybrid orbital silicon supply chain: captive digital from Terafab, merchant analog from TI Richardson and Sherman north of Dallas.

F-35 and the DFW Defense Anchor

Dallas–Fort Worth's defense-industrial identity is anchored by the F-35 program at Lockheed Martin's Fort Worth facility — the final assembly and checkout location for the most advanced tactical aircraft in the US inventory, and the primary fighter platform for a coalition of allied air forces whose combined procurement represents decades of sustained production demand.

The F-35's relevance to the thesis is not its airframe. It is its avionics: the electronic warfare systems, sensor fusion architecture, and communications infrastructure that define the aircraft's capability edge. The F-35 program and the Trusted Foundry program are connected not by organizational relationship but by the supply chain logic a long-duration, high-value defense program creates for the domestic semiconductor infrastructure being built in the same regional geography.

The broader DFW defense electronics cluster — Raytheon, L3Harris, Bell Helicopter, and the dozens of tier-two and tier-three defense electronics suppliers concentrated in the metroplex — extends the F-35 anchor into a defense-industrial base whose aggregate semiconductor demand is large enough to influence the economics of domestic defense-grade fabrication.

Fort Hood and the Autonomy Agenda

Fort Hood sits 30 miles west of Temple. Its modernization agenda connects the defense layer to the commercial autonomy deployment in ways standard defense procurement coverage does not surface.

The US Army's investment in autonomous logistics, autonomous resupply vehicles, autonomous route clearance systems, and autonomous surveillance platforms creates a procurement relationship with the commercial autonomous vehicle ecosystem — one that benefits from proximity to the country's most advanced commercial autonomy deployment geography.

The Army's requirements differ from commercial ones in ways that matter for the supply chain: ruggedization for off-road and adverse-weather operation, electronic warfare resilience, cybersecurity frameworks commercial systems are not designed for, and the logistics supportability requirements that define military procurement. But the underlying autonomy stack — perception, prediction, planning, control, over-the-air (OTA) update infrastructure — is the same stack commercial programs are developing, and the commercial development pace is faster than any defense-only program could sustain. Fort Hood's proximity to Austin's commercial autonomy cluster creates the technology transfer pathway: through defense contractors who serve both markets, through engineers who move between commercial and defense programs, through the shared talent pool the I-35 corridor produces.

Saronic and Port Alpha

The United States Navy faces a shipbuilding capacity gap versus China's People Liberation Army Navy that cannot be closed by the traditional US shipbuilding industrial base on any near-term timeline. Office of Naval Intelligence assessments have put China's aggregate shipbuilding capacity at roughly two hundred times the US figure. The PLA Navy already fields more hulls than the US Navy and is projected to expand past 400 vessels by 2030 against a US fleet of roughly 290. The four remaining major US shipyards capable of building naval combatants — HII Newport News, HII Ingalls, Bath Iron Works, and General Dynamics NASSCO — deliver frigates in five to seven years, destroyers in four to five, and carriers in eight to ten. That baseline capacity is what the current force structure was designed around. It is also what makes closing the numerical gap through conventional shipbuilding structurally impossible: no conceivable capital expansion of the existing yards can compress those timelines by the order of magnitude the strategic problem requires. The Navy has adopted a hybrid fleet doctrine — complementing manned combatants with numerically large fleets of autonomous surface and subsurface vessels — as the only architecturally coherent path to closing the gap. The doctrine requires production infrastructure the existing shipbuilding base does not have.

Saronic Technologies, headquartered in Austin, has a product line that spans four autonomous surface vessel classes: the 6-foot Spyglass scout vessel, the 14-foot Cutlass, the 24-foot Corsair, and the 180-foot Marauder. The vessels operate through Saronic's Echelon command-and-control platform, which lets a single operator direct individual vessels or coordinate swarms. The design principle is autonomy from the keel up rather than autonomy retrofitted onto a manned hull, with the software layer treated as the load-bearing engineering rather than a bolt-on capability.

Port Alpha is the industrial-scale expression of the Saronic thesis, a $3.2 billion next-generation shipyard at the Port of Brownsville designed to produce autonomous vessels at speed and scale not seen since World War II. The buildout is planned as four phases over ten years, with a projected employment target of approximately 10,000. The Rio Grande Valley will host the two most consequential vertically integrated aerospace and defense industrial buildouts in the United States just 19 miles apart along the Gulf Coast geography.

Port Alpha's strategic significance runs past the private capital deployment it represents. If the Navy's hybrid fleet doctrine matures into procurement at the scale the strategic problem requires — thousands of autonomous vessels rather than dozens — Port Alpha is one of the two production facilities the doctrine will need. Anduril at the old Foss Shipyard in Seattle is the other. That makes Port Alpha part of the national defense-industrial infrastructure regardless of who holds the private capital that builds it. The Rio Grande Valley siting places the facility within reach of the same Mexican precision manufacturing base that supports SpaceX's supplier relationships, the same Gulf Coast propellant and industrial gas infrastructure the launch complex draws from, and the same defense-adjacent workforce pipeline up north that the Fort Hood and Austin autonomy clusters produce. The industrial buildout across launch, orbital compute, and analog silicon extends to autonomous naval combatants through the same geographic and workforce logic.

Blue Origin

The corridor's pull may be drawing the second major commercial orbital launch operator into the same geography as the first. Blue Origin, Jeff Bezos's aerospace company headquartered in Kent, Washington, is evaluating a 1.3 million-square-foot manufacturing, R&D, warehouse, and logistics campus in Hutto. The project, referred to publicly as Project Blue Hub, is valued at $650 million to $1 billion and projected to bring 2,000 jobs to a community whose Skybox-Prologis PowerCampus data center campus is partially operational on the same corridor. Hutto as of June, 2026 was the leading Texas candidate location.

Blue Origin's Texas footprint predates the Hutto announcement. Launch Site One at Corn Ranch — 165,000 acres roughly 30 miles north of Van Horn in West Texas — is the operational launch infrastructure from which New Shepard has flown since 2015, with BE-4 engine test stands supporting both New Shepard and New Glenn propulsion development. Van Horn is geographically distant from the Austin corridor but functionally integrated with it — the same relationship McGregor's Raptor test facility has with Starbase and Austin. Texas hosts launch infrastructure for both of the world's leading commercial launch operators at once, at facilities 730 miles apart on opposite sides of the state.

Blue Origin's orbital programs compete with SpaceX's on two fronts at once. Project Sunrise calls for a 50,000-satellite orbital data center constellation, and is the direct competitive response to SpaceX's million-satellite Starmind constellation. Both filings describe essentially the same architecture: networks of satellites performing advanced computation in orbit, powered by continuous solar energy. The orbital compute layer is becoming a competitive industry whose two leading participants are both Texas-rooted.

Fourteen

THE ENERGY MANUFACTURING LAYER

ONE MANUFACTURING LAYER REMAINS — and it is the layer that powers all the others.

The layer has a dominant anchor and a second significant operator. The anchor is Tesla Energy. Tesla is understood, in most coverage, as a vehicle company with an energy business attached — the energy business treated as a secondary segment, a portfolio diversification, a hedge. That framing is now badly out of date. Tesla Energy is a manufacturing operation building grid-scale battery storage, the integrated grid node that replaces the field-built substation, and terrestrial solar generation hardware at a scale that would place Tesla as the largest solar manufacturers in the world. The energy business is not a hedge. It is a manufacturing layer, and its physical center of gravity is a single industrial campus west of Houston.

The second operator is T1 Energy, an unaffiliated company assembling the first fully domestic utility-scale solar manufacturing supply chain, anchored at a cell fab in Milam County and a module plant near Dallas.

Brookshire Campus

The Tesla campus sits in Brookshire, approximately 35 miles west of Houston. It is not in the Austin corridor. It is the first major node of the AI-industrial convergence documented in this book whose pull points toward Houston rather than the Austin metro — and the reasons it points there are specific and instructive.

The campus hosts two co-located Tesla Energy manufacturing operations. The first is the Megapack Factory — a roughly $200 million facility producing the Megapack 3 utility-scale battery and the Megablock grid node, at 50 GWh of annual capacity. The second is the Brookshire Solar Manufacturing Complex, a fully vertically integrated terrestrial solar manufacturing operation under construction. The two operations will share a campus, a logistics footprint, and a supply chain organization. Together they constitute the physical center of Tesla Energy's manufacturing presence in Texas.

Houston offers three advantages the Austin corridor does not. The first is port access: the Port of Houston is a major equipment-import gateway, and a manufacturing program that depends on imported production equipment needs the deep-water port proximity Austin cannot provide. The second is the industrial labor pool: Houston's petrochemical and heavy-industrial base has produced a deep workforce in exactly the construction trades, process operations, and facility engineering manufacturing requires. The third is existing industrial infrastructure: the Empire West Business Park at Brookshire already had the power, the road access, and the industrial zoning a manufacturing campus needs — which compresses the time from lease to production.

The Megapack Factory

The Megapack is the utility-scale battery. The Megapack 3 is the current-generation unit. The Megablock is the integrated grid node: four Megapack 3 units, totaling approximately 20 MWh of storage, combined with a Tesla-built transformer, the Tesla inverter stack, medium-voltage switchgear, integrated thermal management, on-skid controls, and an energy management system — all assembled and pre-certified into a single factory-built skid. Tesla claims this plug-and-play architecture can commission 1 GWh of storage in 20 business days, drastically reducing installation complexity and construction costs.

The data center chapter documented why the Megablock matters for the buildout: it converts the field-built substation — a multi-vendor construction project gated by 12 to 24-month transformer lead times — into a factory product that ships on a truck and is deployed at manufacturing speed. The Megablock forms the backbone of a utility-scale microgrid ideal for gigafactories, fabs, and data centers. The Megablock's deployment advantage exists only because there is a manufacturing operation capable of producing Megapacks and the integrated skids at volume.

The Tesla Solar Factory

The solar half of the Brookshire campus is even more ambitious. Tesla's solar manufacturing history is a decade-long record of uneven commitments and modest restarts, beginning with the 2016 acquisition of SolarCity and the Buffalo, New York factory — promised as the largest solar manufacturing plant in the Western Hemisphere, and never that. For years the Buffalo facility produced Supercharger components and served as a data labeling site rather than a solar plant. Tesla's solar deployment fell so low the company stopped reporting it as a separate metric. A quiet revival began in late 2025 with a new panel produced at Buffalo — but Buffalo's capacity is a rounding error against the program Tesla has now committed to.

The new committed program is 100 GWs of annual domestic solar manufacturing capacity — announced as a target spanning the full supply chain, from raw materials to finished panels. The number is worth pausing on. One hundred gigawatts of annual solar manufacturing capacity is not an incremental expansion of the Buffalo operation. It is a manufacturing program more than two orders of magnitude larger, and it would place Tesla, on its own, as the largest solar manufacturer in the world.

What distinguishes the Brookshire solar operation from most domestic solar manufacturing is that it is fully vertically integrated, and the vertical integration extends to the very beginning of the production chain. Most solar cell manufacturing begins with purchased silicon wafers, sliced and polished by an upstream supplier. The Brookshire complex is being built to grow its own silicon ingots, slice its own wafers, produce its own photovoltaic cells, and assemble its own finished panels — the entire chain under one operation. Ingot growth is a categorically more demanding manufacturing capability than cell production from purchased wafers. It is closer to a semiconductor crystal-growth operation than to a conventional solar panel assembly line, requiring cleanroom-grade manufacturing environments and process control of a kind the solar industry has historically concentrated in a small number of specialized producers, most of them in China.

The supporting equipment reflects that ambition — and contains a precise irony. To build domestic solar manufacturing capability that reduces dependence on Chinese solar panels, Tesla has moved to acquire Chinese solar manufacturing equipment: a procurement program reported at $2.9 billion from Chinese equipment suppliers. Solar manufacturing equipment is one of the few elements of the solar supply chain not subject to US tariffs, which makes the Chinese-equipment route the fastest available path to domestic production capability. The irony is real, and it is also the rational move — the fastest way to build an independent domestic solar supply chain runs, at the equipment-acquisition stage, through the China supply chain Tesla is trying to become independent of.

The Brookshire solar program must be distinguished from two other Texas solar manufacturing programs, because the three are frequently conflated and they are genuinely different. SpaceX's Project Echo Solar facility at Bastrop, documented previously, is a separate program — smaller, on the order of 10 GW, producing radiation-hardened solar cells for orbital deployment rather than terrestrial panels. The other is T1 Energy.

THE T1 ENERGY SOLAR FACTORY

Tesla Energy is moving towards being a dominant operator in the solar energy manufacturing layer, but it won't be the only one. T1 Energy, an unaffiliated company, is assembling what will be the first fully domestic utility-scale solar manufacturing supply chain in the United States, and the two principal nodes of that chain are in Texas. T1 Energy is a different company from Tesla, pursuing a different strategy at a different scale, and the energy manufacturing layer is not fully described without it.

The achievement T1 Energy is pursuing is the closure of the solar supply chain on domestic soil — a chain China has dominated at every stage for two decades: polysilicon, wafers, cells, modules. T1's G2_Austin facility, under construction at Rockdale, is the cell-manufacturing node. Rockdale is on US 79 — the same Milam County stretch where the Sandow Lakes data center campus pulls its power. Its Phase 1 capacity of 2.1 GW of high-efficiency TOPCon cells, targeted for production by the end of 2026, exceeds the entire prior US capacity to manufacture silicon-based solar cells. T1's G1_Dallas facility, a module-manufacturing plant at Wilmer, assembles cells into finished panels. With polysilicon and wafers supplied from a domestic upstream source, the chain runs end to end without leaving the country: domestic polysilicon and wafers, T1 cells at Rockdale, T1 modules near Dallas. The first fully domestic utility-scale solar supply chain is, at both its cell and module stages, a Texas manufacturing operation.

The contrast with the Tesla Brookshire program is instructive. The two operations occupy the same layer but solve the domestic-solar problem differently. Tesla Brookshire pursues vertical integration within a single operator and a single campus, extending all the way upstream to ingot growth, the most demanding stage of the chain. T1 Energy pursues the domestic supply chain as a multi-node, multi-facility system, with a distinct upstream polysilicon and wafer supplier feeding T1's own cell and module plants. Tesla's model concentrates the chain inside one company; T1's distributes it across a domestic supplier network. Tesla's announced terrestrial scale, anchored at the 100-GW target, is an order of magnitude larger than T1's current committed capacity. Both are real manufacturing programs, both are sited in Texas, and both are responses to the same strategic problem: the concentration of the global solar supply chain in China, and the resulting exposure of every downstream solar deployment to a single national supply base.

The presence of two distinct terrestrial solar manufacturing programs, from two unaffiliated operators, alongside the separate SpaceX space-grade solar program, is itself a data point about the convergence. Three solar manufacturing programs, three operators, three scales, three strategies — all sited in Texas. Texas is becoming a solar manufacturing geography at the same time, and for the same underlying reasons that it is becoming a semiconductor, battery, robotics, autonomy, gigafactory, and AI compute geography. The energy manufacturing layer, like every other layer, is concentrating in the same place.

Energy Manufacturing as System

Tesla Energy's products are increasingly sold as integrated. A solar generation source without battery energy storage is non-functional as firm power; a storage system without generation has nothing to store. The two halves of the energy stack are co-dependent, and manufacturing them on one campus, with shared logistics and a shared supply chain, is the production-side expression of a product strategy that already treats them as a single system.

The same integrated stack has an application beyond the terrestrial grid. Generation paired with storage paired with dispatch intelligence is exactly the architecture a lunar South Pole installation or a Mars surface base will require — because a microgrid facing dust storms, eclipse periods, and extreme thermal cycling is the same engineering problem as one on the terrestrial grid, only less forgiving. Brookshire manufactures the terrestrial version. The same product organization, the same vertical integration, the same Megablock architecture are what the outer spheres may eventually draw on.

The energy manufacturing layer is the layer beneath the entire convergence — and it is also the layer that will travel.

PART III

THE CRITICAL DENSITY

Fifteen

THE CAPITAL STACK

THE AI-INDUSTRIAL BUILDOUT IN THE Texas Triangle is not a market phenomenon with federal support layered on top. It is a market phenomenon and a federal strategic infrastructure program operating simultaneously in the same geography, each reinforcing the other in ways that neither would produce independently. Separating the two misreads the structure of what is being built.

The capital that integrates the two flows through five distinct layers.

The structure is not new. Pittsburgh's steel industry had the same five-layer capital structure: federal tariff protection and land grants, state infrastructure investment, county and municipal industrial recruitment, Carnegie and Frick private capital, and the entrepreneur risk capital of the supply chain builders around the steel anchor. Detroit's automotive industry had the same structure with different instruments.

The Texas Triangle has the same structure with the instruments this era provides: CHIPS Act instead of tariffs, TSIF instead of land grants, MUD financing instead of railroad bonds, infrastructure REITs instead of robber baron capital, venture funds instead of Gilded Age entrepreneur risk capital.

The Federal Layer

The federal government has made a deliberate decision to construct a domestic semiconductor and AI infrastructure stack in Texas. The capital committed, the programs funded, the institutional frameworks being stood up — these are not the sum of independent agency responses to grant applications. They express a strategic logic, implemented at a scale the United States has not attempted since the semiconductor industry was born out of the defense procurement relationships of the 1960s.

The logic rests on three premises.

First: the United States cannot hold its military and economic position if the silicon running its weapons, its communications, its AI platforms, and its critical civilian infrastructure is fabricated exclusively in Taiwan and South Korea — jurisdictions whose exposure to Chinese pressure is a supply chain vulnerability no amount of diplomacy erases.

Second: rebuilding domestic leading-edge fabrication from a standing start needs federal capital at a scale private markets will not commit without the demand certainty and risk mitigation only a federal commitment provides.

Third: the AI capability competition with China is not, at bottom, a software or model competition. It is a silicon competition. The side that controls leading-edge fabrication controls the hardware on which frontier AI capability is built.

Hold those three premises together and they produce the CHIPS and Science Act of 2022, and the investment pattern that has followed it into Texas.

The federal decision is not only to rebuild domestic fabrication. It is to concentrate it. A complete semiconductor and AI-industrial stack — raw material inputs, fabrication, process prototyping, IP and design, specialized devices, photomask supply, the technician workforce — is being assembled in one geography rather than spread across the country. The Austin cluster already holds the anchor pieces. Samsung Taylor brings leading-edge process capability. The UT Austin Texas Institute for Electronics brings a

DARPA-funded prototyping mandate. A supplier ring spans photomask, process materials, chip architecture IP, and silicon design. And the community colleges and training schools feed technicians into fabs and equipment suppliers faster than four-year engineering programs can.

Federal capital reaches Texas through four distinct mechanisms, whose conditions and risk profiles differ in ways that determine how private capital responds to them.

Direct facility grants are the most visible mechanism. Their condition structure is the analytical key: restrictions on Chinese capacity expansion, domestic workforce development requirements, technology transfer commitments that build indigenous capability over time.

Research funding — DARPA, NSF, NASA university programs — flows through grant mechanisms whose conditions are oriented toward knowledge production and workforce development rather than facility construction.

Defense procurement — Fort Hood modernization, F-35 avionics, Golden Dome compute, Starshield — flows as demand rather than supply-side capital. It does not fund facilities directly. It creates the sustained demand signal that makes defense-grade domestic semiconductor investment economically viable beyond the CHIPS Act's direct funding horizon. A Trusted Foundry-certified facility with a multi-decade defense procurement demand base has a revenue visibility that changes the private capital risk calculus for construction and operation far more fundamentally than a one-time grant does.

The fourth mechanism is not a flow of money but an effect: the signal. Federal commitment at this scale de-risks every layer below it. When the federal government stakes strategic capital and political credibility on the industrial trajectory, it communicates a demand-durability judgment that private capital incorporates into its own underwriting. The de-risking effect is the federal layer's largest contribution to the stack — and it does not appear on any program's balance sheet.

The Private Institutional Layer

What the federal commitment de-risks, the private institutional layer scales. The most strategically innovative expression of this layer is not a facility investment but a fund structure. Digital Realty Trust — the fifth-largest publicly traded REIT in the United States, headquartered in Austin — closed its first US hyperscale data center fund at $3.25 billion, drawing capital from public pensions, sovereign wealth funds, endowments, foundations, corporate pensions, and family offices across a global investor base. It is the first dedicated US hyperscale data center vehicle of its type: a structure that lets global institutional investors access hyperscale data center economics through a REIT-managed vehicle without developing or operating facilities themselves.

The significance is what it signals. Sovereign wealth funds and public pensions — the most conservative patient pools of capital in the world — are now treating AI data center capacity as sovereign-grade infrastructure, an asset class as fundamental to a long-horizon portfolio as toll roads, airports, and utilities. That reclassification, more than any single facility, is the clearest evidence of how durable the institutional capital layer judges the data center demand to be.

The State Layer

The Texas Semiconductor Innovation Fund is the most strategically designed state instrument in the buildout. Its portfolio composition mirrors the federal CHIPS Act's stack construction logic at the state level, filling gaps federal programs leave and amplifying federal investment where the two target the same facilities.

The Permanent University Fund's distributions to UT Austin provide research infrastructure funding that makes the Cockrell School of Engineering and the Texas Institute for Electronics viable as world-class research institutions without depending entirely on annual appropriations or federal grant cycles. The PUF's sovereign wealth character gives UT Austin a research funding stability most peer institutions lack.

The Texas Enterprise Fund and the governor's economic development programs provide the discretionary incentive layer, deployed at critical moments in major facility location decisions. The TEF is smaller in aggregate than the TSIF or PUF, but performs a specific deal-closing function — the final margin of incentive that converts a favorable site selection assessment into a committed investment announcement — whose value is disproportionate to its capital scale.

The County and Municipal Layer

County and municipal capital provides the site-level infrastructure — roads, utilities, district financing — that makes individual facilities operationally viable, and it often commits before the demand justifying it is fully visible. Williamson County's 2019 Road Bond, approved by voters two years before the Samsung Taylor announcement it inadvertently preceded, is the clearest illustration. The commitments at this layer are typically made on general trajectory confidence rather than specific project knowledge — which is why they so often appear, in retrospect, to have anticipated the catalyst.

The Venture and Growth Capital Layer

The venture and growth capital layer funds the technology companies whose products and services the industrial infrastructure is built to support.

The startups being funded in Austin today — autonomous systems software, AI application layers, robotics software platforms, energy management systems, semiconductor design tools — are building products that will run on or support the AI infrastructure being constructed in the Triangle. The industrial infrastructure is the foundation; the venture-backed companies are the applications layer that runs on it. The venture layer is funding the demand that justifies the infrastructure layer's supply, and when those companies succeed their talent recycles back into the regional engineering labor market and eventually its residential wealth concentration.

Strategic Logic of Concentration

These five layers are not a collection of independent capital flows that happen to intersect in Texas. They express a single strategic logic. The conclusion: the United States needs a geography where the full stack of capability is concentrated densely enough that the components can be integrated, iterated, and deployed at the pace the competition with China demands.

The Texas Triangle is where that convergence is taking shape: a domestic AI-industrial complex whose capability does not depend on foreign-operated facilities, foreign-controlled supply chains, or the geopolitical stability of a foreign jurisdiction. The capital stack is what assembles it.

Sixteen

THE WORKFORCE

THE AI-INDUSTRIAL COMPLEX RUNS ON electricity and silicon and water. It also runs on people — and the people constraint is, in some respects, the most binding of all of them.

Consider the others. Water availability can be engineered around, through brownfield inheritance and aquifer management. Grid capacity can be expanded, through transmission investment and behind-the-meter generation. Land, in Texas, is abundant by any measure that matters. Each of those constraints yields to capital and time. The human one does not. The specific capabilities a leading-edge semiconductor fab, a hyperscale AI training cluster, a humanoid robot factory, and an orbital launch complex all require at once cannot be manufactured on any timeline that capital alone sets. They are produced by educational systems, professional development pipelines, and labor market dynamics whose response to a demand signal is measured in years and decades — not months.

So the Texas Triangle's workforce situation is, at the same time, its most impressive structural achievement and its most binding near-term constraint. The achievement is the depth and diversity of the talent pipeline that UT Austin, Texas A&M, and the broader university system have built across decades of sustained investment. The constraint is that the buildout's demand is arriving faster than that pipeline was designed to supply. And certain critical skill categories are scarce enough globally that no regional pipeline can fully satisfy local demand on the timeline required.

The University Anchor

UT Austin, is at the center of the Texas workforce pipeline. The Cockrell School of Engineering ranks consistently among the top ten engineering schools in the United States by research output, industry placement, and peer assessment. Three of its disciplines map most directly onto the buildout's core workforce requirements — electrical and computer engineering, aerospace engineering, materials science and engineering — and all three are running at or near capacity in graduate enrollment, faculty research funding, and industry partnership activity.

The Texas Institute for Electronics at UT Austin — the DARPA- and state-funded prototyping facility whose $840 million investment is documented earlier — is the research infrastructure that converts engineering talent into semiconductor process engineers whose skills transfer directly to Samsung Taylor and Terafab. TIE's NGMM program is not a commercial production facility. It is a workforce development instrument disguised as a research facility: the environment where the next generation of domestic process engineers learns to operate leading-edge fabrication equipment, develop process variants, and manage the yield learning curve that separates a functioning fab from an expensive building full of idle equipment.

The strategic significance of TIE is a timeline argument. Samsung Taylor's current workforce needs are being met three ways at once — experienced Samsung engineers relocated from Korea, domestic engineers with relevant backgrounds recruited nationally, and UT Austin graduates whose preparation is strong but whose process-specific training has to be finished on the job. That mix works now. It will not scale. As Samsung Taylor moves toward full Fab 1 capacity and Fab 2 construction, and SpaceX Terafab moves from site preparation toward operations, demand for process engineers with leading-edge fab experience will outrun what national recruitment and on-the-job training can supply. TIE is the infrastructure whose output — engineers trained on leading-edge equipment in a research environment before they ever join commercial operations — compresses the gap between hiring and productive contribution in the most skill-intensive roles the fab has.

Texas A&M's engineering programs contribute a different slice of the pipeline, particularly in chemical engineering, nuclear engineering, and materials science. The Cyclotron Institute at College Station is at once a research facility and a workforce development environment — for the specialized nuclear and radiation physics engineering that commercial orbital and defense semiconductor qualification demands. The Texas A&M Semiconductor Institute is the most significant state-funded semiconductor R&D base outside UT Austin's TIE program, providing process development, advanced packaging research, and workforce training infrastructure.

Rice University in Houston contributes at the graduate research level, particularly in materials science and nanotechnology — disciplines directly relevant to semiconductor process development at leading-edge nodes and to the battery chemistry Giga Texas's cell production requires. Rice's proximity to Houston's energy finance and engineering ecosystem creates a pipeline whose output flows two ways: into the AI-industrial complex's technical roles, and into the energy finance roles that fund the buildout's energy infrastructure layer.

The Engineering Labor Market

Between the technician trades and the PhD-level researchers sits the corridor's largest and most diverse workforce segment: the engineering labor market itself. Electrical engineers, mechanical engineers, software engineers, systems engineers, and the interdisciplinary practitioners whose work falls in the seams between those categories. This is the segment whose depth is Austin's single most significant competitive advantage over every other comparable geography in the United States — and whose continued growth is the variable that decides whether the corridor's industrial ambition can be staffed at the pace required.

Austin's engineering labor market has had a decade of help. Technology company expansion drew engineering talent out of the Bay Area, Seattle, Boston, and New York — markets where cost of living, state income tax, and housing economics made Austin look better every year. The pandemic accelerated it: remote work normalized, then became permanent for a substantial fraction of the technology workforce. The result is an Austin engineering labor market whose depth, measured by the count of qualified engineers available to hire, is far greater than the city's university output alone would produce — a structural advantage built from years of in-migration compounding on top of the local pipeline.

The Domain in north Austin is the demand anchor that justifies the in-migration. Engineers move to Austin because the jobs are there; the jobs are there because the employers followed the engineers who were already arriving. That positive feedback between employer and talent concentration is the labor market's version of the self-reinforcing dynamic described elsewhere — a dynamic whose momentum shows up in the sustained pace of both corporate expansion announcements and residential buildout across the corridor.

One segment is scarcer than all the others: the intersection of semiconductor process engineering and AI systems development — engineers who understand both the chip hardware and the AI stack that runs on it. This is where the chip-and-fab co-design work happens, where AI training infrastructure is tuned to specific silicon architectures, where the autonomy stack is matched to the inference characteristics of specific chip families. The number of engineers worldwide genuinely competent at that intersection is small — thousands, not tens of thousands. And Texas is competing for every one of them against every other AI-industrial geography on Earth: TSMC's Hsinchu operations, Samsung's Korean fabs, Intel's Oregon and Arizona facilities, the Bay Area's AI research ecosystem.

THE FOUNDER FLYWHEEL

The Austin engineering labor market produces a kind of output the workforce numbers do not capture: founders. The anchor employers – Tesla, Samsung, Apple, Dell, IBM, the defense contractors in Austin and serving Fort Hood – are at once the corridor's largest employers and its most productive founder incubators. An engineer who develops domain expertise at Giga Texas and then leaves to start a robotics or autonomy company carries out the door the specific technical knowledge, the supplier relationships, and the understanding of the anchor's unsolved problems that make an adjacent startup viable. Process engineers from Samsung Taylor spin out semiconductor equipment maintenance companies. AI infrastructure engineers from the Tesla Cortex program start inference-optimization companies. Defense-tech founders come out of Capital Factory's defense accelerator with Fort Hood procurement relationships already in hand.

Capital Factory is the institutional infrastructure that captures this recycling and turns it into investable ventures. The combined anchor-to-founder mechanism is what makes the corridor's talent system self-compounding rather than merely self-sustaining.

THE DEMOGRAPHIC BASE

The workforce story has a demographic dimension the engineering-labor-market analysis alone does not capture. Texas is the second most populous state in the United States, and it has the youngest median age of any large state – a profile that reflects both the high birth rate of the state's population and the in-migration of young professionals drawn by the technology economy. The pipeline is not just the current cohort of engineers and construction workers. It is the K-12 students in Williamson County's rapidly expanding school districts, the students in Austin Community College's advanced manufacturing programs, and the current undergraduate engineering students at UT Austin and Texas A&M whose career choices will be shaped by the industrial landscape they graduate into.

Liberty Hill's school district is among the fastest-growing in Texas — a direct consequence of the residential buildout the industrial anchor has generated. The district's capital planning problem is acute: build schools fast enough to absorb enrollment that arrives faster than bond-funded construction can finish. It is the educational system's version of the same buildout dynamic straining the water infrastructure, the road network, and the construction trades labor market all at once. The students entering Liberty Hill ISD's kindergartens today will graduate into a labor market whose character will be fully established rather than still emerging — and whose demand for the specific skills the corridor's industrial base requires will be the defining feature of their professional landscape.

The STEM and Training Schools

The K-12 response to the corridor's workforce demand is not only the rapid expansion of conventional school districts struggling to absorb enrollment. A parallel pattern is taking shape across the corridor's residential communities: specialized STEM and science academies whose curriculum is explicitly oriented toward the technical workforce requirements of the industrial buildout assembling around them. Cedar Park's STEM programs and Leander ISD's emerging science and technology academies are visible expressions of school districts responding to a labor market signal its own residents are sending. Parents who work in semiconductor fabs, gigafactories, and technology companies know what skills their children's professional landscape will require, and they are advocating for school programs that develop those skills earlier than the conventional curriculum does.

The pattern compounds. Communities that attract technology workers generate school districts whose STEM orientation reflects the professional identity of the parent population. STEM-oriented districts produce graduates better prepared for the technical workforce. A better-prepared local cohort reduces the corridor's dependence on in-migration to fill entry-level technical roles. Reduced dependence on in-migration steadies the housing and infrastructure demand that in-migration creates. The educational system is not simply responding to the buildout. It is becoming part of it.

Tesla's manufacturing technician training, structured around the specific equipment, processes, and quality systems Giga Texas's production lines require, is the corporate equivalent of the apprenticeship model the construction trades have run for generations: skill development embedded in the production environment, evaluated against actual job performance, producing a competency the classroom alone cannot certify. The CHIPS Act workforce development conditions attached to Samsung Taylor's federal funding require a similar commitment — structured training programs whose output is measured in certified process technicians rather than enrollment numbers.

The Technician Pipeline

The semiconductor workforce pipeline is being addressed at the technician level through a coordinated cluster of TSIF-funded programs whose proximity within 40 miles of Samsung Taylor is not chance. Multiple institutions have received TSIF grants for semiconductor workforce development in the Austin I-35 corridor. Austin Community College District for semiconductor training at its Round Rock campus. Temple College to build a Central Texas Chips Hub. UT Austin's Texas Quantum Institute and Prairie View A&M's Roy G. Perry College of Engineering for research and engineering pipeline programs. Texas State Technical College's campus in Hutto for its Accelerated Semiconductor Technician Training Program — a ten-week modular curriculum producing equipment technicians, process technicians, and gas control system specialists, trained inside a mock cleanroom that replicates Samsung Taylor's fab environment.

The five institutions form a complete educational stack. The community colleges and technical programs at ACC and TSTC produce the equipment and process technicians who run the tools — the fab's largest workforce category by headcount, outnumbering engineers roughly five to one. Temple College and Texas A&M Central Texas produce associate's and bachelor's degree graduates for operator and supervisor roles.

The Construction Trades

The buildout's most immediate workforce constraint is not process engineers or AI researchers. It is ironworkers, electricians, pipefitters, concrete masons, welders, millwrights — the full range of construction trades whose labor the simultaneous buildout of dozens of large industrial facilities requires. A semiconductor fab under construction needs thousands of craft workers, not for weeks but for years; for a facility of Samsung Taylor's scale, the timeline from groundbreaking to full campus buildout may run across a decade of continuous construction. A data center campus needs electricians who can install medium-voltage switchgear and the specialized power distribution hyperscale compute requires. A gigafactory needs the full range of industrial construction trades plus the specialized mechanical and electrical contractors who install the manufacturing and test equipment.

And it is all happening at once. The construction of Samsung Taylor, Giga Texas's ongoing expansion, the data center clusters, the Austin–San Antonio megaregion's industrial buildout, and the master-planned residential communities absorbing the workforce all of this generates — every one of these is competing for the same pool of construction trades labor, in the same geography, at the same time. The competition is not theoretical. It is visible in construction labor cost inflation, in project schedule delays attributed to craft worker availability, and in the recruiting activity of the major construction contractors operating in the corridor.

The workforce system described here is not failing. It is operating at the limits of its designed capacity under demand conditions it was not designed for.

Seventeen

THE TRICKLE-DOWN ENGINE

A REPRESENTATIVE PROPERTY IN WEST LAKE HILLS today can have an energy infrastructure that 20 years ago would have been mistaken for a utility substation. Rooftop solar across the main house, guest house, and pool house provide 85 kW of generation. Ground-mount solar on the south-facing slope below the pool deck adds another 40 kW. A microgrid provides 130 kWh of battery storage, backed by a natural gas generator tucked behind the detached garage that delivers 100 kW. The microgrid controller coordinates all of it across 8,400 square feet of conditioned space, the four-car garage's EV charging, pool equipment, multiple HVAC systems. Total load that rivals that of a small commercial building.

This sovereign estate harvests its own water, generates its own power, maintains its own communications, and runs its own physical management infrastructure — through a closed loop whose human oversight is supervisory, not operational. No component is speculative. Each exists in deployable form today. Integrating them into a closed-loop system is what the first-adoption market is currently funding.

The closed-loop estate is the most complete residential expression of the convergence. Every layer — silicon from Samsung, connectivity from Starlink, energy from Tesla's solar and storage manufacturing, physical AI in the near future from Optimus and

Figure, autonomous mobility from robotaxis, AI orchestration from the same multi-agent coordination that runs the Giga Texas floor — converges at the individual property. The result is a system whose operational self-sufficiency is the personal-scale expression of the same vertical integration logic that makes the AI-industrial complex distinctive.

The installation is not a green lifestyle statement. It is an engineering decision — made by someone who rode out Winter Storm Uri in a West Lake Hills home whose grid connection failed, and who responded by building the infrastructure the grid did not provide.

The western Travis County luxury enclaves — Rollingwood, West Lake Hills, Barton Creek, Spanish Oaks — and the Lake Travis waterfront communities concentrate a demographic whose professional identity and consumer behavior are inseparable from the AI-industrial buildout. These are not generically wealthy people who happen to live near a technology corridor. They are the engineers, executives, founders, and investors building it — and their personal exposure to its success, through equity compensation, venture positions, and company ownership, makes their consumer behavior a direct expression of their professional conviction.

The Fleet Opportunity

Two products may soon turn the sovereign estate into a sovereign business: Tesla Cybercab and Optimus. Their economics are structurally identical, and for this demographic both are instances of the same move — early acquisition of an asset whose cost is trivial and whose revenue potential is immediate.

Start with the Cybercab. Its most significant characteristic for this demographic is not its autonomy; it is its price. At the $30,000 cost Tesla has publicly targeted — comparable to a low- to mid-range ICE vehicle — a Cybercab running 16 hours a day in the Tesla robotaxi network returns its purchase cost within months. An owner who acquires five units — one or two for personal and estate use, three earning fleet revenue — has created a pilot transport business whose capital requirement is a rounding error on a Lake Travis portfolio, and whose operating model needs no employees, no scheduling, and no management beyond the Tesla app.

Optimus can be the same opportunity, commercially larger. At Tesla's targeted $30,000 unit price — well below the annual cost of a full-time household employee — a unit's cost is recoverable within a single year. And the estate adds something a purely commercial operator cannot easily replicate: a controlled, task-rich training environment. An Optimus unit accumulates domain capability — groundskeeping, pool maintenance, security patrol, package handling, vehicle management — before commercial deployment in contexts whose tasks overlap. A unit trained on a Lakeway estate for six months is a more capable commercial asset than one shipped straight from the factory to a job site. The estate is not just the unit's home base. It is its apprenticeship.

The financing structures just beginning to take shape mirror the residential rental playbook: debt-financed acquisition, debt service covered by fleet revenue, equity building as each FSD or capability update compounds the asset's productivity. A five-unit Optimus fleet at target prices is a $150,000 commitment — two or three units for estate operations, two or three in a commercial leasing pool earning $10 to $20 per hour during off-peak estate hours. The Tesla fleet network handles dispatch, insurance, maintenance, and revenue collection. The owner supplies the capital and absorbs the early-adopter risk.

This is the Roadster-to-Model-S arc compressed into a single product cycle: early high-net-worth acquisition funds the production volume that drives unit cost down to the next demographic tier. The economics work at estate scale first and at mass-market scale eventually. The sequence is the one this demographic has run for every prior technology transition. The first private commercial Cybercab and Optimus fleet operators in Texas may almost certainly be these households — their estate infrastructure supplied the training ground, their financial profile supplied the capital, and their proximity to Giga Texas supplied the service relationship early fleet operators require. The sovereign home is not just where the fleet lives. It is where the fleet business begins.

The Trickle-Down Adoption Cascade

The first-adoption foundation matters not because wealthy estates are a large market, but because they are the leading indicator of an adoption cycle whose mass-market phase follows the high-net-worth phase at a predictable lag — a lag set by production scale and cost reduction.

The technology a West Lake Hills estate deploys in 2028 — Optimus fleet, Cybercab fleet, drone patrol network, AI orchestration, hybrid microgrid with solar and BESS — is the technology a Georgetown master-planned community deploys across its shared amenity infrastructure in 2032. The cascade above is not abstract. The shared Optimus fleet managing a 6,500-home community's common areas in 2032 is the estate-scale technology the western Travis County household funded in 2028. By 2038 it is a San Marcos apartment complex's management company. The individual household follows, at the mass-market price point that estate-funded production scale makes possible.

Every modern technology cycle has run this cascade. Home battery energy storage, solar generation, smart home systems, personal computing — all began as leading-edge products for the wealthy and reached the mass market the same way. Wealthy first adopters fund production scaling, generate the usage data that drives capability improvement, surface the failure modes that need engineering out, and provide the aspirational demonstration that makes the technology legible before it is affordable.

Austin's geographic closed loop compresses that cascade. The Optimus unit working a Spanish Oaks estate is 45 minutes from the Giga Texas floor assembling the next generation. The operational training data it generates flows to Cortex the same day. The next OTA update addresses the failure modes the estate surfaced. The cycle that took personal computing 20 years to run from wealthy first adopters to the mass market will take autonomous physical AI considerably less — in an environment where the production campus, the training infrastructure, the first-adoption customer base, and the feedback loop all sit within a 45-minute radius.

Eighteen

THE SOLO-BUILDERS

THE CHAPTERS PRECEDING THIS ONE have documented the physical substrate of the AI-industrial complex — the fabs, the compute clusters, the transmission upgrades, the water rights, the campus footprints, the capital stack. The substrate has a second layer that follows from the first mechanically. Every gigawatt provisioned for AI training is also, at a fractional cost, provisioned for AI inference. Every hour of inference consumed by a large operator is also, at commodity pricing, available to any individual with a subscription. The physical infrastructure that trains Cortex is the same type of infrastructure that trains Claude, ChatGPT, and Cursor. The inference generated in the Cortex cluster is the same class of inference generated by ChatGPT at Stargate in Abilene and served in a browser window at a home office north of Austin.

That coupling — industrial-scale AI training producing consumer-accessible AI inference as a byproduct — is the phenomenon this chapter documents at the individual scale. The physical buildout has generated a cognitive amplification layer whose consumers are not primarily enterprises. They are individuals, working alone, running businesses whose entire operational stack fits on one desk. The layer exists because the substrate exists. The two are the same buildout observed at different scales. The industrial chapters described what the buildout is producing at gigawatt-and-square-foot scale. This chapter describes what the same buildout is producing at laptop-and-home-office scale. The infrastructure is one thing. Its consequences are two.

Drive through the master-planned communities north of Austin on a weekday morning. The streets are quieter than the residential density would suggest. The construction crews are pouring foundations next street over. But the driveways tell the story the quiet streets obscure.

Custom wraps on pickup trucks. A domain name on a tailgate. Google Maps business pins inside master-planned communities. These are not roofing contractors or plumbers, though those decals are here too. These are one-person media companies, analytics practices, AI-assisted research operations — run from home offices behind the garage doors, serving audiences and clients whose geographic distribution has nothing to do with Liberty Hill or Georgetown. The person inside is not commuting to The Domain or downtown Austin. They are already at work, with a laptop, a Mac Mini, a camera, a microphone, a 3D printer, a CNC machine, and a suite of AI tools whose combined capability has made them, functionally, a one-person institution.

These are the solo-builders — the part of the convergence that reaches not into the fab or the data center or the gigafactory, but into the home office, the spare bedroom, the kitchen table at six in the morning before the kids wake up. The same inference compute running on the same power feeding the same data centers coalescing 30 minutes east is what these individuals are consuming when they open Claude Code, Cursor, or ChatGPT and begin the day's work. The physical infrastructure and the cognitive infrastructure share a geography. The buildout produces both at once.

The Productive Unit That Did Not Exist

Five years ago the solo-builder did not exist as a viable productive unit. The capability existed in fragments — individuals had always been able to write, to research, to create videos — but what did not exist was the tooling that collapses the productivity gap between an individual and an institution across every cognitive function at once. The solo-builder is the individual who deploys AI tooling as a cognitive amplifier to produce business operation outputs that previously required teams.

The LLM changed that. Not by making individuals smarter — the solo-builder's domain expertise is the human input the LLM amplifies rather than generates — but by making every cognitive task whose execution previously required specialist labor executable by a single individual with sufficient focus, persistence, and domain expertise to direct the tool and evaluate and incorporate its output. Ideation, research synthesis, structured writing, code generation, image production, video editing, data analysis, customer communication, legal document drafting, financial modeling: all of it, by one person. The LLM is not the solo-builder. It is the solo-builder's team, operating at a subscription cost whose monthly total is less than one hour of any of those specialists' billing rates.

The productivity multiple this tooling provides is not linear. It is not that the solo-builder now does the work of two people. It is that the solo-builder now does the work of a team whose coordination overhead, management cost, and institutional friction have been eliminated entirely. A team of eight specialists producing a research report requires scheduling, briefing, drafts, review cycles, revision, integration, and the communication overhead any multi-person production process involves. A solo-builder with equivalent domain expertise and LLM tooling produces the same report in a fraction of the elapsed time — no coordination overhead, no briefing cost, no revision cycles between specialists, no integration work. The output is different in some dimensions: the team's specialist depth in each domain may exceed the solo-builder's LLM-assisted depth. But for the majority of digital production tasks whose market value the knowledge economy generates, the solo-builder's output is commercially equivalent at a small fraction of the cost and time.

The Solo-Builder Capital

The solo builder documented here and the startup founder documented earlier are two expressions of the same underlying dynamic: the conversion of domain expertise with AI tooling into productive autonomy that does not require institutional employment to sustain. The distinction is organizational scale. The solo builder operates as a one-person institution indefinitely, compounding output without needing to grow. The startup founder uses the same domain expertise and AI tooling as the initial productive base, then builds organizational infrastructure around it. Austin concentrates both expressions at once — the home office in the northern suburbs and the Capital Factory cohort on Brazos Street — because the same conditions that enable solo-building enable startup formation: deep domain expertise, low overhead, and cultural permission to build rather than join.

Austin's emergence as one of the most concentrated startup and solo-builder geographies did not just happen. It is the product of the same convergent factors that make it the most concentrated AI-industrial geography applied at the individual rather than the institutional level. Three of those factors are worth drawing out.

The first is human capital. Austin's educational attainment is among the highest of any major US metro, reflecting both the UT Austin pipeline and a decade of in-migration by credentialed technology professionals from the Bay Area, Seattle, Boston, and New York. Austin's concentration of domain expertise — in software engineering, semiconductor design, AI systems, energy technology, aerospace, financial analysis — is the human capital that AI tooling amplifies. A city full of domain experts with daily LLM usage and expertise is a city full of potential solo-builders whose productive ceiling has been raised by an order of magnitude.

The second is cultural permission. Austin's entrepreneurial culture — the Domain's startup scene, the South by Southwest ecosystem, the technology company alumni networks whose members have both domain expertise and risk tolerance, the general social permission to be building something rather than employed by someone — creates the disposition solo-building requires. The outdoor cafe at The Domain where everyone appears to be working on a startup is not a demographic accident. It is the cultural expression of a city that has rewarded entrepreneurial initiative consistently enough that individual productive autonomy has become the aspirational norm rather than the exception.

The third is residential economics. The master-planned communities provide the low-overhead residential environment that makes solo-building financially viable. A home office in Liberty Hill costs a small fraction of equivalent commercial office space in Austin proper and comes with the quality-of-life infrastructure, fiber-optic broadband, schools, amenities, and community, that sustains the domestic environment from which solo-building operates. The wrap on the Cybertruck in the driveway is the brand. The home behind it is the production facility.

On a Wednesday morning at the local H-E-B supermarket, I recognized someone I had been listening to for more than a year — a YouTube creator with a substantial following, one of several Austin-based creators who have built large audiences around Tesla, autonomy, AI, and the technology transition underway. We talked briefly, in the way neighbors do when they discover they live in the same area.

He produces content at a pace and scale that would have required a production team three years ago. The AI tools amplifying his research, scripting, and video editing run on inference compute drawing power from data centers emerging 30 minutes east of where we were standing in the aisle. His channel generates the kind of annual revenue that redefines what individual productive autonomy means — not supplemental income, not a side project, but a primary economic output whose scale would have required an institution to achieve a decade ago.

He is not unusual in Austin. He is representative. The corridor produces this demographic the way Samsung Taylor produces wafers — not one at a time, but at scale, as the natural output of a geography that concentrates domain expertise and AI infrastructure simultaneously. They are the cognitive layer's most visible expression — individuals who have converted domain knowledge using AI tooling into businesses that reach audiences larger than most regional newspapers, from home offices in master-planned communities 30 miles from the Texas Capitol.

The semiconductor fab worker and the YouTube creator in the H-E-B are both outputs of the same underlying conditions. The fab worker requires the physical infrastructure. The YouTube creator requires the inference computing. Both are here because the geography assembled the foundation that makes both possible. The grocery store is where they share a checkout line.

The Democratization Argument

The productive threshold — the output level at which the solo-builder's work is commercially viable, reaches an audience, and generates returns that sustain continued production — has been lowered by at least an order of magnitude. The former Apple engineer generating seven figures annually from a YouTube channel built with AI tooling assistance is operating well above that threshold. The suburban home office operator whose SaaS tool serves 500 paying customers is operating above it. The author whose self-published book reaches the institutional audience whose analytical needs it addresses is operating above it. The threshold crossing — from employment-dependent to productively autonomous — is now achievable by a broader fraction of the knowledge worker population than at any prior point in time.

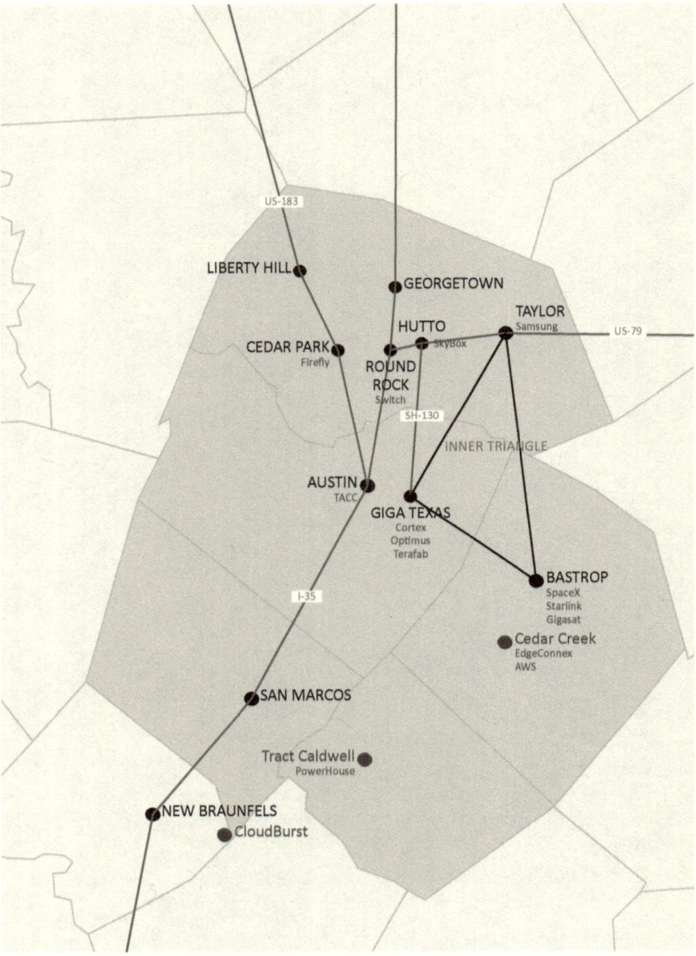

Inside the Singularity

AUSTIN SUPERCLUSTER · 2026

Nineteen

THE AUSTIN SUPERCLUSTER

FIVE INDUSTRIAL SUPERNODES SIT within 45 miles of each other inside the Austin metropolitan region: the Tesla Giga Texas complex, the Samsung Taylor semiconductor campus, the SpaceX Bastrop orbital compute compound, the Williamson County US 79 AI compute corridor, and the San Marcos–Uhland AI compute corridor. Each is a supernode in its own right. These five are so tightly co-located, so deeply integrated through shared workforce and corridor infrastructure, and so interlocked with the academic, design, and institutional layers underneath them, that they collectively constitute a supercluster.

Definitions: a supernode is a concentration of nodes (facilities) operating as a coherent system. A supercluster is a concentration of supernodes sharing common substrate at a scale that makes the collection itself structurally and geographically distinct.

The Austin Supercluster is the only supercluster mapped in the Texas AI-industrial graph. No other US region hosts more than two AI-industrial supernodes in close proximity. By the supercluster standard, the United States currently has exactly one.

THE FIVE SUPERNODES

The five differ in structural signature and complement rather than overlap:

Tesla Giga Texas is the gravitational center — the manufacturing and intelligence vertex where vehicle and robotaxi production, Cortex AI training, Optimus robotics, and Terafab Research operate (or will operate) on a single campus at 1 Tesla Road. Four different AI-industrial programs under one operator on one site is the highest concentration in the United States.

Samsung Taylor is the silicon vertex — the Fab 1 and Fab 2 complex 30 miles northeast along SH 130 and US 79, with a ten-fab long-term vision, the captive Tesla foundry relationship at Fab 1, and its semiconductor supply chain anchoring Williamson County. Together with Giga Texas it forms the supercluster's Austin–Taylor semiconductor axis.

The SpaceX Bastrop Complex is structurally distinct: a single-operator captive vertical-integration compound with six industrial functions under one operator — Starlink terminal manufacturing, space-grade solar cell production, AI1 orbital compute satellite manufacturing, rad-tolerant semiconductor packaging, Boring Company tunneling operations, and the Snailbrook worker housing development. Together with Giga Texas it forms the supercluster's SpaceX–Tesla axis. The emerging Cedar Creek three-operator data center cluster on SH 21 sits just south but on a separate vector from Austin.

The US 79 AI compute corridor in Williamson County is the most inheritance-based of the five. Its substrate — Hutto's industrial infrastructure, Round Rock's Dell-anchored data center cluster, Taylor's freight rail — was assembled before the AI-industrial buildout arrived. The hyperscale operators chose Williamson County because the stage was already built. Hutto itself is a microcosm: the Skybox / Prologis PowerCampus, the Hutto Mega Tech Center, Blue Origin's Project Blue Hub in site evaluation, a semiconductor workforce training facility, and growth that has made it one of the fastest-growing cities in the United States. ASML is there too.

The San Marcos–Uhland compute corridor is the youngest and only pre-operational of the five: CloudBurst Evolve, Sabey, Edged Energy, Prime Data Centers Project Snow, and Tract Caldwell Valley along I-35 south of Austin — the supercluster's southward extension toward San Antonio.

The supercluster is asymmetric. The five supernodes are not equivalent in density, maturity, or function. Their integration is what produces the supercluster's coherence.

THE INNER TRIANGLE

The Texas Triangle is the macro geographic frame: Dallas–Fort Worth, Houston, and the San Antonio–Austin megaregion — more than 22 million people, the outer boundary within which the buildout is occurring.

The Inner Triangle within the Austin MSA is the densest concentration in the Texas AI-industrial graph: three vertices whose corridor edges have compounded into the most analytically consequential concentration of strategic industrial capital in the United States. Giga Texas is the manufacturing and intelligence vertex. Samsung Taylor is the silicon vertex. The SpaceX Bastrop Complex is the defense and orbital infrastructure manufacturing vertex.

The connection corridors are not incidental. SH 130 ties the silicon vertex to the manufacturing vertex — a 35-mile state highway through Central Texas farmland along which workforce, supplies, and institutional relationships flow at 85 mph. SH 71 is the logistics spine along which Starlink terminals move from the Bastrop compound toward Austin distribution, and along which SpaceX engineering teams commute between the Austin engineering presence and the Bastrop compound. US 79 is the data center spine along which the Williamson County data center corridor consolidates from Round Rock through Hutto to Taylor.

Austin proper sits at both the Inner Triangle's western vertex and the larger Texas Triangle's interior anchor at the same time — the city whose geography and institutions were the stage the emergence required, now physically surrounded by the manufacturing infrastructure of the assembly itself. The concentric geometry begins at a specific address: 1 Tesla Road.

The manufacturing cluster generates the data center demand. The data center cluster generates the energy demand. The energy demand generates the power infrastructure investment. The power infrastructure investment makes the next manufacturing anchor viable. And the cycle compounds. The Inner Triangle is where that compounding is most dense, most visible, and most consequential

The Institutional Substrate

The supernodes did not grow on empty land. The substrate predates the buildout by decades — the framing Part I established.

UT Austin anchors the academic layer. Its semiconductor research programs, AI research, and STEM workforce pipeline have shaped Austin as a technology hub for more than four decades. AMD's headquarters, NXP's design footprint, Silicon Labs' founding — these are first-order consequences. Without UT Austin, the AI-industrial supernodes may have chosen another geography.

The Texas Institute for Electronics is the deliberate semiconductor-industrial policy bet. TIE coordinates state, federal, and industry funding into a consortium spanning UT Austin, Texas A&M, multiple operators, and DARPA.

The Texas Advanced Computing Center on UT Austin's campus operates Frontera, Stampede3, and Vista — one of the largest academic supercomputing footprints in the United States, with Frontera the largest academic supercomputer in the world when commissioned in 2019. Horizon at Sabey Round Rock adds a second academic compute anchor under NSF's Leadership-Class Computing Facility program. The supercluster's compute capacity is, in this sense, three-layered: commercial hyperscale, commercial frontier AI training, and academic supercomputing.

Approximately two dozen key semiconductor-related companies operate in the Austin core — Applied Materials, NI, AMD, NXP, Silicon Labs, Cirrus Logic, Arm, Apple silicon, Intel design operations, and a long tail of fabless and design-services firms. This is the design substrate complementing the manufacturing supernodes.

The argument extends beyond the commercial-centric gigafactory, semiconductor fab, and AI data center domains.

The Aerospace Concentration

The supercluster's orbital and aerospace manufacturing capability is already at supernode density at the Bastrop Complex — Starlink terminal manufacturing, upcoming orbital solar cell production, and upcoming orbital AI1 satellite manufacturing produce the ground hardware, photovoltaic infrastructure, and orbital platforms for SpaceX's Starmind constellation.

A second aerospace concentration is forming in the northern half of the metro region in Williamson County, separate from Bastrop both geographically and operationally. The emerging cluster — Firefly Aerospace at Cedar Park; the CTSDC Spaceport and Defense Innovation Campus also at Cedar Park; and potentially Blue Origin's massive Project Blue Hub at Hutto — is not yet at supernode density.

The trajectory is clear. The northern concentration is broadening rather than duplicating Bastrop's specialization: launch, lunar, and defense aerospace domains adjacent to but distinct from Starlink terminal manufacturing and orbital compute satellite production. If the northern cluster densifies as the trajectory suggests, the supercluster will operate two distinct aerospace supernodes inside one metropolitan region: Bastrop as the SpaceX captive vertical-integration compound, and the northern cluster as the multi-operator launch, lunar, and defense aerospace concentration.

The Defense Concentration

If aerospace anchors itself in two geographic clusters, defense technology in Austin organizes along an institutional axis. The structural anchor is the US Army Transformation and Training Command (T2COM) as the merger of Army Futures Command and the Training and Doctrine Command. T2COM consolidates force design, force development, and force generation under one four-star headquarters — the Army's modernization apparatus made into a single unified command — headquartered in downtown Austin with a Resource Operations Center on the Dell Technologies campus in Round Rock.

T2COM operates downtown alongside the Army Software Factory at Austin Community College — the first soldier-led software factory in the US Army — and inside an innovation ecosystem where Capital Factory has hosted Defense Innovation Unit (DIU) and AFWERX offices for years. Project Convergence, the Army's annual capstone integration experiment, is anchored institutionally to this command. No other US metropolitan region houses an Army four-star command of this scope alongside a soldier-led innovation campus, a top-tier research university with a frontier-AI talent base, and a venture-coordinated startup ecosystem inside a single labor market.

The commercial layer has thickened around that institutional anchor. Saronic Technologies is the apex case: an autonomy-first shipbuilder producing multiple unmanned surface vessels, with Port Alpha as the next-generation shipyard at ten times the capacity of its existing Franklin, Louisiana site. Brownsville is the selected location for Port Alpha. Allen Control Systems, another Austin unicorn at $2.2 billion, builds the Bullfrog AI-powered autonomous counter-drone weapon station under Army Applications Lab contract.

The structural argument for the supercluster's defense concentration has a single defining test case: Golden Dome. Golden Dome is the largest single homeland-defense program initiated in a generation sitting at $1.2 trillion, with initial capability targeted for 2028 and the full architecture for the mid-2030s.

The Austin Supercluster holds direct production roles in the program rather than peripheral ones. SpaceX is set to receive a roughly $6.45 billion contract for a 600-satellite constellation supporting missile targeting — built on the Starshield platform whose terminals and satellite hardware are manufactured at the Bastrop Complex.

Anduril is also jointly developing the program's command-and-control software — the "glue layer" that fuses radars, sensors, and missile batteries across services — with Palantir, both of which run substantial Austin operations. Live-flight validation for Anduril's Fury Collaborative Combat Aircraft takes place at the company's massive West Texas test site. The supercluster does not merely host vendors to Golden Dome; it manufactures the satellite layer, co-develops the command-and-control software, and tests the airpower platforms on which the program depends.

The Comparative Case

The claim that no other region matches this density deserves examination.

The test is cross-domain integration. The Austin Supercluster combines seven industrial domains at supernode density — semiconductor fabrication, frontier AI training compute, electric vehicle manufacturing, upcoming humanoid robotics, upcoming orbital infrastructure manufacturing, AI and hyperscale data centers, and defense technology manufacturing — plus two domains at substrate density — semiconductor design and academic supercomputing.

Silicon Valley (with Southern California). Design dominance (Intel design, Nvidia HQ, Apple silicon, AMD, Qualcomm San Diego), AI software (OpenAI, Anthropic, Google AI), capital concentration, and aerospace-defense (SpaceX HQ Hawthorne, Anduril Costa Mesa, Shield AI San Diego, General Atomics, Northrop Grumman Space Systems in Redondo Beach, Lockheed Skunk Works in Palmdale, plus Skydio and Vannevar Labs in the Bay). But Intel's manufacturing is in Oregon and Arizona, Tesla's HQ moved to Austin in December 2021, major new program investment concentrated in Texas rather than the Bay Area, and there is no equivalent to Samsung Taylor, Cortex, Bastrop's captive vertical integration, Optimus-scale humanoid robotics, or T2COM-scale defense command. A half-supercluster: design and software strong, manufacturing half elsewhere. Three of nine at strict counting; the coastal California composite reaches four if the SoCal aerospace-defense weight is folded in generously.

Phoenix. The strongest US semiconductor manufacturing concentration currently — TSMC Fab 21, Intel's Ocotillo Fab 52 (operational on 18A) and Fab 62 (targeted for 2028). Real and significant. But no AI training at frontier scale, no EV manufacturing, no humanoid robotics, no orbital, no defense-tech concentration at scale. Three of nine.

Northern Virginia. The largest US data center geography today, plus the legacy defense industrial base — Pentagon adjacency, all five primes (Lockheed Martin, RTX, Northrop Grumman, Boeing, General Dynamics) at headquarters scale, and the intelligence community contractor density. No fabs, no design at scale, no vehicles, no robotics, no orbital, no academic supercomputing at frontier scale, no AI training at frontier scale. The defense concentration is real but legacy-prime in character rather than autonomy-and-AI native, and cross-domain integration with manufacturing is absent. Two of nine, both at world-class scale.

Seattle. Software (Microsoft, Amazon HQ), aerospace and orbital manufacturing (Boeing; SpaceX Starlink manufacturing in Redmond producing roughly 70 satellites per week; Amazon Leo satellite manufacturing in Kirkland at up to five per day; Blue Origin in Kent — Washington State has produced roughly two-thirds of all operational satellites in orbit), hyperscale data centers, modest defense-tech (Boeing defense, some Microsoft defense contracts). No fabs in the Pacific Northwest beyond Intel's Hillsboro 170 miles south, no vehicle manufacturing at scale, no humanoid robotics, no equivalent defense command, no AI training at frontier scale in-region. Three of nine — with orbital manufacturing arguably the closest single-domain challenger to Bastrop in the United States, though the specialization is satellite communications rather than orbital compute, and the six-function captive vertical integration under one operator at Bastrop has no parallel.

Detroit. Vehicle manufacturing legacy, with the EV transition happening primarily elsewhere. The Detroit Arsenal at Warren provides a ground-vehicle defense connection at sub-supernode density. No fabs, no AI training, no robotics, no orbital, no comparable research layers. One of nine.

Research Triangle and Boston. Research-strong concentrations with biotech and AI research; Boston adds Draper Laboratory, MIT Lincoln Laboratory, and a modest defense-tech base. Both lack the manufacturing depth. Two of nine each.

Hsinchu. Taiwan's Hsinchu Science Park has the world's deepest semiconductor fabrication concentration via TSMC and a supplier ring deeper and longer-established than Austin's — but lacks frontier AI training, EV manufacturing, humanoid robotics, orbital, and US-aligned defense-tech under a unified command structure. Three of nine.

Shenzhen. China's Shenzhen has manufacturing density exceeding Austin's across many product categories — including the parallel PLA-aligned defense industrial base and the world's deepest commercial drone supply chain — but operates under US export controls constraining its frontier AI compute. Three to four of nine in equivalent-domain terms.

Tokyo, Seoul, Munich. Multi-domain industrial regions strong in semiconductor manufacturing and automotive but lacking frontier AI training and the new-wave autonomous defense base. None score above four.

Austin scores seven of nine at supernode density plus two at substrate density. No other US region exceeds four at confirmed density across these domains — and reaching four requires treating coastal California (Bay Area plus Southern California) as a single composite. No region in the world exceeds four. The structural singularity holds — and arguably hardens with defense technology folded in, because a four-star modernization command co-located with frontier-AI talent has no peer outside the Pentagon corridor, while the autonomous-systems commercial base has no peer anywhere.

The Center of the Trajectory

The Austin Supercluster is the operating center of the AI-industrial convergence: five supernodes whose integration density makes the collection structurally distinct, the institutional substrate that predates and feeds the buildout, the aerospace concentration, and the recursive architecture closing more densely here than anywhere else. As of mid-2026, this is the densest concentration of AI-industrial supernodes in the United States and, by the cross-domain integration test, the world.

The title of this book promises a trajectory from Austin to Alpha Centauri. This chapter has summarized what "from Austin" means now structurally.

Twenty

THE STEP FUNCTION

THERE IS A MOMENT IN THE DEVELOPMENT of every industrial convergence when the components stop being a collection of related assets and become a system. It is the moment when the interactions between the components begin producing outcomes none of them could produce in isolation — and when the system's momentum becomes self-sustaining rather than dependent on continued external inputs. Multiplicative, not additive.

Pittsburgh reached that moment sometime in the 1880s, when the concentration of steel production, railroad infrastructure, coal supply, and financial capital had compounded to the point where the system's own gravity was pulling in additional capital and talent faster than any individual actor's decisions could redirect. Detroit reached it in the 1910s, when Ford's moving assembly line, the supplier ecosystem that came together around it, and the labor market that concentrated in southeastern Michigan had created an automotive industrial complex whose trajectory was no longer determined by Henry Ford's choices but by the system's own compounding logic.

The step function is not an announcement. It is a condition: the state of a system whose compounding has reached the threshold where its own gravity becomes the primary determinant of its trajectory. The Texas Triangle's AI-industrial complex is approaching that threshold from below. Specific observable indicators mark the approach. The convergence has already propagated beyond the industrial layer into the consumer, cognitive, and institutional layers that make it self-sustaining in a way prior industrial convergences were not.

The builders arriving today are not betting on what Texas could become. They are responding to what Texas already is.

Self-Sustaining Momentum

The diagnostic for self-sustaining industrial momentum is not the size of the assets present. It is the rate at which the system attracts additional assets without external recruitment — the degree to which each new facility's site selection decision is determined by what is already in the geography rather than by incentives that compensate for what the geography lacks. Three commitments show the diagnostic met.

Samsung Taylor did not locate in Taylor because of federal funding alone. It located there because Samsung's existing Austin operations since 1997, the available power infrastructure, the University of Texas talent pipeline, and the prior technology company base had already made Central Texas a viable semiconductor manufacturing geography before the federal funding was committed. The CHIPS Act funding accelerated the decision and increased the facility's scale. It did not create the conditions that made the decision possible. The conditions were already there.

The Tesla–Samsung supply relationship required no external incentive to produce. It is the market outcome of two operators whose site selection decisions, made independently for independent reasons at different points in time, placed them 40 minutes apart along the same highway corridor. The geographic proximity made the supply relationship's economics obvious. That is the self-reinforcing dynamic that characterizes a system approaching self-sustaining momentum: each new commitment makes prior commitments more valuable, which makes additional commitments more attractive, which accelerates the compounding.

THE BUBBLE QUESTION

The self-sustaining momentum argument invites the obvious objection: the same pattern was visible in fiber during the 1990s, in housing during the 2000s, and in the private-market technology cycle of the late 2010s. Formation events look inevitable while they are occurring. They also frequently collapse. The AI-industrial buildout at Texas Triangle scale — measured in tens of gigawatts of announced compute capacity, hundreds of billions of dollars of committed capital, and multi-year construction commitments made against demand curves that no one has seen sustained through a full cycle — is not exempt from that history. The demand-side risk deserves serious treatment rather than dismissal.

The case for concern is real. AI training capacity is being built against operator projections whose validation depends on continued frontier-model scaling economics, whose scaling laws are contested at the current parameter counts, and whose enterprise adoption curves are still forming. The public discourse's implicit demand model — every enterprise will train frontier models, every consumer will run AI-powered agents, inference demand will grow at the rate compute capacity is being provisioned — is a projection, not a measurement. If model economics harden into commodity margins faster than the buildout completes, some fraction of the announced capacity ends up as stranded assets whose depreciation schedules do not match their utilization curves. The 70 GW Texas dataset projection for 2030 is not a floor. It is a ceiling that assumes the demand curve continues to justify the supply curve.

Two distinctions separate this cycle's exposure from the fiber and housing analogs.

The first is that the substrate has multiple uses. Fiber optic capacity built for late-1990s traffic assumptions was single-purpose: it either carried the projected traffic or sat dark. Housing built for 2005 mortgage assumptions was single-purpose in a different way: the market required buyers at the projected price points or the units did not clear. Data center capacity, by contrast, has fungible workloads. A gigawatt provisioned for frontier AI training can be reallocated to inference, to enterprise cloud, to scientific computing, to media rendering, or to any other high-density compute workload if the training-specific demand curve softens. The elasticity is not perfect — some hardware is training-specific and some campus architectures are optimized for particular workload profiles — but it is materially higher than fiber or housing offered. Excess capacity in the compute layer is more likely to reprice than to strand.

The second is that the buildout is not being financed primarily by the equity markets whose sentiment shifts drive bubble collapses. The federal layer, the sovereign wealth pools, the hyperscaler balance sheets, and the private institutional capital documented in Chapter 15 are patient capital operating on multi-decade horizons. Digital Realty's $3.25 billion fund closed with sovereign wealth funds and public pensions specifically because those investors are underwriting the demand curve as sovereign-grade infrastructure rather than as a growth trade. Patient capital does not liquidate on a quarterly earnings miss. It absorbs. The buildout's financing structure is closer to the mid-twentieth-century utility investment pattern than to any recent technology cycle, and that difference determines what happens if demand softens: repricing, not collapse.

None of this eliminates the risk. Any facility whose capital return schedule assumes 24/7 utilization at frontier-training rates is exposed to a demand curve that may not deliver those utilization rates on that timeline. Any operator whose site selection assumed a specific tenant profile is exposed to the tenant profile shifting under it. Some fraction of announced capacity will underperform its projections. Some fraction of announced facilities will be delayed, scaled back, or repositioned. That is the ordinary friction of a formation event and does not invalidate the structural argument.

What would invalidate the structural argument is a demand collapse so complete that both the fungibility and the patient-capital financing fail simultaneously — no reallocation absorbs the capacity, and the sovereign, hyperscaler, and institutional pools reprice to distressed exit rather than long-horizon hold. That scenario is not zero-probability. But it requires assuming that AI capability at 2026 levels of demonstrated utility fails to compound into further demand, and that the most conservative long-horizon capital pools in the world all decide simultaneously that the compounding is not real. Neither assumption is inconsistent with the empirical record of the past three years. Both are inconsistent with the direction of the past three years, whose trend has been demand outrunning capacity rather than capacity outrunning demand.

The bubble question, honestly stated, is not whether some overbuilt capacity will emerge. Some will. It is whether the average outcome across the buildout is closer to the utility investment cycle — patient capital, long-horizon absorption, gradual return realization — or to the fiber cycle's stranded-asset collapse. The evidence available in 2026 supports the utility framing. The evidence will look different in 2028, and the honest analyst will read it then.

The Five Loops

Five feedback loops are operating in the Texas Triangle right now, each at a different stage of closure. Each connects a physical input to an intelligence output that improves the next iteration of the input. Each closes along a traceable physical path.

Loop 1 — Energy to Compute to Intelligence. West Texas wind and solar feed ERCOT; ERCOT feeds Cortex, Stargate, and the other AI data centers; trained AI models improve grid management. The loop closes across the data center concentrations within the Austin metro: Williamson County hyperscale, Cortex inside Giga Texas, and the San Marcos–Uhland campuses. Closed and compounding.

Loop 2 — Vehicle to Data to Intelligence. Every Tesla from Giga Texas is a sensor platform feeding video and telemetry to Cortex; improved FSD deploys back to the fleet over-the-air; the fleet grows and the data compounds. The loop closes most cleanly at Giga Texas itself — data ingestion and model training on the same campus as the vehicle production line, a physical compression no other supernode replicates. Operating at 11 billion FSD miles and accelerating.

Loop 3 — Manufacturing to Robotics to Intelligence. Pre-production Optimus units on the Giga Texas floor generate the training data that improves the efficiency of the robots that will eventually run the floor. The loop closes across Giga Texas (Optimus pre-production) and Bastrop (The Boring Company's tunneling automation). Not yet closed — closing.

Loop 4 — Orbital Infrastructure to Inference to Update. Starlink supplies the connectivity that carries Tesla OTA updates globally; the orbital compute layer — D3 and the space-grade silicon for inference — is still in development. The loop anchors at Bastrop's Starlink terminal manufacturing, the upcoming Project Echo orbital-grade solar cell facility, and the upcoming Gigasat advanced manufacturing buildout. It closes globally, but its orbital infrastructure originates inside the Austin cluster. Optimistically 2030, when Terafab reaches yield on D3 and Starship is populating the Starmind constellation.

Loop 5 — Capital to Vertical Integration to Moat. The Samsung Taylor agreement, Terafab, the Intel partnership, and a five-source capital stack are deploying at once; each integration step captures margin and widens the moat, which compounds capital access. The loop anchors at Giga Texas, where vehicle production, AI training, the upcoming humanoid robotics, and the Terafab Research Fab share one operator, one campus, and one capital stack. Deploying at once.

Two walkthrough paths show the interlocking at work.

Path A — chip to vehicle to Cortex to fleet. A Tesla AI5 inference chip fabricated at Samsung Taylor travels 30 miles down US 79 and SH 130 to be installed in a Cybercab at Giga Texas. The vehicle starts, generating telemetry that travels 500 feet across the same campus to Cortex. Trained models deploy over-the-air to the global fleet and to the Optimus units on the manufacturing floor where the next vehicle is being built. Loops 2, 3, 4, and 5 close inside a 30-mile radius.

Path B — terminal to constellation to fleet. A Starlink Mini terminal manufactured at Bastrop ships to a customer who mounts it to their vehicle, then connects to the constellation. That same constellation carries OTA updates back to the Tesla fleet, which generated the data that trained the model that improved the FSD that the next Tesla off Giga Texas deploys with. Loops 2 and 4 interlock across two supernodes.

No other supernode group hosts all five loops. Abilene closes Loop 1 cleanly but participates only secondarily in the rest. The El Paso–Borderplex anchors Loop 1 at frontier scale but lacks Loops 2 through 5. Houston–Brookshire anchors Loop 5 strongly but participates secondarily in the rest. Only the Austin Supercluster hosts all five.

These are not theoretical loops. They are physical paths traversed every day. Five loops together compound faster than any one alone, and the Triangle is the geography close enough for all five to physically interlock. The compounding is the whole argument.

The Critical Density

The operators who chose Texas in 2021 — Tesla, Samsung, the first wave of data center developers — made first-principles location decisions based on underlying base analysis. They evaluated what Texas could become and bet on it. Those bets required conviction about a future state that did not yet exist.

The operators arriving five years later in 2026 are making a different kind of decision. These operators are not betting on what Texas could become. They are responding to what Texas already is. The decision calculus has changed from conviction to observation — and observation-based decisions arrive faster, in larger quantities, and with less friction than conviction-based ones.

The acceleration is the most observable characteristic of the phase transition. The announcement cadence has compressed from quarterly to weekly. Samsung Taylor took four years of planning before its 2021 announcement. Amazon's Cedar Creek acquisition in Bastrop County closed in days — land identified, diligenced, and purchased without a public site search process, without a competitive incentive auction, without the years of negotiation that characterized the anchor operators' location decisions. Operators arrive because other operators are already here, and because the infrastructure, the labor market, and the institutional relationships the anchor operators built have made Texas the path of least resistance for the next wave.

PART IV

THE OUTWARD ARC

Twenty-One

THE ORBITAL COMPUTE SHELL

THE BOUNDARY BETWEEN THE TERRESTRIAL and the orbital infrastructures is not a conceptual boundary. It is a physical one: the Karman line, 100 km above sea level, where Earth's atmosphere becomes too thin to support aerodynamic lift and the domain of orbital mechanics begins. Below that line are the silicon fabs, the compute clusters, the autonomy deployments, the energy base, the capital stack, the workforce pipeline. Everything above it is what those terrestrial systems are building toward.

The orbital compute layer is not a single technology program. It is three co-dependent infrastructure programs whose simultaneous readiness determines when meaningful orbital AI inference becomes operational — and whose development timelines, assessed honestly against current site preparation status and production ramp realities, point toward 2030 as an optimistic threshold rather than a near-term milestone. The three are the compute layer, the power layer, and the thermal management layer. Each is a major industrial development effort in its own right. Their co-dependence means the system's deployment timeline is determined by the slowest program's readiness, not the fastest.

And before any of the three can deploy at meaningful scale, a fourth dependency must be satisfied: Starship.

The Starship Dependency

Every program documented here is contingent on Starship achieving the launch cadence and economics its architecture promises. The orbital compute shell and buildout toward the Starmind constellation the SpaceX FCC filing describes, Golden Dome's space layer, sun-synchronous orbital deployments, lunar logistics, Mars colonization — all of it requires launch economics that only Starship's full reusability and rapid turnaround architecture can provide. Without Starship reaching its designed cadence, every downstream program's timeline extends by decades, and the economics of most do not close at all. Starship is not one component of the orbital infrastructure argument. It is the load-bearing platform on which every other component rests.

Starship's development methodology — iterative hardware testing that accepts early failures as information rather than program-threatening events — has produced a vehicle substantially more capable today than it was two years ago, and substantially less capable than its designers intend it to be at full operational maturity. The Falcon 9's development followed the same pattern across a longer timeline, eventually producing the most-launched orbital rocket in history, with a reliability record the launch industry had not previously achieved. Starship's trajectory is consistent with that pattern at larger scale and higher complexity. Consistency with a historical pattern is not a guarantee. It is the best available evidence for a program whose outcome is not yet determined.

The Starship launch cadence is itself dependent on a supplier ring whose constraints are as real as any of the three orbital layer programs. Raptor engine production at McGregor is the primary cadence constraint. The Raptor V3 production rate determines how many Super Heavy boosters and Starship upper stages can be built per month, which determines the maximum launch cadence the vehicle fleet can sustain, which determines the constellation deployment rate every orbital program depends on.

The propellant supply chain is the second constraint, less discussed but equally binding. Liquid methane and liquid oxygen at the volumes high-cadence Starship operations require are not commodity purchases at Starbase's scale. Methane sourcing, liquefaction, and cryogenic transport infrastructure; LOX production at volumes sufficient for multiple launches per week at full operational tempo — these are industrial production programs whose development pace constrains launch rate independently of engine production and vehicle refurbishment. A launch vehicle without propellant does not launch, regardless of its technical readiness.

The water deluge system at Starbase consumes volumes per launch that make water supply, storage, and replenishment a non-trivial logistics constraint in the coastal South Texas environment. At the launch cadences Starship's economic model requires — dozens of launches per month at full operational tempo — the deluge water infrastructure is an industrial water management program in its own right.

Boca Chica is not SpaceX's only Starship launch facility. Launch Complex 39A at Kennedy Space Center in Florida — the same pad that launched Apollo 11 and the final Space Shuttle missions — has been modified for Starship operations, giving SpaceX a second orbital launch site whose geographic separation from Boca Chica provides both operational redundancy and launch azimuth flexibility a single Gulf Coast site cannot offer. Florida's eastern seaboard location enables direct-ascent trajectories to equatorial and mid-inclination orbits that Boca Chica's southern Texas location constrains. The dual-site architecture is the launch infrastructure expression of the same supply chain risk logic that dual-sources AI5 between Samsung Taylor and TSMC Arizona: no single point of failure for a program whose launch cadence is the gating variable for every downstream program.

The Orbital Architecture

The case for orbital AI compute leans heavily on continuous solar illumination — the argument that compute in space avoids the energy constraints of terrestrial data centers because the satellite is "always in the sun." That argument is true for some orbits and false for others, and the difference matters for whether the constellation can deliver what its program targets are stated to deliver. SpaceX's FCC filing of January 2026 for the Starmind constellation discloses how the architectural choice has been made.

The filing proposes up to one million satellites in narrow orbital shells 50 km thick, distributed across altitudes from 500 to 2,000 km — all within low Earth orbit (LEO) — and split between two inclination types. Sun-synchronous orbits (SSO) at the higher altitudes carry satellites that remain in sunlight more than 99 percent of the time, eliminating the eclipse problem at the lower level. These satellites handle the workloads requiring constant computing capacity — AI training, baseload inference, the workloads whose interruption is operationally costly. Thirty-degree inclination orbits at the lower altitudes carry satellites that face standard LEO eclipse periods of roughly 30 minutes per orbit, but they provide better coverage of populated mid-latitudes and serve demand peaks rather than constant baseload. The 30-degree shells are the burst capacity layer that responds to demand surges; the SSO shells are the continuous compute substrate. SpaceX has confirmed that different satellite versions will be designed for different orbital shells, with hardware tuned to the workload profile each shell serves.

The architectural specialization resolves the constraint triangle the orbital compute layer faces. Continuous solar power, low-latency LEO altitude, and global constellation scale are not simultaneously achievable in any single orbital regime — but they are simultaneously achievable in a constellation that allocates different orbital regimes to different workload classes. The SSO satellites get continuous solar power and LEO latency at the cost of restricted orbital volume and limited mid-latitude coverage. The 30-degree satellites get LEO latency, mid-latitude coverage, and scalable orbital volume at the cost of eclipse interruption. Both classes ride at LEO altitudes — 500 to 2,000 km, with round-trip latency under 15 milliseconds at any altitude in that range — so neither pays the latency penalty of higher orbits. Inter-satellite optical links route workloads between the two classes based on which satellites are currently illuminated and which workloads can tolerate eclipse periods. The constellation's ground connectivity flows through existing Starlink satellites as the relay backbone.

What the filing has not disclosed is the per-satellite specification: battery energy storage for the 30-degree shells, peak compute capacity by shell class, the operational protocols for routing workloads between SSO and 30-degree satellites, the relative satellite counts in each class. The 30-degree satellites either carry substantial onboard battery storage to bridge eclipse periods or operate in workload classes that tolerate eclipse interruption — most likely both, with batteries sized to maintain critical inference during shadow passes and intermittent workloads queued to be processed when sunlight returns. The architectural framing the filing establishes is that orbital specialization — not a single uniform satellite design — is how the program reconciles continuous solar with low-latency LEO operation at million-satellite scale. The trade-offs are real. The resolution is hybrid.

The AI1 Satellite Platform

The AI1 satellite is large. A 70-meter wingspan dominated by deployable solar arrays, a 20-meter overall height, and a power-to-mass density of 70 kW of compute payload per ton of total satellite mass. At the disclosed peak compute capacity of 150 kW and average payload of 120 kW, each AI1 satellite masses approximately two tons in operational configuration. The platform is larger than the current Starlink V3 satellites and is sized for Starship-class launch cadence — the cost-per-kilogram-to-orbit assumption that makes the Starmind constellation economics work.

The Compute Architecture

The honest timeline for D3 in operational LEO deployment is 2030 by optimistic estimate. The final gates are satellite integration and launch aboard Starship at a cadence that can populate a meaningful constellation. Each gate is a program in its own right. None can be compressed below its physical minimum timeline by capital deployment alone.

The compute architecture is a centralized module that can run interchangeable processors from multiple chipmakers, configured for specific AI workloads. At 120 kW of average compute payload, each AI1 satellite carries roughly the compute capacity of a single terrestrial high-density server rack. The interchangeability of the processor module is the design feature that lets SpaceX upgrade compute capability without redesigning the satellite around it; D3 silicon from Terafab can replace earlier-generation chips in the same module as the silicon trajectory advances. The constellation's compute capability is therefore not fixed at the satellite hardware generation it launches with. It tracks the silicon roadmap of whichever foundry is producing the current generation.

The AI1 bus, solar array, and thermal envelope can accept different compute modules across multiple chip generations and vendors. The modularity also hedges against Terafab's own production ramp, which itself is a multi-year construction buildout whose fab construction, radiation qualification, and yield optimization at leading-edge nodes are hard engineering programs whose outcomes are all in the future.

The primary alternative compute platform today is Nvidia's Space-1 Vera Rubin Module, announced at GTC 2026 in March as the flagship of a three-tier orbital product family. Space-1 Vera Rubin targets data center-class orbital compute directly, combining two Rubin GPUs with a Vera CPU on a SWaP-optimized module with claimed 25x the AI compute of an H100. Below it in the family, Nvidia's IGX Thor targets mission-critical edge, and Jetson Orin targets SWaP-constrained satellite payloads that are already reaching orbit — Kepler Communications has deployed forty Jetson Orin modules across ten satellites in its network. Nvidia's launch partners for the space computing platform include Aetherflux, Axiom Space, Kepler, Planet Labs, Sophia Space, and Starcloud. Outside of Terafab's future D3 program, Nvidia is the only orbital-grade AI datacenter-class silicon supplier in near production.

The Power System

The power system is built around a 150 kW deployable solar array operating at approximately 250W per square meter — a solar density consistent with high-efficiency space-grade photovoltaic technology, manufactured at the Project Echo solar cell facility on the Gigasat campus in Bastrop. The array is sized to deliver peak compute capacity directly from solar power when illuminated. The excess capacity during illumination phases supports thermal management, attitude control, the inter-satellite optical link power budget, and the battery charging that any satellite in the 30-degree-inclination shell needs to bridge eclipse periods.

The radiation tolerance requirements for space-grade PV panels are distinct from, but parallel to, D3's process requirements. Silicon and multi-junction III-V solar cells degrade under sustained particle bombardment. The qualification program for space-grade PV, at the required radiation tolerance and production scale, is a multi-year effort running in parallel with D3's chip qualification — and its timeline, like D3's, is gated by physical minimums that capital deployment cannot compress.

Thermal Management System

The thermal management system is the engineering achievement most easily overlooked in the public discussion of orbital compute. Every watt of compute is a watt of heat, and in vacuum the only way to dissipate heat is radiation. The AI1 design carries a 110-square meter deployable liquid radiator system with redundant pumping loops and micrometeoroid shielding — a structure approaching the radiator area of the International Space Station, fitted to a single small satellite. The radiator is what makes high-density compute possible in orbit; without it, the payload would overheat within minutes of running at full power draw. Terrestrial data centers move heat through water, refrigerants, and evaporative cooling. Orbital data centers can only radiate it, and the radiator area is the structural price the architecture pays for operating compute in vacuum. Of every visible square meter on an AI1 satellite, the radiator and the solar array are the two surfaces that dominate the design; the compute module itself is small, the structure that powers it and cools it is most of what the satellite physically is.

The thermal cycling environment in LEO adds a materials challenge terrestrial thermal management does not face. A satellite in a 90-minute LEO orbit experiences approximately 16 sunrise-sunset cycles per day, each transition imposing a rapid temperature swing on exposed surfaces between the solar-illuminated maximum and the eclipse minimum. Over an operational lifetime of five to ten years, the cumulative thermal cycle count reaches tens of thousands — a fatigue loading regime whose materials implications require design margins and materials selections terrestrial thermal management practice does not routinely address.

The Communications Layer

D3 compute, space-grade power, and thermal management are the hardware an orbital compute satellite requires to operate. The communications layer that connects orbital compute satellites to each other and to ground-based infrastructure is Starlink's inter-satellite link (ISL) mesh — the laser-based optical communications network Starlink's Generation 2 satellites carry, which makes the constellation a continuous global mesh rather than a series of ground-contact windows.

The coverage explains why inter-satellite link matters for orbital compute. A single LEO satellite at 550 km altitude has a ground station in view for approximately 10 minutes per 90-minute orbit. Without ISL, the satellite can only transmit data to the ground during those windows — roughly 11 percent of its operational time. An AI1 satellite running inference continuously generates outputs, and requires model update ingestion, that cannot wait for the next ground station pass. With ISL, the satellite relays its data continuously through the Starlink mesh to whichever satellite currently has a ground station in view, eliminating the coverage gap entirely. The mesh is what makes continuous orbital compute operationally viable rather than intermittently connected.

The ISL mesh also enables coordination between orbital compute nodes — the multi-satellite inference distribution the architecture implies. A compute workload too large for a single satellite's inference capacity can be distributed across adjacent satellites in the mesh, with results aggregated through the ISL links before ground transmission. The same multi-agent coordination architecture that will operate at the Giga Texas production floor — multiple Optimus units coordinating tasks through shared situational awareness — will operate at the orbital layer through Starlink's ISL mesh.

The Radiation Gradient

Low Earth orbit is the easiest place to operate hardware beyond Earth's surface. The planet's magnetic field still deflects most galactic cosmic radiation and the bulk of solar particle flux at LEO altitudes, and spacecraft pass through the inner edges of the Van Allen belts only briefly at high-latitude crossings. Radiation hardening for LEO is the entry rung on the qualification ladder, not the end of it.

The ladder is steep and non-linear. Each outward regime represents a step change — not a percentage — in environmental severity. Sun-synchronous and polar orbits add modest particle exposure as magnetic shielding thins toward the poles. Medium Earth orbit (MEO), home to GPS and the other navigation constellations, sits inside the Van Allen belts and exposes spacecraft to some of the highest sustained particle fluxes in near-Earth space. Geostationary orbit lies outside most of the magnetosphere's protection and absorbs continuous energetic electrons, solar storms, and cosmic radiation. Beyond Earth orbit, the full galactic cosmic ray environment arrives undiluted: lunar missions encounter it directly, and Mars trajectories add years of cumulative dose on top. The difference between LEO and deep space is not measured in percentages but in multiples, often orders of magnitude, each requiring new approaches to shielding, packaging, fault tolerance, and process qualification.

This gradient maps directly onto the orbital compute architectures now under construction. SpaceX's AI1 Starmind satellites are designed for LEO and SSO, where radiation exceeds terrestrial data centers but remains partially mitigated by Earth's magnetic field. The multi-layer architectures proposed for Golden Dome distribute assets across LEO, MEO, and GEO simultaneously, meaning a single program must qualify hardware against three different radiation regimes at once. A constellation that crosses orbital boundaries is not one engineering problem but several, stacked.

The qualification challenge is also heterogeneous within each spacecraft. Different component classes fail in different ways. Advanced logic — the kind of 2-nanometer accelerators D3 is targeting for orbital inference — is dominated by single-event upsets and transient faults, energetic particles striking transistor structures small enough that a single ionization track can flip a bit. Analog and mixed-signal devices suffer cumulative dose degradation and parameter drift. Power electronics face displacement damage in their semiconductor lattices. Memory has its own upset modes; sensors, optics, and photovoltaics each degrade on their own clocks. A space-qualified compute platform is therefore not a hardened chip but a hardened supply chain: AI accelerators, power management, memory, networking, sensors, solar arrays, and packaging must each be qualified against the regime where they will operate, and the qualification has to compose at the system level.

The Moon and Mars push every parameter further. Lunar orbit and the lunar surface have neither a global magnetic field nor a meaningful atmosphere; hardware is exposed directly to galactic cosmic rays and solar energetic particle events with nothing in between. Mars adds a very thin atmosphere — enough to attenuate some primary particles but also enough to generate secondary radiation through cosmic-ray spallation — on top of pervasive dust intrusion and thermal cycling. Each outward step adds requirements that compound rather than substitute. Radiation tolerance, thermal management, power density, packaging, and reliability all tighten together.

This is the broader strategic point, and it is where Terafab's two-node architecture earns its position. The value is not a single radiation-hardened part. It is the manufacturing knowledge, process control, and qualification capability needed to produce successive generations of hardware for successively harsher environments. The chips that fly in LEO this decade are the simplest version of the systems that lunar surface infrastructure, of Mars, and eventually what deep-space platforms will require. Whoever can iterate up the ladder — LEO to MEO to GEO to cislunar to Mars — holds the manufacturing position that defines the next several decades of off-Earth compute. The radiation gradient is the engineering specification. The fab capable of climbing it is the strategic asset.

The Golden Dome Queue

Before the SpaceX orbital compute shell serves commercial AI inference at scale, it serves Golden Dome. The missile defense architecture's sensor and compute layer has procurement priority over commercial orbital compute.

The physics of hypersonic defense dictates why the space layer matters for Golden Dome specifically. Ground-based radar systems have a geometric limitation: the Earth's curvature limits the horizon distance at which a ground-based sensor can detect and track a low-altitude maneuvering target. A hypersonic glide vehicle at Mach 20 flying a depressed trajectory — specifically to exploit that limitation — can reduce the discrimination window available to ground-based interceptors below the minimum needed for a reliable intercept solution, which is measured in seconds.

The chips that run Golden Dome's sensor and compute layer require a separate qualification program running in parallel with D3's certification, whose completion gates Golden Dome's space layer deployment.

Therefore, Terafab's first certified output will most likely serve Golden Dome's requirements before it serves SpaceX's commercial compute. This sequencing is strategically logical and commercially consequential. The 2030 projection for meaningful commercial LEO compute deployment should be understood as contingent on Golden Dome's most urgent procurement requirements having been substantially satisfied, and on Terafab's production yield having reached the scale commercial constellation deployment requires.

Orbital Shell as System

The three AI1 infrastructure programs — D3 compute, space-grade PV power, thermal management — are not independent development efforts that happen to be needed by the same application. They are a co-dependent system, whose integration requirements create design constraints each program must satisfy relative to the others. The compute density D3 enables determines the waste heat load the thermal system must reject; that determines the radiator area the satellite bus must accommodate; that determines the satellite's mass budget; that determines the power generation capacity the PV array must provide within the satellite's size and mass constraints — which in turn constrains the compute density D3 can operate at within the available power envelope.

The orbital compute shell is a Texas industrial program distributed across the state's geography — from Terafab Gibbons Creek to College Station's Cyclotron Institute to Bastrop Gigasat to Mcgregor's test stands to the Starbase's launch complex.

The Orbital Compute Industry

The orbital compute programs documented here were, until recently, a SpaceX-specific thesis. They are becoming an industry.

Google's Project Suncatcher was announced by CEO Sundar Pichai as the subject of active launch partnership talks with SpaceX. It envisions 81 satellites spanning a 1-km radius, forming an in-orbit compute cluster that carries Google's Trillium TPUs, communicates via free-space optical inter-satellite links, and runs on continuous solar energy, with two prototype satellites targeted for early 2027. Planet Labs is the satellite manufacturer. Google owns approximately 6.1 percent of SpaceX. The talks are not between strangers.

Blue Origin's Project Sunrise, for 50,000 satellites, is the second competing orbital compute constellation program — from a company whose satellite manufacturing campus may come to Hutto. SpaceX's own orbital data center filing targets one million satellites. Three operators with confirmed or near-confirmed programs are treating orbital compute as a planning reality at the same time. That serious operator attention — Google, Blue Origin, SpaceX — is the single most important validation of the orbital compute thesis: not because any one program has proven the economics, but because three of the most analytically rigorous operators in the technology industry are committing prototype capital to the same architectural bet at the same time.

The economics require honest treatment. Google's own analysis pins the financial equilibrium for space data centers at approximately $200 per kilogram launch cost. SpaceX's current rideshare pricing is $7,000 per kilogram. The gap is 35x. Sam Altman called orbital data centers "ridiculous for now." The orbital compute thesis does not depend on denying that gap. It depends on Starship's cost reduction trajectory closing the gap within the three-to-five-year window the prototype programs are using to validate the architecture before committing to full constellation deployment. If Starship achieves its designed economics, the $200 per kilogram equilibrium becomes reachable. If it does not, the orbital compute shell remains a compelling architecture whose economics do not yet support commercial deployment at the scale its proponents envision.

What is already in evidence: SpaceXAI has leased the Colossus 1 data center in Memphis to Anthropic, the AI safety company, for $1.25 billion per month for its inference workloads, with potential future collaboration on orbital compute. Google has followed suit, and will pay $920 million per month to rent Colossus 2 capacity for both their AI training and inference workloads. Reflection AI is also leasing Colossus 2 capacity for their frontier open-weight model at $150 million per month through 2029. SpaceX is now a computing infrastructure provider to third-party AI companies across both terrestrial and orbital layers — before a single orbital compute satellite has been deployed. The infrastructure operator's commercial relationships are assembling ahead of the infrastructure. That sequencing — customers before deployment, revenue relationships before operational capability — is the strongest available signal that the orbital compute shell's economics are being taken seriously by operators whose capital allocation decisions are grounded in financial analysis rather than aspiration.

Twenty-Two

THE MOON

THE MOON IS 384,400 KM FROM EARTH. At the speed of the Apollo 11 command module, that transit takes about three days. At the speed of a radio signal, the one-way communication delay is 1.3 seconds — long enough to make real-time teleoperation of surface equipment from Earth unreliable, short enough that a human operator with situational awareness can manage it given the right system design.

That distance is the whole problem, stated in one number. The Moon is close enough that its resources, its surface area, and its position in orbit make it the logical first industrial node beyond Earth's atmosphere. It is far enough that getting there, operating there, and returning require engineering solutions whose difficulty the Apollo program's success has caused a generation of observers to systematically underestimate.

Apollo went to the Moon six times between 1969 and 1972. It landed 12 humans on the surface for stays of 21 to 75 hours, and returned every one of them safely. The achievement was real and extraordinary. But notice what Apollo did not do, and was never designed to do: establish a sustained human presence, develop lunar resources, or build the industrial infrastructure that makes the Moon economically productive rather than scientifically interesting. Apollo was an exploration program. What follows is an industrialization program. The two are related the way Lewis and Clark's expedition is related to the transcontinental railroad. The expedition established that the destination existed and could be reached. The industrialization is a different undertaking entirely — different technology, different economics, and a different relationship between the people doing it and the place they are going.

The Return Operators

NASA's Artemis program is the federal framework for returning humans to the lunar surface. Artemis IV, currently targeting 2028 for the first crewed lunar landing since Apollo 17, uses the Space Launch System as the crew launch vehicle and either Blue Origin's or SpaceX's Human Landing System as the surface descent and ascent vehicle. It is a federal-commercial partnership whose operational execution runs through SpaceX's Starship program — which makes Artemis's timeline contingent on Starship's development maturation, the load-bearing variable for every cislunar program.

The use of Starship as the potential Human Landing System matters beyond the mission itself. It means the vehicle whose economics are the key to lunar industrialization is being flight-qualified in the most demanding context possible: crewed lunar surface landing and ascent, under a NASA contract that supplies both the funding and the regulatory framework for the qualification program. SpaceX is not developing Starship's lunar capability speculatively and then offering it to NASA. NASA is funding that capability as a procurement — which means the qualification milestones, the safety review processes, and the operational procedures crewed lunar landing requires are being developed inside a federal contract framework that supplies institutional rigor alongside commercial development speed.

NASA's Moon Base program— a continuous-operations human presence on the lunar South Pole — is realistically a 2035 to 2040 program, given Artemis IV's current timeline and the infrastructure that sustained surface operations require beyond the initial landing.

The gap between the first one-week-duration crewed landing and a continuous base is not one that additional Artemis missions alone can close. It requires three things: cargo delivery at the scale base construction and sustained operations demand, power infrastructure that supports more than short-duration surface stays, and the in-situ resource utilization (ISRU) capability that removes the most operationally limiting dependency on Earth-sourced consumables. All three point back to Starship — cargo Starship variants for bulk delivery, nuclear or solar power systems, lunar terrain vehicles, Starlink satellites deployed in orbit, construction equipment, Optimus robots, and the other hardware Starship delivers to begin producing power, water, oxygen, and eventually propellant from lunar resources.

The Artemis architecture and the SpaceX Human Landing System are the headline programs of the lunar return. They are not the only ones. NASA's Commercial Lunar Payload Services program — CLPS — has contracted a roster of commercial providers to deliver science instruments, technology demonstrations, and robotic precursor hardware to the lunar surface on vehicles that are neither SLS nor Starship. CLPS is the commercial lunar services layer operating beneath the crewed Artemis missions, delivering the surface characterization data and technology validation that sustained human operations require before they can be safely attempted.

Firefly Aerospace is a CLPS provider, and its headquarters in Cedar Park, Texas, makes it the most locally grounded expression of the lunar economy taking shape around the Texas AI-industrial complex. Firefly's Blue Ghost lander achieved a successful upright lunar landing in 2025, delivering NASA payloads to the surface aboard a Falcon 9 and operating for a full lunar day — roughly 14 Earth days — before the lunar night's thermal environment ended the mission.

Firefly's lunar program illustrates something the Artemis and SpaceX narrative tends to obscure: the commercial lunar services layer is not a monolith. It is an ecosystem of providers at different scales, with different vehicle architectures and different mission profiles, whose aggregate activity is building the database and the operational experience base that sustained human operations require. Blue Ghost's landing, Intuitive Machines' Nova-C landings, and the broader CLPS roster are the commercial lunar economy's early formation — the equivalent of the first commercial satellite operators, whose missions in the 1960s and 1970s built the operational experience the mature commercial space industry now rests on.

The Cedar Park connection is worth holding onto precisely because it is easy to miss in the standard coverage. The Moon is 384,400 km away. Firefly Aerospace is in Cedar Park, Texas. The convergence extends from an Austin suburb to the lunar surface — not metaphorically, but through a small company whose engineering teams work here and whose hardware has operated on the Moon.

THE SOUTH POLE

Every serious lunar base architecture targets the same place: the lunar South Pole. The reasons are scientific, operational, and economic at once. The South Pole's permanently shadowed craters hold water ice, confirmed by multiple orbital missions, in quantities that are sufficient to support sustained human operations and potentially propellant production at meaningful scale. The crater rims next to those shadowed regions receive near-continuous solar illumination. These are the "peaks of eternal light," whose solar exposure approaches 90 percent of the lunar year in some locations — a power generation base for a surface base, without the two-week darkness that equatorial and mid-latitude sites endure every lunar night.

The thermal environment compounds the advantage. Equatorial lunar surface temperatures swing from 127 degrees Celsius in direct sunlight to minus 173 degrees in the two-week night — a 300-degree excursion that imposes severe thermal cycling on surface hardware. The permanently shadowed craters at the South Pole sit near minus 200 degrees Celsius continuously: extremely cold, but stable. Pluto has warmer temperatures. The illuminated rim sites swing far less than equatorial sites, because the solar angle is low and nearly continuous rather than high and alternating with darkness.

The water ice is the resource that changes the lunar industrialization economics most fundamentally. Water is the consumable a human surface presence needs in the largest quantities: for drinking, for cleaning, for oxygen production through electrolysis, and for radiation shielding whose mass requirements make Earth-sourced water prohibitively expensive to deliver. Water is also the feedstock for propellant. Lunar oxygen, produced by electrolysis, is the oxidizer half of Starship's methane-oxygen propellant — and the larger half by mass. Whether lunar propellant production reaches operational scale determines whether Starship and other cislunar vehicles refuel at the Moon — the architectural requirement that makes the Moon a logistics node in the cislunar system rather than a terminal destination requiring round-trip propellant from Earth.

The Mass Driver

The mass driver is the technology that turns the Moon from a science destination and human outpost into an industrial node — one capable of contributing material output to the broader cislunar economy. The operating principle is straightforward. An electromagnetic launcher accelerates payloads along a track using linear motor technology, imparting enough velocity to reach lunar orbit — or, with the right trajectory design, to send payloads directly to Earth-Moon Lagrange points, to the Earth orbital compute shell, or onto interplanetary trajectories, with no chemical propellant at all. The energy comes from the lunar power system infrastructure. The propellant cost is zero.

Its economic significance comes from the shallowness of the lunar gravity well. Escaping Earth's gravity requires reaching roughly 11.2 km/s — an escape velocity whose achievement demands the enormous propellant fractions that make Earth-to-orbit launch so expensive. Lunar escape velocity is 2.38 km/s, less than a quarter of Earth's, and the Moon has no atmosphere to impose drag during ascent. An electromagnetic launcher on the lunar surface can accelerate a payload to escape velocity using electrical energy, at a cost per kilogram orders of magnitude below any chemical propulsion system operating from Earth.

The materials a mass driver can launch are exactly the ones the cislunar economy needs in the largest quantities: metals extracted from regolith through in-situ processing, and water ice processed into propellant. Delivering that propellant to orbital depots eliminates the need to lift propellant from Earth's deep gravity well for every cislunar mission. A lunar mass driver supplying orbital propellant depots transforms the economics of every cislunar operation. Spacecraft refuel in orbit from lunar-sourced propellant rather than carrying Earth-launched propellant for the whole mission — which cuts the Earth-to-orbit launch mass fraction for every cislunar mission, and cuts its cost correspondingly.

The mass driver is also what makes the orbital compute shell's construction economics viable at scale. The satellites, radiators, solar panels, and structural components a large orbital compute constellation requires represent enormous mass — mass currently assumed to be launched from Earth. If in the next three decades lunar-sourced materials can instead be delivered to orbital construction sites by mass driver, at a fraction of the cost of Earth-launched equivalent mass, the shell's construction economics improve by the ratio of lunar launch cost to Earth launch cost per kilogram. Once a mass driver is operational, that ratio is large enough to change the economic feasibility of orbital infrastructure programs whose Earth-launch mass budget currently leaves them marginal. The engineering challenges are hard.

The Nuclear Power Question

Moon Base's sustained operations — human habitation, ISRU processing, mass driver operation, communications, and the compute infrastructure that manages all of it — require more power than photovoltaic generation alone can provide at the South Pole.

Solar power at the rim sites is more reliable than at equatorial sites, but the low solar elevation angle reduces the power density per unit of panel area.

Fission surface power — compact nuclear reactors designed for lunar surface deployment — is the architecture NASA and the Department of Energy have been developing for the Artemis program's sustained surface operations phase. The Kilopower and subsequent fission surface power programs target reactor systems in the 10 kW range: enough for early base operations, but requiring multiple units or much larger successor systems for the industrial-scale power that mass driver operation and large-scale ISRU processing demand. The development timeline for fission surface power capable of supporting industrial-scale lunar operations runs parallel to and beyond the Artemis crewed landing.

The Cislunar Economy

Moon Base, if it develops along the trajectory the current program implies, is not the end state of lunar industrialization. It is the infrastructure whose existence makes the next phase's economics viable — the same relationship the transcontinental railroad's completion had to the settlement of the American West, or the interstate highway system's completion had to the logistics industry that reorganized around it.

The cislunar economy that Moon Base enables has four layers, and the sequence of their development is set by the infrastructure each layer requires from the one before it.

The first layer is science and exploration: the human and robotic presence Artemis establishes, generating the surface data that industrial planning requires.

The second is resource extraction: water ice mining at the South Pole, regolith processing for metals and construction aggregate, the ISRU capability that produces propellant from local resources.

The third is industrial production: manufacturing on the lunar surface, using lunar-sourced materials and lunar power, producing components whose delivery to orbit by mass driver feeds the orbital infrastructure programs.

The fourth is the mass driver economy itself: the sustained industrial output whose orbital delivery transforms the economics of every cislunar program and makes the orbital shell's construction viable at the scale a planetary AI-industrial complex eventually requires.

The timeline from the Artemis first crewed landing to a mature mass driver economy is not a decade's work. It is a multigenerational industrial program. Its first phase — the crewed landing and initial surface operations — is scheduled within the current decade. Its second — sustained base operations and initial ISRU — is a 2030s program. Its third and fourth phases are 2040s and beyond. The timeline cannot be compressed by urgency or capital availability alone. It is set by the physics of what each phase requires from the phase before it, by the engineering development timelines hardware qualification imposes, and by the political and institutional continuity a multigenerational program demands from the organizations responsible for it.

Twenty-Three

MARS

TWO TESLA PERFORMANCE AWARDS anchor the corporate governance frame this chapter relies on. The 2018 award granted Elon Musk options on approximately 304 million Tesla shares (split-adjusted), valued at approximately $56 billion at the time of the 2024 Chancery decision, vesting against twelve market-capitalization milestones paired with revenue and adjusted-EBITDA operational targets. The Delaware Court of Chancery voided the award in January 2024; Tesla shareholders ratified it at the June 13, 2024 annual meeting — the same meeting at which they approved the company's redomicile from Delaware to Texas; and the Delaware Supreme Court reversed the rescission in December 2025, restoring the award while explicitly declining to address the trial court's underlying findings on control, fairness, and ratification.

The 2025 award, approved by Tesla shareholders in November 2025, runs alongside it: twelve market-capitalization tranches rising from $2 trillion to $8.5 trillion, paired with twelve operational milestones — four product and deployment targets (20 million vehicles delivered, 10 million active FSD subscriptions, one million bots delivered, one million robotaxis in commercial operation) plus eight adjusted-EBITDA tiers ascending to $400 billion. Potential value at full vesting approaches $1 trillion. It is the largest executive compensation structure in corporate history.

The milestone structure is what matters analytically.

Neither award contains Mars colonization as a contractual vesting milestone. The vesting tranches of both are financial and operational. What the awards do establish is the multi-decade fiduciary horizon over which Musk is contractually retained as Tesla's CEO. An $8.5 trillion market capitalization is not a destination Tesla reaches inside a single product cycle or a single decade. A million humanoid robots delivered and a million robotaxis in commercial operation are not milestones achievable on a five-year planning horizon. The compensation architecture is, in effect, a board-approved, shareholder-ratified, court-tested commitment that Musk operates Tesla on a generational timescale toward outcomes whose achievement would in itself reorganize the industrial landscape.

The Mars commitment sits adjacent to that framework rather than inside it. SpaceX, not Tesla, is the entity whose stated corporate mission explicitly includes Mars colonization. Musk is its CEO and majority owner. The integration of Tesla, SpaceX, the Boring Company, and Neuralink under shared leadership, shared capital, shared technology stacks, shared workforce, and increasingly shared geography is what converts Mars colonization from a personal aspiration into a corporate program executed across multiple legal entities under unified strategic direction. Tesla's Master Plan IV — the strategic document Tesla's board has cited as the context for the 2025 award — articulates the long-horizon vision within which the compensation framework operates. The fiduciary instruments that compensate Musk for Tesla outcomes are one of several instruments that hold this integrated structure in place.

Treating the empire-level Mars trajectory as a corporate program rather than a personal dream changes every analytical question. The question is not whether Elon Musk believes in Mars colonization. His belief is well documented and not analytically interesting. The question is whether the capital allocation, the engineering trajectory, and the operational programs of the multiple entities under unified Musk leadership are consistent with the trajectory's eventual achievement. That is an engineering and economics question, not a psychology question — and it has an engineering and economics answer.

The consequence is a category shift. Mars colonization does not appear as a line item in either Tesla award — the vesting milestones are Tesla-specific product and financial targets. What the awards do is bind Musk to Tesla, and Tesla to Musk, across the exact multi-decade horizon that the SpaceX mission requires. Combined with his majority ownership of SpaceX and his operational command of the integrated Tesla–SpaceX–Boring–Neuralink apparatus, the fiduciary architecture around Tesla effectively armors the timeline over which the SpaceX mission is executed. Mars colonization is no longer a founder's aspiration. It is the animating mission of an integrated corporate program whose operational continuity shareholders have funded, whose board has engineered, and whose courts have now armored against unwinding. Serious analysis of the next quarter-century of American industrial policy has to start from that fact.

The Mars Argument

The Mars question — why Mars specifically, why now, why through a corporate structure rather than a national program — has a first-principles answer the personality-focused coverage of Musk's Mars ambition consistently obscures.

Earth is a single point of failure for human civilization. Every extinction-level risk humanity faces — asteroid impact, supervolcanic eruption, engineered pandemic, nuclear exchange — is a risk whose consequences are total if humanity exists on only one planet. The risk is not imminent on any human planning horizon. It is certain on a geological one. A civilization confined to a single planetary body is a civilization whose long-term survival probability, integrated over geological time, approaches zero — regardless of how well it manages its near-term risks.

Mars is the only other body in the solar system that currently presents a path to self-sustaining human civilization within the engineering and economic constraints of foreseeable technology. The Moon is too small, too resource-limited, and too gravitationally close to Earth to function as an independent civilization rather than an Earth-dependent outpost. Venus's surface conditions — 465 degrees Celsius, 92 atmospheres of pressure, sulfuric acid clouds — are more hostile to human habitation than anywhere in the solar system. The outer solar system's distance from the Sun imposes energy constraints that put self-sustaining civilization there generations of technology away. Mars has a 24.6-hour day, a very thin but present atmosphere, water ice at its poles and in its subsurface, and temperatures that are extreme but within the range engineered habitats can manage. Its gravity, at 38 percent of Earth's, might be enough to prevent the long-term physiological degradation that zero-gravity exposure produces — though the specific health implications of 0.38g over multi-year timescales remain incompletely characterized.

Mars is not a comfortable destination. It is the least hostile available option for a second planetary human civilization — which is a different thing.

Starship Architecture for Mars

Starship is the specific propulsion and payload architecture that makes Mars colonization economically tractable rather than prohibitively expensive. Its key innovation for Mars — beyond full reusability — is the methane-oxygen propellant combination, whose components can be produced from Mars's available resources through in-situ resource utilization chemistry that is well understood and has been demonstrated at laboratory scale.

The chemistry is not exotic. Mars's atmosphere is 95 percent carbon dioxide. Its subsurface holds water ice, confirmed by orbital observation and surface measurement across multiple landing sites. The Sabatier reaction — carbon dioxide plus hydrogen heated yields methane plus water, with the hydrogen produced by electrolysis of water — is the ISRU pathway that produces Starship propellant from Martian atmospheric carbon dioxide and subsurface water ice, using electrical power from solar or nuclear generation on the surface. It is nineteenth-century industrial chemistry applied to a twenty-first-century propellant production problem on a planet 225 million km from Earth.

So the Mars ISRU propellant plant is not a convenience. It is the load-bearing economic element of the whole colonization program. Without it, either every Mars mission is a one-way trip, or every mission has to launch its return propellant from Earth, imposing a mass multiplier that makes large-scale colonization economics impossible within any foreseeable launch cost structure. With it, Starship becomes a reusable interplanetary transport whose per-mission cost is dominated by vehicle amortization and operations rather than by propellant mass launched from Earth.

The Convergence of the Empire

The standard frame treats the Musk empire as disparate ventures whose only connection is one man's attention and capital: an electric vehicle company, a battery storage manufacturing division, a solar panel manufacturing division, a rocket company, an AI company, a tunneling company, a satellite internet provider. The frame is wrong in its most important dimension. The components are not disparate. They are the layers of a single integrated planetary colonization program — an architecture that becomes legible only when all the components are held at once and their Mars application is made explicit.

Starship is the primary transportation layer: the vehicle that moves people and cargo from Earth to Moon to Mars, and eventually between planetary surfaces. Without Starship's economics, every other layer's Mars application is stranded on Earth by launch costs no colonization program can sustain.

Starlink is the communications layer — planetary communications for Moon and Mars colonies, orbital data relay between Earth and surface operations, the OTA update infrastructure that keeps autonomous systems current across interplanetary distances. The 3-to-22-minute one-way light travel time between Earth and Mars, varying with orbital position, makes real-time Earth-Mars teleoperation impossible and forces Mars surface systems to operate with substantially higher on-device autonomy than Earth operations require. Starlink's architecture, extended to Mars relay satellites in areosynchronous orbit, is the communications infrastructure that makes that autonomous operation possible while holding the Earth-Mars data link that training updates, mission planning, and emergency consultation require.

The Bastrop solar manufacturing program is the power generation layer for both orbital compute and potential Mars surface operations. Mars receives 43 percent of Earth's solar irradiance at its average orbital distance, enough for meaningful surface power generation across the relatively dust-storm-free periods that make up much of the Martian year. The same rad-tolerant PV panels the orbital compute program requires serve Mars surface power generation. The specification overlap between LEO radiation tolerance and Mars surface radiation tolerance is not complete, but it is substantial enough that a single development program serves both applications with appropriate process variants.

Megapack and the grid-scale Megablock battery storage architecture are the energy storage layer. Mars's dust storms — regional events lasting weeks, global events lasting months, recurring just frequently enough that surface power system design needs to treat them as a baseline condition rather than an exception — impose energy storage requirements whose scale, relative to installed generation, substantially exceeds what renewable variability drives on Earth. The same battery chemistry, the same power electronics, and the same grid management AI that optimizes dispatch for terrestrial customers will manage Mars surface microgrid dispatch for a colony whose energy security is as consequential as any grid reliability requirement on Earth.

Tesla Cybertruck variants, both autonomous and human-driven, are a possible surface mobility layer. The Martian surface requires pressurized vehicles for any transit beyond the immediate habitat perimeter, and electric drivetrains are the only viable propulsion architecture for them: combustion engines need an oxidizer Mars's thin atmosphere cannot supply. The same battery chemistry that powers Cybertruck on Earth powers a pressurized Mars rover in 0.38g, with equivalent energy density and with thermal management adapted for Martian temperature ranges.

Optimus is the most-likely labor layer — and the most strategically significant single component of the Mars colonization program currently in active development. The architecture the foundation stack analysis implies requires substantial base infrastructure in place before the first human colonists arrive: habitat modules assembled, ISRU propellant plants constructed and operating, power systems deployed, food production systems initialized, a minimum viable industrial manufacturing capability established. Human labor performing that pre-arrival construction requires human life support for the construction workforce — a recursive dependency that multiplies the mission mass requirement for every human worker deployed. Optimus units performing the same tasks require no life support, no pressurized habitat, no food production, no medical capability. They require only power, which the solar and storage infrastructure provides, heat rejection, and the compute and connectivity that Starlink and on-hardware AI supply.

So the pre-human Optimus deployment on Mars is not a feature of the colonization program. It is a requirement whose absence makes the program's mass budget unworkable. The cargo Starships that deliver Optimus units, ISRU hardware, power systems, and habitat modules to Mars before the first crewed missions arrive are building the minimum viable base that makes human Mars residence feasible. The ratio of Optimus to human in early Mars operations may exceed ten to one because the cost of human life support makes Optimus the economically correct choice for every task it can perform without human presence.

Terafab's chip roadmap — D3 for LEO, its successors for the lunar and Martian surfaces — is the intelligence layer that runs every autonomous system in the program. The training infrastructure at

Cortex that improves FSD on Earth may well produce the models FSD's Moon and Mars surface adaptation runs in pressurized rovers. The training data Optimus accumulates on the Giga Texas production floor is the foundation on which Optimus's Mars construction capability is built. The OTA update loop that closes at Earth distances closes via Starlink relay at Mars distances, with the autonomy level adapted for the communication delay. The intelligence layer is the same layer at every sphere of the concentric geometry. Its process requirements become more demanding with each outward sphere, but its architectural logic is continuous from Austin to Arcadia Planitia.

The Boring Company is the infrastructure layer whose Mars application is more strategically significant than any of its terrestrial deployments. Underground pressurized tunnels on Mars are not an engineering convenience. They are the radiation shielding solution that makes multi-year human surface residence physically acceptable. Mars's thin atmosphere and absent global magnetic field expose surface inhabitants to galactic cosmic ray flux and solar particle events at doses that exceed acceptable limits for above-surface habitation. Underground construction, using Martian regolith as radiation shielding mass, reduces surface dose rates to acceptable levels — the same physics that makes submarine and underground bunker habitation effective for radiation protection on Earth, applied on Mars. The Boring Company's tunneling technology, adapted for Martian subsurface geology and operating conditions, is the construction tool whose Mars application was implicit in its development from the beginning, whether or not that application was ever publicly articulated. The same logic applies to the lunar South Pole: tunneling into crater walls or subsurface lava tubes provides radiation shielding and thermal stability no surface habitat structure can match.

THE MARS SUBSTRATE STACK

The one-million-colonist threshold Musk has stated as the target for a self-sustaining Mars civilization is not achievable through transportation alone. It requires a complete industrial substrate on Mars — and the individual layers of that substrate stack must be developed in sequence, each providing the foundation the next one requires. The substrate stack is the honest analytical frame for what the Mars milestone actually demands: not the headline number, but the industrial program that makes the headline number physically possible.

The first layer is transportation infrastructure: Mars and Earth reach favorable launch alignment — opposition — roughly every 26 months. Each opposition window is the primary opportunity for efficient Earth-Mars transit, its duration and energy requirement varying with the specific orbital geometry. High-cadence colonization requires launching many Starship missions per window — which makes the launch infrastructure, the propellant production, and the rocket fleet the gating variables for every program below.

The second layer is habitation infrastructure: pressurized volume sufficient for human occupation, radiation shielding adequate for surface activities, and life support systems whose reliability over multi-year operational periods is the specific engineering. The Boring Company's tunneling technology may be the construction tool that makes buried habitat construction operationally tractable rather than manually prohibitive.

The third layer is food production infrastructure: closed-loop agricultural systems capable of producing enough caloric and nutritional output to support the population without dependence on Earth resupply for food. Colonization at the scale the compensation framework's milestone implies requires self-sufficient food production operating reliably in a pressurized habitat whose atmosphere, water, lighting, and thermal conditions are all engineered.

The fourth layer is energy infrastructure: power generation and storage sufficient to support habitation, food production, ISRU propellant production, computing, communications, and the industrial manufacturing self-sufficiency requires. Solar power at

Mars receives about 43 percent of the irradiance available at Earth's distance from the Sun — enough for meaningful generation, but requiring larger panel areas per unit of output than equivalent Earth or lunar installations. Fission surface power is the generation architecture that provides dust-storm resilience and the power density industrial-scale ISRU and manufacturing require.

The fifth layer is industrial manufacturing infrastructure: the capability to produce on Mars the components, tools, structures, and systems the base requires for maintenance, expansion, and eventual self-sufficiency from Earth supply. A Mars base that cannot manufacture replacement parts for its own critical systems is a base whose survival is contingent on Earth resupply timelines measured in years. Industrial manufacturing on Mars, using Mars-sourced materials and Mars-generated power, is the layer that converts a research outpost into a self-sustaining settlement.

The One Million Colonists Number

That one-million-colonist milestone is not an arbitrary round number. It is the population threshold the SpaceX technical team has determined as sufficient for a self-sustaining Mars civilization — one whose genetic diversity, skill diversity, and institutional knowledge base are large enough to maintain and develop the industrial and social infrastructure civilization requires, without depending on continuous input from Earth.

The population genetics argument is straightforward. A founding population of a million individuals drawn from Earth's diverse gene pool provides enough genetic diversity to avoid the inbreeding depression and genetic drift that would threaten a smaller founding population's long-term viability. The skills argument is more complex. A self-sustaining industrial civilization requires practitioners of thousands of distinct technical and social specializations — not just engineers and scientists, but physicians, farmers, teachers, administrators, artists, the full range of human capability that maintains the social fabric technical civilization depends on. A population of a million provides the demographic depth to staff that full range at the minimum viable level for each.

The timeline for reaching a million Mars colonists from the current state of zero is not a decade's program or a generation's. It is multigenerational. Its early phases — the first crewed Mars landings, the first permanent surface base, the first self-sustaining food production — are the current generation's contribution to a trajectory whose completion belongs to their grandchildren or great-grandchildren.

The Multigenerational Question

Institutional continuity across generations is not guaranteed by the quality of the founding vision or the capital invested in its pursuit. The organizations, the engineering programs, and the cultural commitment to Mars colonization the current generation is building will require successors whose commitment to the directional goal is as genuine as the founders'. The most likely source of that continuity is the generation now growing up inside the aerospace workforce — the children of the SpaceX engineers at Starbase, of the NASA personnel running Artemis at Johnson Space Center and Kennedy Space Center, of the Firefly engineers in Cedar Park, of the broader space-industrial workforce. Their parents are doing the foundational work, and their dinner-table conversations are the cultural transmission mechanism no corporate succession plan can substitute for. Whether that workforce-generation continuity is sufficient is genuinely uncertain. What is not uncertain is the requirement. A Mars colony of a million self-sustaining inhabitants requires the work of multiple human generations. Each generation inherits the bedrock the prior generation built and adds the layer the next one requires.

The engineers at McGregor producing Raptor V3 engines today are doing work whose ultimate significance will be legible to their grandchildren in a way it cannot be fully legible to them. The researchers who are qualifying D3's radiation tolerance at Texas A&M are building the institutional capability that AI9's lunar qualification and AI12's Mars qualification will require from their successors. The students in Cedar Park's STEM programs, in Georgetown's science academies, in UT Austin's engineering programs are the generation that may plan out the first Mars base — not build it from scratch. Their preparation is the human foundation the current generation's industrial buildout is building toward.

The Texas buildout is the first chapter of a story whose subsequent chapters will be written on the Moon, on Mars, and eventually in the spaces in between.

Twenty-Four

ALPHA CENTAURI

THE NEAREST STAR SYSTEM TO EARTH is 4.37 light-years away. At the speed of the Apollo command module – roughly 40,000 km/h – a spacecraft would take about 115,000 years to get there. At the speed of the fastest object humanity has ever launched, the Parker Solar Probe at its closest solar approach, roughly 690,000 km/h, the transit falls to around 7,000 years. At 10 percent of the speed of light – which no human-made object has approached, but which the physics of foreseeable propulsion concepts does not prohibit – the transit takes 43 years. At 20 percent of light speed, which the Breakthrough Starshot initiative proposes for gram-scale lightsail probes, it takes about 20 years, plus the 4.37 years for the signal confirming arrival to return to Earth.

Alpha Centauri is not a destination on any timeframe relevant to any current programs. It is the horizon the subtitle holds in view – the directional endpoint of a geometric progression that begins at Terafab, Bastrop, and Starbase and extends outward through the concentric spheres. Alpha Centauri is not reachable on any foreseeable timeline. But the trajectory is pointed there. The same compounding logic that is building the terrestrial AI-industrial complex, pushing compute to LEO, establishing the first lunar industrial node, and encoding Mars colonization as a corporate fiduciary commitment is a trajectory whose extension, followed far enough, reaches the stars.

The Physics of Getting There

Alpha Centauri is a triple star system. Alpha Centauri A and Alpha Centauri B are a binary pair — G-type and K-type main sequence stars orbiting each other with a period of approximately 79 years, separated at their closest approach by a distance comparable to the Saturn-Sun distance. Proxima Centauri is a red dwarf gravitationally bound to the binary pair at a distance of approximately 0.2 light-years, the closest of the three to Earth. It is the host of Proxima Centauri b, a potentially rocky planet, with a magnetic field, orbiting within the star's habitable zone. Its atmospheric and surface conditions remain unknown, but its existence as an Earth-mass planet at the right orbital distance has made it the most discussed candidate for an interstellar destination in the current scientific literature.

Reaching Alpha Centauri within any timeframe a human institution can plan for — a century, two centuries — requires propulsion whose specific impulse and exhaust velocity are orders of magnitude beyond chemical rockets, and substantially beyond any propulsion system currently operational. Three concepts have serious physics behind them, as distinct from speculative engineering.

The first is laser-driven lightsails. Breakthrough Starshot — the $100 million initiative announced in 2016 by Stephen Hawking, Yuri Milner, and Mark Zuckerberg — proposes a ground-based laser array of approximately 100 GW focused on a gram-scale reflective sail, accelerating it to 20 percent of light speed over a burn lasting minutes. The physics is sound. A 100-GW ground array is an enormous but not physically impossible undertaking, comparable in power to roughly 100 nuclear power plants operating simultaneously for the duration of the burn. The gram-scale probe carries cameras, sensors, and a laser communication system whose signal, transmitted at light speed, takes 4.37 years to return to Earth after the probe arrives. The engineering is extremely hard. The materials science of a sail that survives 20-percent-light-speed acceleration without ablating, the pointing stability of a 100-GW array focused on a gram-scale target, and the miniaturization of functional scientific instruments to gram-scale mass are all unsolved problems whose solutions require advances beyond the current state of the art. Not to mention the hundreds of billions of dollars it will take to construct and operate a precision 100-GW laser array. But none of this violates known physics.

The second is nuclear pulse propulsion — the Orion concept, whose development was halted by the Partial Nuclear Test Ban Treaty in 1963. A spacecraft propelled by sequential nuclear detonations behind a pusher plate achieves specific impulses thousands of times higher than chemical rockets, putting transit times to Alpha Centauri in the range of centuries rather than millennia. The physics was validated in the original Orion studies. The political and environmental barriers to atmospheric nuclear detonation for propulsion have not diminished. Nuclear pulse propulsion in deep space, away from Earth's biosphere, faces fewer treaty constraints — but it requires a political and institutional framework for authorizing nuclear detonations in the solar system that does not currently exist.

The third is fusion propulsion — spacecraft powered by controlled nuclear fusion, whose energy density and specific impulse place transit times to Alpha Centauri within the range of centuries. Fusion propulsion requires controlled fusion, which remains the famously receding horizon of energy technology. But the recent progress is genuine: Commonwealth Fusion Systems' high-temperature superconducting magnet results, the National Ignition Facility's ignition achievement, the commercial fusion development ecosystem that has attracted substantial private capital. Together they suggest the "always 30 years away" characterization may be less accurate in 2026 than it was in 1976.

THE VON NEUMANN PROBE

The most analytically interesting interstellar concept is not the crewed interstellar mission or the lightsail probe to Proxima Centauri. It is the Von Neumann self-replicating probe — the concept that connects the terrestrial AI-industrial complex documented to the interstellar horizon most directly.

John Von Neumann's original formulation — machines that can reproduce themselves from available raw materials — describes a class of system whose interstellar implications were first analyzed seriously by Frank Tipler and subsequently developed across the astrobiology and SETI literature. A self-replicating probe that arrives at a star system, mines local raw materials, builds copies of itself, and dispatches the copies to adjacent star systems creates an exponentially expanding wave of exploration — a wave whose speed of advance is set by the probe's transit speed between stars and its replication time at each system.

The mathematics are striking. An ultra-rugged, ultra-space-hardened, ultra-ablative-resistant, nuclear-powered Von Neumann probe traveling at 10 percent of light speed, with a replication time of 500 years at each destination — conservative assumptions by the concept's standards — could reach every star in the Milky Way within roughly 10 million years of a single launch. The galaxy contains 200 to 400 billion stars. Ten million years is less than 0.1 percent of the galaxy's age. A civilization that launches a single self-replicating probe could, in principle, have its technological presence at every star in the galaxy within a timeframe that is very long by human standards and very brief by galactic ones.

THE TRAJECTORY

The question is not whether Alpha Centauri is reachable. It is whether the trajectory that begins in Texas points in that direction — whether the compounding logic building semiconductor fabs at Gibbons Creek and launch complexes at Starbase is the same logic, extended far enough, that eventually produces interstellar probes.

The honest answer is that the logic is continuous but the timescales are not. The Texas Triangle's AI-industrial buildout is a decades-scale program. The lunar industrial node is a mid-century program. Mars colonization is a multigenerational program. The interstellar horizon is a civilizational-scale program whose timescale is measured in centuries at the optimistic end and millennia at the realistic one. The trajectory is real. The compression of that trajectory into a single human lifetime, or a single corporate planning horizon, is not.

The universe is 13.8 billion years old. The physics permitting interstellar travel does not violate any known law. A civilization that solves the intermediate problems — sustainable Mars colonization, self-sufficient orbital infrastructure, reliable autonomous systems operating across communication delays measured in minutes — has removed the primary obstacles between its current state and the interstellar horizon. Those intermediate problems are what the Texas Triangle and Starbase are working on right now. The interstellar horizon is what comes after them, not instead of them.

The early steps of a geometric progression do not look like the later ones. The first Raptor engine at McGregor does not look like a Von Neumann probe. The first D3 chip does not look like a mature orbital compute shell. But the progression from the first to the second is a matter of compounding rather than a change in direction — and the direction is set by the same logic that built the semiconductor fab in Taylor.

The launch complex at Starbase that will lift that chip to orbit, to the Moon, and eventually toward Mars is six hours south, on the road that passes through the brush country and the chaparral of South Texas before pointing toward the Gulf Coast and the launch towers visible from the highway on a clear day.

Alpha Centauri is farther. But there is a trajectory that begins here, extends through the orbital compute shell, passes through the lunar mass driver, continues to Mars, and runs forward through the Von Neumann probe concept that the AI systems being developed at SpaceX may eventually embed intelligence in.

Epilogue

THE VIEW FROM INSIDE

ONE MORNING LAST FALL THE POWER went out in my neighborhood. A local event — the foundation dependency made briefly visible, in the form of a blank screen and a silent house. I drove to the coffee shop in Leander and set up at the communal table. Others had done the same: a woman on a Google Meet, an Amazon sprint reviewer, an Apple engineer whose name I never learned, all of us displaced from our home offices by the same interruption. The table was full of people whose work depended on infrastructure they had not thought about until it stopped working. Around noon the power came back on. The Apple engineer's phone showed his home online. He packed up and left. The others followed. Within 20 minutes the table was empty and the infrastructure was invisible again.

That is the condition: the moment when the buildout of the AI-industrial revolution is still new enough to be seen as infrastructure, before it becomes simply the way things are.

WHAT HAS BEEN BUILT

The Texas Triangle's AI-industrial complex, assessed at the system level through mid-2026, is the most consequential concentration of strategic industrial capital in the United States since the automotive complex assembled in southeastern Michigan in the first two decades of the twentieth century. The comparison is not rhetorical. Pittsburgh built steel. Detroit built automobiles. Houston built petrochemical infrastructure. The Texas Triangle is building the physical base of artificial intelligence — the silicon that thinks, the compute that trains, the energy that powers both, the autonomy systems that deploy the resulting capability into the physical world, and the launch infrastructure that carries it beyond Earth's atmosphere.

It is visible from the ground — in the construction cranes, the Teslas on the highways, the permit filings, and the corporate announcements that, taken together, are what assembling an AI-industrial supercluster looks like at the scale of a metro region.

What the Constraints Are

The honest epilogue acknowledges the constraints with the same specificity applied to the assets. There are four.

Water is the binding physical constraint whose ceiling is hardest to engineer around at the pace the buildout requires. Brownfield inheritance is the market's partial workaround, not a complete solution. The greenfield capacity ceiling is real, and its approach is not distant.

The grid is the second. The many gigawatts of disclosed power capacity represent a load addition whose full transmission infrastructure implications are still being assessed. The energy sovereignty architecture — behind-the-meter generation, nuclear power, BESS integration — is the market's response to that failure mode risk. It is spreading, but not yet universal.

The workforce is the third. The capital is deploying faster than the people can be trained. The construction trades shortage is real. The semiconductor process engineering shortage at leading-edge nodes is more structurally challenging.

Political continuity is the fourth — the constraint the investment thesis literature discusses least and the corridor's trajectory depends on most at the program level. The CHIPS Act conditions, the TSIF allocations, the DoD certification frameworks, and the federal procurement relationships are products of a specific political moment.

The constraints are real. They are not fatal to the trajectory. They are the variables whose management determines whether the buildout completes at the pace its momentum implies, or completes more slowly as binding constraints impose friction on a system whose direction is not in doubt.

What Comes Next

The formation has crossed a threshold whose significance will only be fully legible in retrospect. The builders arriving now are not making first-principles bets on what Texas could become. They are responding to what Texas already is. The gravity well is deep enough to pull without persuading.

The first D3 chips qualified for the Starmind orbital compute deployment will mark the point at which the Texas Triangle's silicon supply chain extends fully from the terrestrial complex to the orbital layer.

The crewed lunar landing, targeted for 2028, will mark the point at which the concentric sphere progression moves from theoretical to operational at the lunar layer. The engineers who help design that mission are in Austin and Houston. The engines that power the vehicle that executes it are tested at McGregor. The launch complex that sends it is at Starbase, Texas, and Kennedy Space Center, Florida.

The first cargo Starship missions to Mars, targeting the early-to-mid 2030s in the most optimistic scenarios, will mark the point at which the Mars foundation stack's transportation layer becomes operational. The Optimus units aboard those early cargo missions will be the first physical AI agents to operate autonomously on the Martian surface — the furthest expression of the autonomy deployment whose terrestrial origins have been documented on the Giga Texas production floor.

Each of these next events is a formation in its own right: the visible surface of a system compounding below the threshold of daily attention until a specific observable milestone makes the compounding undeniable. Formations are most legible in retrospect. The retrospective legibility of the Texas Triangle's formation will be clearest to the generation that inherits the infrastructure whose construction the current generation is living inside.

What It has done

I write this book using LLMs as cognitive infrastructure. Not as a search engine. Not as a grammar checker. Not as a ghostwriter. As a working assistant across the documentation of a network spanning six industrial-AI knowledge-graph sites, a 24-chapter manuscript, a notes and ideation apparatus, an entity-and-edge taxonomy, and the cross-domain reasoning required to hold semiconductor fabrication, gigafactory production, data center buildouts, autonomous vehicle deployments, orbital launch programs, AI compliance regimes, and capital flows in a single coherent frame. A frame that one person, in one home office north of Austin, could not have held three years ago.

This is the application-layer answer to the question this book has so far ignored to address: what is the buildout for. The substrate has been documented. What runs on the substrate is what those gigawatts are being consumed in service of, and the current public discourse on what runs on the substrate is concentrated on the failure modes: AI as surveillance instrument, AI as slop video generator, AI as search-replacement that hallucinates citations, AI as the engine of a labor market disruption whose distribution of pain is unaddressed. The failure modes are real. They are also a small fraction of the use cases that exist among users who have integrated the technology into knowledge work.

The version I work with daily is cognitive amplification in the older and more precise sense of the term. It is the capacity to hold more variables simultaneously than unassisted cognition can manage, to test more reasoning paths against more sources of evidence than I could test alone, to do in a working session what would have taken a research team a week to do in 2021. It is not magic. It is engineering applied to the substrate of cognition itself, and like every prior application of engineering to a previously unautomated domain, it produces a step change in what one person can accomplish.

The transcontinental railroad created use cases over four decades. The personal computer created use cases over three. The web created use cases over two. The current transition is creating use cases faster than that, and the people building the substrate cannot tell you, in 2026, what the application landscape will look like in 2036, because the application landscape is being created by the users, in interaction with the technology, on a timeline measured in months rather than decades. The substrate is built before the use cases are legible.

I am among the early integrators. I am a sample of one, and my testimony is not a substitute for systemic analysis. What I can say is that the benefits of every prior infrastructure transition accrued first to its builders and early integrators and reached the broader population across a generation. The current transition is not exempt from this pattern, only from its timeline. The trickle-down will happen the way it always happens, through the gradual diffusion of the early integrators' work product into the experience of the broader population, on a clock the substrate has compressed.

The buildout documented in this book is for this. For the cognitive amplification of human capability across domains the unamplified individual could not previously span. For the autonomy systems that will eventually relieve human workers from the physically dangerous and cognitively numbing tasks industrial production currently requires. For the research instruments that will compress the timeline of medical, materials, and energy breakthroughs whose human consequences will outweigh, eventually, every concern about resource consumption the present discourse is raising. The trickle-down has begun. It has reached, among others, a one-person operation north of Austin attempting to document the formation that made the operation possible.

I have sipped from the fire hose. There is no going back. It is the beginning of the first inning.

What I Believe

This is what I believe, as distinct from what I have documented.

The AI transition will proceed faster than any prior industrial transformation humanity has experienced. The step changes — in productive capability, in what becomes technically possible, in the reach of human agency — will arrive before the social, regulatory, and institutional frameworks designed to absorb them are ready. That gap between technological capability and institutional readiness is real. It is already visible in the Austin metro corridor's infrastructure systems straining to absorb a formation they were not designed for, and it will widen at the national and global scale before it narrows. It will produce disruption, displacement, and genuine human cost whose honest acknowledgment is the price of intellectual credibility on this subject.

The catastrophist scenarios — AI systems pursuing goals misaligned with human welfare, autonomous weapons operating beyond meaningful human control, labor displacement outpacing economic adaptation — are not zero-probability outcomes. They are the failure modes of a transition whose management requires exactly the kind of serious analytical attention that catastrophist framing often prevents, by substituting dramatic narrative for rigorous engineering and governance work. The people building the systems are aware of these risks. The autonomous systems safety architectures, the human-in-the-loop design requirements, the AI governance frameworks that serious practitioners impose on their own programs are not marketing gestures. They are engineering responses to engineering problems whose difficulty is real and whose resolution is not guaranteed.

The abundance narrative deserves equal skepticism. Its current expression — Step 1: wake up; Step 2: AI has accomplished something remarkable; Step 3: free services, free energy, nobody needs to work, all disease eradicated — omits precisely the steps that matter most. The human failure modes are not edge cases. Regulatory capture by incumbents threatened by the transition is historically consistent behavior, not a speculative risk. The concentration of transition benefits among builders and early adopters, before broader distribution mechanisms develop, is what every prior industrial transition produced. The Pittsburgh steelworker's standard of living did not improve until decades after Carnegie's did — and the mechanism of improvement was not technological inevitability but labor organizing, antitrust enforcement, and political contestation whose outcomes were not predetermined. Greed, corruption, institutional inertia, and the impulse toward greater control are not bugs in the human system that better technology will fix. They are features of human organization that every technological transition must navigate rather than assume away.

What I believe — held after years of helping design and document the infrastructure to see this formation clearly, and after living inside it long enough to observe what it actually produces rather than what its critics fear or its evangelists promise — is more specific, and more modest, than either frame allows.

The transition is positive in its tendency. The compounding logic is building the Texas Triangle's AI-industrial complex. That logic is also producing the cognitive amplification and the autonomy technology that will eventually relieve human workers from the physically dangerous and cognitively numbing tasks industrial production currently requires of them. These are genuinely good outcomes, whose tendency is toward expanded human capability and reduced human suffering — even as their distribution is initially concentrated and their pace of broader diffusion is uncertain.

The transition is also faster than prior transformations. The industrial revolutions that preceded it — steam, electrification, computing — each required decades for their most consequential effects to propagate through the systems they transformed. The current transition is compounding on a foundation of prior transitions whose infrastructure — global broadband communications, digital finance, computational ubiquity — compresses the diffusion timeline in ways that make the pace genuinely unprecedented. Faster is not automatically better. Faster means less time for the institutional adaptation that distributes benefits and manages failure modes. The pace is the variable that makes the human failure modes most consequential — not because the failure modes are new, but because the timeline for addressing them is shorter than prior transitions allowed.

The analysis settles the direction; the distribution, the governance, and the human cost of the pace belong to the people now responsible for them.

The direction is forward. The pace is remarkable. The outcome is not guaranteed. But the emergence is real, and it is happening now, and it matters enormously.

What I See

I drive toward Taylor when I want to see Old Texas and New Texas held in superposition. The Samsung Taylor construction cranes are visible from the highway — not dramatic from a distance, just present, the way large industrial facilities are present when you are accustomed to looking at them across a cornfield and know what they represent.

The Giga Texas campus on the southeast side of Austin is a presence that grows measurably between drives. It is not the raw construction pace of the early years, but the steady buildout of a campus whose original programs are operational and whose new programs are in early construction.

The buildout is not complete. It is underway. The groundwork is being built in the specific geography — in the meeting rooms and the home offices, in the classrooms, under the construction cranes, in the shadow of the launch towers. The intelligence being trained will eventually run in orbit. The chips being designed will eventually operate on the lunar surface and on Mars. The engineers whose children are in Cedar Park's STEM classrooms today will be the generation that operates the first lunar base.

The Hill Country sky is wide to the west, and full of stars on clear nights — light that has been traveling longer than human civilization has existed. The Milky Way runs above the southern horizon. Alpha Centauri is just below it from this latitude, a southern hemisphere object Central Texas observers cannot see directly. But its direction is known precisely. It is there, 4.37 light-years out, in the direction the trajectory is pointed.

The singularity is not a future event. It is this: the AI-industrial convergence happening now, in this geography, observed from this vantage point inside it — before it becomes simply the way things are.

From Austin.

To Alpha Centauri.

Note on the Living Layer

This book was written during the buildout it documents. That is its primary authority claim and its primary limitation at the same time. The authority claim is that the analysis is grounded in direct observation from inside the geography where the convergence is occurring. The limitation is that the formation event does not stop at the publication date. Facilities documented here will reach operational milestones after the manuscript is locked. New facilities, expansions, and capital funding will be announced. A printed book cannot track any of that.

The argument does not go obsolete; the evidence does. The analytical contributions of this book are fixed claims whose value compounds as the phenomena they describe become more historically significant. But the specific facts those claims rest on continue to move, because the buildout continues.

Readers who want to track the buildout progression past the publication date can find that at AustinIO.com. The book is the fixed artifact. The site is the living layer. The book makes the argument; the living layer maintains the evidence.

Note on the Book Structure

This book is built in two integrated parts.

The chapters develop the argument that the Texas Triangle, and specifically the Austin metro region within it, has become the first US region to reach supercluster density across seven of nine AI-industrial and substrate domains. The prose is deliberately compressed. Where longer treatments of this material would circle back and reinforce, this one moves forward once and trusts the reader to hold the thread.

The five appendices that follow enumerate the evidence base on which that argument rests. Appendix A documents the geographic concentrations. Appendix B documents the anchor facilities within them. Appendix C documents the brownfield inheritances the buildout is siting itself on. Appendix D documents the Tesla silicon roadmap. Appendix E documents the Musk entity cluster as the only structurally integrated multi-layer operator in the buildout.

The chapters cite the appendices; the appendices ground the chapters. Neither stands alone. The chapters are meant to be read. The appendices are the receipts. Together they are the case.

Appendix A — AI-Industrial Supernodes

The Texas AI-industrial buildout is most accurately understood not as a collection of independent projects in separate cities, but as a graph: nodes (individual facilities), edges (the corridors of power, fiber, road, rail, supply chain, and labor that connect them), and supernodes (concentrations of nodes whose integration density makes them strategically distinct from the surrounding geography).

This appendix maps the supernodes operating, emerging, and announced as of mid-2026, with scope defined by the AI-industrial convergence thesis of this book — data centers, fabs, and gigafactories as the three archetypes whose mutual reinforcement drives the buildout.

The first five listed are tightly co-located, deeply integrated through shared workforce and corridor infrastructure, and interlocked with the academic, design, and institutional layers underneath them. As such, they collectively constitute a supernode-of-supernodes. These belong to the Austin Supercluster.

Operating Supernodes — integrated concentrations where multiple facilities are operational or under active construction, the corridor edges are functioning, and the supernode's analytical coherence is observable in real time.

Emerging Supernodes — concentrations where the facility base is forming and the corridor integration is in progress, but the supernode quality is anticipated rather than fully realized.

Potential Supernodes — Major announced facilities whose surrounding cluster has not yet formed, but whose scale and strategic role suggest a supernode trajectory.

***Note:** Appendix A documents the state of the buildout as of manuscript completion in July, 2026. Announced capacities, capital commitments, construction timelines, and operational status are dynamic across every facility documented. Capacities scale, funding tranches evolve, timelines shift, and phase transitions occur continuously across the buildout. Readers checking specific facility details against later reporting should treat the appendix as the structural baseline.*

Austin Supercluster

The Austin MSA metropolitan region is the only AI-Industrial Supercluster identified in this appendix. It comprises five complementary supernodes and corridors that share a common workforce, transportation network, transmission infrastructure, supplier ecosystem, and institutional substrate. All within a 45-mile radius from Austin. At its strategic core is the Inner Triangle formed by Tesla Giga Texas, Samsung Taylor, and SpaceX Bastrop.

Member Supernodes: (1) Tesla Giga Texas Campus; (2) Samsung Taylor Fab Complex; (3) SpaceX Bastrop Compound; (4) US 79 Compute Corridor; (5) San Marcos–Uhland Compute Corridor.

Tesla Giga Texas Campus

Type: Foundational member of the Austin Supercluster, strategic vertex of the Inner Triangle and center of the AI-industrial convergence singularity.

Geographic anchor: Austin, Travis County.

Key nodes: *Vehicle and robotaxi* manufacturing facility; ***Cortex*** 1 and 2 AI training cluster; ***Optimus*** factory (in construction); ***Terafab Research*** facility (in construction). Also the University of Texas at Austin, Texas Institute for Electronics, and the broader Austin semiconductor ecosystem 30 miles to the west.

Corridor edges: SH 130 linking to Hutto-Samsung Taylor and SH 71 toward Bastrop. Vertex of the Inner Triangle.

Supernode density: Four tightly integrated industrial programs on one campus create the highest concentration of AI-industrial production in the United States. Three structural features produce the density: co-located fabrication of the silicon, the vehicles that consume it, and the AI training infrastructure that uses both — vertical integration across three layers of the same industrial stack within a single labor market.

Comparable: Detroit in the 1910s for the integrated industrial stack co-developing the dominant industry of its era.

Samsung Taylor Fab Complex

Type: Semiconductor member of the Austin Supercluster. Vertex of the Inner Triangle

Geographic anchor: Taylor, Williamson County.

Key nodes: Fab 1 (Tesla-captive); Fab 2 (merchant); long-term ten-fab vision; supplier ecosystem; adjacent AI data center campus.

Corridor edges: US 79 and SH 130. Vertex of the Inner Triangle

Supernode density: America's largest planned leading-edge semiconductor manufacturing campus.

Comparable: Hsinchu Science Park in Taiwan housing the TSMC fab complex and the TSMC fab complex in Arizona.

SpaceX Bastrop Compound

Type: Orbital compute infrastructure member of the Austin Supercluster. SpaceX industrial concentration; orbital infrastructure manufacturing; captive aerospace and semiconductor vertical integration under one operator.

Geographic anchor: Bastrop County near Bastrop and along SH 71 east of Austin.

Key nodes: ***Starlink*** terminal manufacturing facility (largest US satellite-component manufacturing operation by unit count); ***Project Echo Solar*** (aerospace-grade rad-hardened photovoltaic cells); ***Gigasat*** (AI1 orbital data center satellite development and manufacturing factory); SpaceX advanced semiconductor research and chip packaging; ***The Boring Company*** operations; ***Snailbrook*** worker housing development.

Corridor edges: SH 71 as the spine to Austin and Giga Texas; SH 95 as the 35-mile corridor to Samsung Taylor; labor flows between Bastrop and Austin proper that have made the corridor a commute zone.

Supernode density: The Bastrop Compound is among the densest single-operator vertical integration in the Texas dataset. SpaceX runs satellite terminal manufacturing, orbital-grade solar cell production, AI1 orbital compute satellite manufacturing, aerospace semiconductor advanced packaging, tunneling automation development, and worker housing — six distinct industrial functions under one corporate operator. The structural signature is captive vertical integration.

Comparable: Ford's River Rouge complex in the 1920s for the captive vertical-integration pattern — multiple normally separate industrial functions co-located under one operator at one site. The closest current analog within the Texas dataset is Giga Texas itself, which runs the same model.

US 79 Compute Corridor (Williamson County)

Type: Member of the Austin Supercluster. AI and hyperscale compute concentration — data center buildout for hyperscale cloud, enterprise infrastructure, and AI inference workloads.

Geographic anchor: Williamson County north of Austin, spanning Round Rock to Taylor, with Hutto in between where the corridor connects to SH 130 to the Giga Texas supernode.

Key nodes: ***Skybox PowerCampus*** (Hutto); ***Iron Mountain Hutto***; Blueprint Data Centers (Georgetown and Taylor sites); ***Project Comal*** (Taylor); the broader hyperscale concentration across Round Rock representing ***Sabey*** (home of TACC's Horizon supercomputer), ***Switch "The Rock"***, Skybox, and Amazon AWS campuses and unnamed tenants.

Corridor edges: I-35 north of Austin as the workforce and logistics backbone; SH 130 as the eastern bypass connecting the corridor to Hutto-Taylor and to the Giga Texas Complex; the ERCOT transmission upgrades supporting Williamson County load growth as a primary case study for Texas grid management.

Supernode density: The densest concentration of hyperscale and AI workload data center siting in Texas outside the Abilene Supernode, and the only one embedded inside a metropolitan workforce rather than positioned for energy-substrate proximity. Adjacent to the Sandow Lakes Cluster.

Comparable: Northern Virginia's Loudoun County concentration in pattern, but Texas-pattern in execution — more land, faster permitting, ERCOT rather than PJM, and a workforce drawn from the Austin metropolitan area rather than from federal-contractor flows.

San Marcos-Uhland Compute Corridor

Type: Newest member of the Austin Supercluster. Pre-operational multi-anchor data center concentration — 2 GW-scale campuses under early construction with distinct energy strategies, in the I-35 corridor south of Austin.

Geographic anchor: The San Marcos / New Braunfels / Uhland triangle, straddling Hays, Guadalupe, and Caldwell Counties.

Key nodes: ***CloudBurst*** San Marcos DC1 (1.2 GW master plan, 706 acres on Francis Harris Lane in Hays / Guadalupe, ~$14.5 billion committed capital, behind-the-meter natural gas via Energy Transfer's Oasis Pipeline under a ten-year supply agreement, first 50 MW phase Q4 2026, three 400 MW phases at full build-out); ***Evolve Holdings*** as CloudBurst's design-build partner. ***Edged Energy*** to build a $7.3 billion, 462 MW campus outside of San Marcos using a zero-water ThermalWorks cooling technology. ***Prime Data***

Centers Project Snow to build on 786 acres outside of San Marcos; Phase 1 at 144 MW, up to 8 buildings covering 2 million sq ft

Tract Caldwell Valley Technology Park (multi-tenant master-planned park, ~3,000 contiguous acres in Caldwell County near Uhland, up to 4 GW at full build-out, grid-connected via Bluebonnet Electric Cooperative and LCRA partnership with a 1.6 GW on-site substation, first 360 MW phase Q1 2028, supplemental on-site generation provisioned); first tenant is ***PowerHouse Data Centers*** on 500 acres at 700 MW.

Corridor edges: I-35 as the workforce and logistics backbone, with the cluster positioned 35 miles south of Austin and 40 - 50 miles north of San Antonio; the Kinder Morgan Permian Highway and Energy Transfer Oasis Pipelines as dual gas supply edges; the LCRA and Bluebonnet Electric grid corridors as the transmission edges; the long-haul fiber routes intersecting the corridor.

Supernode density: The cluster forms through structurally distinct operators committing to gigawatt-scale buildouts in adjacent counties within twelve months of each other, sharing the same workforce and corridor infrastructure but pursuing different energy strategies — CloudBurst behind-the-meter natural gas primary, Tract grid-primary with supplemental on-site generation. The pattern mirrors the way the Williamson County Compute Corridor emerged: multiple operators co-locating into a workforce-adjacent stretch of I-35. Whether the two anchors integrate operationally over the decade, or remain parallel campuses serving distinct tenants, is the open question.

Operating Supernodes

Integrated concentrations where multiple facilities are operational or under active construction, the corridor edges are functioning, and the supernode's analytical coherence is observable in real time.

Brownsville-Starbase Space-Defense Corridor

Type: Aerospace specialization — launch operations, vehicle production, satellite manufacturing, logistics.

Geographic anchor: Boca Chica / ***Starbase*** (Cameron County) and the Brownsville metropolitan area.

Key nodes: ***Starship*** production facilities; Starship orbital launch complex; ***Gigabay*** vehicle assembly; on-site propellant production infrastructure (air separation for liquid oxygen, methane liquefaction for liquid methane); the orbital launch mount water deluge system; ***Saronic Port Alpha*** location at Port of Brownsville; SpaceX expansion projects on adjacent acreage.

Corridor edges: The Gulf coastline launch trajectory as the geographic constraint anchoring the location; US 77 and US 281 connecting Brownsville to the broader Texas Triangle; the Port of Brownsville logistics network for heavy hardware ingress and egress; SH 4 as the local Brownsville-to-Boca Chica spine; the regional natural gas pipeline network supplying methane feedstock for propellant liquefaction; the Cameron County water supply feeding deluge and industrial cooling loads; the Texas Gulf Coast cryogenic and industrial gas industry as the regional supplier base, originally built for petrochemical and LNG export and now supplying surge capacity for orbital propellant demand; the launch azimuth complement provided by SpaceX's Florida Launch Complex 39A at Kennedy Space Center — the dual-site architecture as launch supply chain redundancy.

Supernode density: The supernode quality comes from the integration of two normally separate aerospace functions — vehicle production and launch operations — at one geographic anchor. This is no longer a launch site with manufacturing support. It is an aerospace-industrial city whose output gates every downstream program documented in Part IV of this book. Without Starbase reaching the launch cadence its architecture promises, the orbital compute shell, the lunar return, and the Mars trajectory all stall.

Comparable: No precise historical analog. The closest parallel is the role Cape Canaveral played in the Apollo era — but the production-launch-manufacturing integration at this scale, under one operator, is structurally new.

TI Semiconductor Corridor

Type: Semiconductor concentration with analog specialization — fab cluster anchored by Texas Instruments at single-operator multi-fab scale.

Geographic anchor: Sherman (Grayson County), 60 miles north of downtown Dallas and 48 miles north of Richardson fabs along the US 75 corridor.

Key nodes: ***TI SM1*** (operational, 300mm analog and embedded processing); ***TI SM2*** (operational); TI SM3 and SM4 (future expansion within the existing master plan, sequenced against analog demand); the supplier ring forming around the TI anchor in Grayson and Fannin Counties.

Corridor edges: US 75 north of DFW as the primary north-south corridor; the DFW metropolitan supplier and labor base 60 miles south as the structural feeder; the BNSF and Union Pacific rail infrastructure serving the broader DFW–Sherman manufacturing geography; the natural gas pipeline network supplying utility load to the fab cluster; Grayson College and the regional community college network as the workforce training pipeline.

Supernode density: The world's largest analog and embedded processing semiconductor manufacturing concentration once the existing master plan is fully built out, and already among the largest in operational capacity. The supernode quality comes from fab-cluster scale rather than from industrial diversification. Sherman is the inverse of Austin–Taylor in structural terms — depth in one sector at one operator, rather than integration across many sectors and many operators. The analytical signature is concentration density of a single specialization at a scale that makes the cluster strategically significant on its own, without requiring adjacent industries to produce supernode coherence.

Comparable: Hsinchu Science Park's TSMC concentration for the fab-cluster pattern at a single dominant operator — though Sherman's specialization is analog and embedded processing rather than leading-edge logic.

Abilene AI Cluster

Type: AI compute concentration — data center buildout for frontier AI training and inference at gigawatt scale.

Geographic anchor: Abilene (Taylor County) and adjacent Shackelford County.

Key nodes: ***Stargate Abilene*** (OpenAI anchor program, 1,200 MW Phase 1, generational capital scale); ***Vantage Frontier*** (frontier-scale AI training mega-campus, 1,400 MW full buildout); ***Crusoe*** AI Factory; ***Oracle Cloud*** infrastructure; ***Microsoft*** expansion; Lancium Clean Campus.

Corridor edges: I-20 as the east-west backbone; ERCOT transmission corridors connecting West Texas wind and solar generation to the compute load; the West Texas natural gas pipeline network supplying behind-the-meter generation where grid interconnection timelines exceed buildout urgency; the regional labor pipeline drawing from Abilene Christian University, Hardin-Simmons, McMurry, and Texas State Technical College.

Supernode density: Roughly 4-plus GW of disclosed compute capacity within approximately 30 miles, integrated through a shared power infrastructure and a shared labor market that did not exist for AI workloads three years ago. The supernode quality is concentration density itself: multiple operators at gigawatt scale, co-located by the same underlying physics — abundant West Texas wind and solar generation, natural gas supply for ride-through, land economics that support multi-thousand-acre campuses, and a regulatory environment permissive to fast buildout.

Comparable: Northern Virginia's Loudoun County data center concentration, but purpose-built for frontier AI training rather than enterprise cloud, and with onsite generation closing the energy-compute loop rather than depending on conventional grid supply.

Emerging Supernodes

Concentrations where the facility base is forming and the corridor integration is in progress, but the supernode quality is anticipated rather than fully realized.

Permian Basin Gas-AI Cluster

Type: Multi-anchor behind-the-meter AI concentration in the Permian Basin oil-and-gas substrate — 3 GW-scale operators using off-grid gas-fired generation, integrated with Permian midstream gas infrastructure, anchored by the first energy-major hyperscaler pairing in the appendix graph.

Geographic anchor: Pecos and Reeves Counties in West Texas, anchored on the Permian Basin oil-and-gas production fairway. Pecos County hosts GW Ranch approximately 17 miles north of Fort Stockton and the Chevron–Microsoft–Engine No. 1 campus in the same Fort Stockton corridor; Reeves County hosts Alpha Digital Campus near the Waha natural gas hub. The two counties are adjacent and together form the densest behind-the-meter data center concentration in Texas.

Key nodes: ***GW Ranch*** (Pecos County, Pacifico Energy Group as developer with CEO Nate Franklin, 8,000-plus-acre site approximately 17 miles north of Fort Stockton, 7.65 GW of gas-fired generation permitted by TCEQ in January 2026 — the largest data center power air permit ever granted in the United States, paired with 1.8 GW of battery storage and 750 MW of solar, fully off-grid private-grid architecture, first 1 GW targeted 2028 and full 5-plus GW by 2031);

Chevron–Microsoft–Engine No. 1 (Pecos County, partnership between Chevron Corporation, Microsoft Corporation, and energy investment firm Engine No. 1, approximately $7 billion estimated for the initial phase, 2.5 GW initial capacity expandable to 5 GW for a single hyperscale customer, 7 GE Vernova combined-cycle gas turbines on order, behind-the-meter Chevron-supplied Permian gas via existing midstream gathering and processing infrastructure, FID targeted early 2026 with Microsoft entering exclusivity April 2026, first power 2027);

LandBridge–PowerBridge Alpha Digital Campus (Reeves County, PowerBridge LLC as developer — Five Point Infrastructure portfolio company with $1 billion equity commitment, led by CEO Alex Hernandez previously founder of Cumulus Data and CEO of Talen Energy Corp; LandBridge Company LLC as surface-rights lessor; up to ~3,400 acres optioned in April 2026; 2 GW initial co-located behind-the-meter generation; GINR filed; first power targeted 2027; direct Waha hub gas supply via LandBridge surface rights).

Corridor edges: I-10 east-west via Fort Stockton and I-20 east-west via Pecos as the dual interstate backbones connecting to El Paso, San Antonio, and DFW; US 285 as the Pecos–Carlsbad north-south spine; the Waha natural gas hub as the central energy substrate edge, with Whistler, Permian Highway, and Gulf Coast Express pipelines as the takeaway infrastructure layer; in-basin Permian gas production as the upstream substrate edge; the planned 765-kV ERCOT transmission corridor expected to reach West Texas later this decade as the future grid edge, though all three anchors are designed to operate independent of it; the Permian oil-and-gas workforce shed spanning Odessa, Midland, Fort Stockton, and Pecos as the integrating labor geography.

Supernode density: Approximately 12 GW of permitted or announced behind-the-meter capacity across 3 GW-scale operators within two adjacent counties, rising to roughly 15 GW if the Chevron–Microsoft campus expands to its 5 GW option. The supernode quality parallels El Paso–Borderplex structurally in the multi-anchor multi-operator pattern, but with four structural firsts for the appendix graph: energy-major-anchored AI, with Chevron as the first oil-and-gas major to anchor a Permian campus and the first hyperscaler pairing in the basin; surface-rights leasing as a data center business model, with LandBridge's royalty-income structure giving each gigawatt of capacity $10–20 million annually in recurring lease revenue against the Permian's existing oil-and-gas tenant base; universal off-grid private-grid architecture across all three anchors, designed to bypass ERCOT interconnect queues entirely rather than supplement grid supply; and the densest county-level data center concentration in Texas, with two adjacent counties accounting for capacity that exceeds any single existing Texas supernode. Pre-operational across all three anchors and likely moves to Operating tier once the first GW Ranch and Alpha Digital phases come online in 2027–2028.

Comparable: El Paso–Borderplex AI Corridor for the multi-anchor gigawatt-scale pattern with distinctive non-Triangle geography; Abilene AI Corridor for the behind-the-meter gas-anchored multi-operator pattern. Distinguishing characteristics versus both: energy-major participation, surface-rights leasing model, universal off-grid architecture, and the largest combined permitted capacity in any single Texas supernode.

El Paso Borderplex AI-Defense Cluster

Type: Multi-anchor frontier-AI training concentration spanning the Texas–New Mexico border — 3 GW-scale operators with distinct anchor types (hyperscaler, AI-developer joint venture, defense-private), all using behind-the-meter or self-contained generation, all AI-purpose-built.

Geographic anchor: The El Paso metropolitan region, spanning El Paso (El Paso County, Texas), Doña Ana County (New Mexico, including Santa Teresa), and Fort Bliss in far East El Paso. The first cross-border supernode in the appendix graph.

Key nodes: ***Meta Seafox*** AI Data Center (Northeast El Paso, approximately 1,039 acres, $10 billion investment scaled from an initial $1.5 billion commitment, 1 GW capacity target 2028, 366 MW behind-the-meter modular natural gas array via 813 modular units in an initial five-year bridge period, AI-optimized closed-loop liquid cooling, October 2025 groundbreaking, JE Dunn and Hensel Phelps as general contractors);

Project Jupiter (Santa Teresa, New Mexico, in Doña Ana County, OpenAI and Oracle as principal tenants, ***BorderPlex Digital Assets*** of Austin as initial developer with Stack Infrastructure as Colorado-based partner, $165 billion total commitment over 30 years with $50 billion in the first five years, four data centers for AI training, approximately 1 GW initial scaling to multi-GW, isolated natural-gas generation system, part of OpenAI Stargate);

Fort Bliss / Carlyle Group (1,384 acres of unused Fort Bliss land in far East El Paso near Montana Avenue and Loop 375, US Army anchor tenant with Carlyle as development partner, approximately 100 MW initial operating capacity targeted 2027 scaling to 3 GW by 2029, AI compute for Department of Defense applications).

Corridor edges: I-10 as the east-west backbone connecting El Paso to the broader Texas grid and to Phoenix to the west; the El Paso–Las Cruces–Ciudad Juárez Borderplex workforce shed as the integrating labor geography; El Paso Electric as the regional grid utility; the El Paso Natural Gas (Kinder Morgan) interstate pipeline and ONEOK intrastate pipeline as the energy substrate edges; multiple I-10 fiber providers including Cogent, Lumen, TransTelco, and Windstream.

Supernode density: Roughly 5 GW of frontier-AI training capacity announced across 3 GW-scale operators within approximately 30 miles of each other, integrated through shared utilities, shared fiber routes, and the Borderplex workforce shed. The supernode quality parallels Abilene structurally — multiple operators at gigawatt scale, co-located by shared energy and labor substrate — with three structural firsts for the appendix graph: cross-border geography, with Project Jupiter in New Mexico as the integrating node across the state line; DoD-anchored AI buildout, with Fort Bliss / Carlyle as the first DoD-anchored AI supernode; and frontier-training concentration outside the Triangle. Pre-operational across all three anchors and likely moves to Operating tier once Meta comes online in 2028.

Comparable: Abilene AI Corridor for the multi-operator gigawatt-scale AI training pattern, but with the cross-border geography, defense-tenant anchor, and natural-gas-primary energy substrate distinguishing it structurally.

Comanche Peak AI Cluster

Type: Multi-anchor nuclear-adjacent AI concentration anchored by the existing 2.4 GW Comanche Peak Nuclear Power Plant — the first nuclear-anchored multi-operator AI buildout in Texas, combining hyperscaler-PPA (AWS) and co-located gas-supplemented nuclear-adjacent (Sailfish) models at gigawatt scale.

Geographic anchor: Hood County (Granbury) and adjacent Somervell County (Glen Rose), approximately 50 miles southwest of Fort Worth. Comanche Peak Nuclear Power Plant sits in Somervell County, with data center buildout consolidating in Hood County across the county line.

Key nodes: ***AWS Project Spectrum*** (Hood County, 1,265 acres at 2300 Coates Rd in Granbury, $5 billion AWS hyperscale campus, 20-year PPA with ***Vistra*** subsidiary Luminant for up to 1.2 GW of carbon-free nuclear electricity, initial 600 MW phase in ERCOT queue February 2026, first power late 2027); ***Sailfish Comanche Circle*** (Hood County, Sailfish Investors as developer, 2,600 acres, 5 GW planned across nine campuses, first 600 MW within 24 months via on-site power islands, subsequent phases through 2028 via additional islands and possible grid integration with Comanche Peak output); ***Pacifico Energy*** Hood County (off-grid behind-the-meter gas project supporting an additional data center development, capacity not yet disclosed); MARA cryptomine (existing Bitcoin mining operation in Hood County). The underlying energy anchor is Vistra's 2.4 GW Comanche Peak Nuclear Power Plant — two pressurized water reactors operational since 1990, NRC license extended through 2053 by the July 2024 approval, operated by Vistra subsidiary Luminant Generation.

Corridor edges: US 67 and US 377 as the regional spine connecting Granbury to the Fort Worth metro; the Comanche Peak transmission interconnection as the grid-baseload edge; the Brazos River and Squaw Creek Reservoir providing cooling water for the nuclear plant and water-resource context for the adjacent data center buildout; the Fort Worth metro workforce shed at 50-mile commuting distance; Vistra's broader industry-leading ~3,800 MW nuclear PPA portfolio with AWS spanning multiple Vistra nuclear sites, making Project Spectrum one node in a multi-site Vistra-AWS national strategy disclosed in Vistra's FY2026 guidance.

Supernode density: Combined planned capacity approaching 7 GW across 2 to 3 operators within 10 to 20 miles, integrated by shared access to Comanche Peak Nuclear and the Brazos River geography. The supernode quality has three structural firsts for the appendix graph: the first nuclear-anchored multi-operator AI concentration in Texas; the first hyperscaler-

PPA-anchored Texas concentration (AWS rather than the Crusoe-Stargate-Vantage operator profile of Abilene or the Pacifico-Chevron-LandBridge profile of Permian); and the first concentration where existing operational nuclear capacity (rather than newbuild like Fermi's planned reactors) provides the energy substrate. Pre-operational across all data center anchors; Sailfish first 600 MW within 24 months, AWS Project Spectrum first power late 2027.

Comparable: Talen Energy / Susquehanna AWS co-location in Pennsylvania as the operational template AWS is applying at Texas scale; Constellation Three Mile Island restart for Microsoft as the other major US nuclear-AI pairing; Fermi America in the Panhandle for the nuclear-AI thesis at scale, but with newbuild rather than existing nuclear and a gas-first staged architecture rather than nuclear-immediate.

Sandow Lakes AI-Industrial Cluster

Type: Multi-function emerging supernode along the US 79 corridor east of Taylor — master-planned brownfield manufacturing megasite paired with greenfield AI data center campus under one county geography.

Geographic anchor: Milam County along the US 79 corridor, with two distinct anchor clusters approximately 15 miles apart: the Sandow Lakes / Xebec advanced manufacturing campus at Rockdale (former Alcoa industrial footprint) and the Stargate Milam County (Freebird) AI data center campus at Burlington. Adjacent to the US 79 Compute Corridor.

Key nodes: *Sandow Lakes / Xebec* Advanced Manufacturing Campus (33,000-acre Xebec-master-planned megasite at Rockdale, brownfield-inherited from Alcoa, with ***T1 Energy's G2_Austin*** solar cell factory as a confirmed tenant and multiple additional tenants across advanced manufacturing categories); Stargate Milam County / ***Freebird*** (OpenAI and SB Energy partnership, 1.2 GW campus on 595 acres in Burlington, 548,950-square-foot Phase 1 building with four data halls, $470 million Phase 1 capital with $3 billion full project value, October 2026 Phase 1 completion target, OpenAI's third confirmed Stargate site after Abilene and Shackelford County). Other nodes include ***Bitdeer***, ***Riot Platforms/AMD***, and Softbank.

Corridor edges: US 79 as the workforce and logistics backbone connecting the two anchor clusters and integrating Milam County with Taylor (Samsung anchor) to the southwest and with Bryan–College Station to the east; the regional Texas Triangle workforce shed; the brownfield power and water infrastructure inheritance from the former Sandow Power Plant and Alcoa operations.

Supernode density: Two gigawatt-scale anchors under active construction within 15 miles on the same US 79 corridor in the same county, each pursuing a structurally distinct industrial function — multi-tenant brownfield manufacturing megasite at Sandow Lakes, single-tenant greenfield AI data center at Freebird. The supernode quality is multi-function diversification within one county and one corridor, anchored by the brownfield platform inheritance at Sandow Lakes and by the Stargate national capital program at Freebird. Whether the two anchors integrate operationally over the decade — through shared workforce, shared US 79 logistics, or supplier-ring overlap — or remain parallel campuses serving distinct industrial functions, is the open structural question.

Comparable: No precise analog. The closest structural parallel within the dataset is the SpaceX Bastrop Compound for multi-function industrial concentration within one county; but Milam County operates multi-operator rather than single-operator captive vertical integration.

Temple–Fort Hood AI–Defense Corridor

Type: Defense and workforce concentration with emerging compute layer — military installation anchor, veteran workforce pipeline, and arriving hyperscale capacity.

Geographic anchor: Bell County, spanning Temple, Belton, Killeen, and Fort Hood along the I-35 corridor between Austin and DFW.

Key nodes: ***Meta Temple*** data center campus; ***Rowan Digital*** Infrastructure; ***Fort Hood*** as the demographic and workforce anchor; additional hyperscale operator interest in the corridor as power and land economics drive site selection.

Corridor edges: I-35 as the workforce and logistics backbone connecting the corridor to both Austin (60 miles south) and DFW (130 miles north); the Fort Hood veteran workforce pipeline as a strategic asset for both defense and commercial technical roles; the regional natural gas and grid infrastructure supplying the arriving data center buildout.

Supernode density: Three parallel concentrations co-located but not yet integrated. The defense base has operated for decades. The hyperscale arrival is recent and driven by power and land economics rather than by defense-commercial coupling. Whether the supernode integrates the two layers — through veteran workforce flows into commercial technology roles, through defense AI applications running on adjacent commercial capacity, through dual-use research — or remains two parallel operations is the open structural question.

Comparable: Huntsville, Alabama, as a defense-to-commercial technology transfer precedent.

DFW Defense-Semiconductor Corridor

Type: Multi-specialization industrial corridor where advanced defense manufacturing, analog and power semiconductor production, legacy enterprise cloud infrastructure, and an emerging generation of hyperscale AI campuses converge. Unlike Austin's integrated AI-Industrial Supercluster, DFW is a collection of complementary industrial ecosystems whose proximity creates a diversified technology region rather than a single AI-centric concentration.

Geographic anchor: The Dallas–Fort Worth metroplex, extending from Fort Worth through Arlington and Grand Prairie to Dallas, Richardson, Plano, and northward to Sherman. The corridor is organized around three primary pillars: the Fort Worth defense complex, the Dallas–Richardson-Sherman semiconductor corridor, and the Dallas enterprise cloud and emerging AI infrastructure belt.

Key nodes: Defense: ***Lockheed Martin*** F-35 production, Bell Textron, RTX, and the aerospace-defense supplier ecosystem centered in Fort Worth.

Semiconductors: ***Texas Instruments*** Richardson and the new Sherman wafer fabrication complex, forming one of the world's largest concentrations of analog, mixed-signal, and power semiconductor manufacturing.

AI infrastructure: ***Provident–PowerHouse*** in Grand Prairie. Digital infrastructure: ***CyrusOne***, QTS, Equinix, Digital Realty, NTT Global Data Centers, and other enterprise colocation providers.

Corridor edges: DFW International Airport; Interstate 20, Interstate 30, Interstate 35E/35W, U.S. 75, the Sam Rayburn Tollway, major transcontinental fiber routes, electric transmission infrastructure, and the Sherman–Dallas–Fort Worth manufacturing axis.

Supernode density: Unlike Abilene or the Austin Supercluster, DFW combines several mature industrial ecosystems rather than one tightly integrated AI stack. Defense manufacturing, foundational analog and power semiconductors, one of the nation's largest enterprise cloud markets, and a new generation of gigawatt-scale AI campuses are converging. The result is an industrial convergence corridor composed of specialized subregions rather than a single frontier-AI hub.

Comparable: Comparable to a diversified industrial convergence corridor rather than a traditional AI supernode. Its distinguishing strength is the combination of defense manufacturing, semiconductor production, enterprise digital infrastructure, and emerging AI factories within one metropolitan region.

San Antonio Compute Cluster

Type: Enterprise cloud and colocation concentration with emerging AI layer — established hyperscale buildout and large AI-driven planning pipeline, no frontier-training anchor.

Geographic anchor: The San Antonio metropolitan area (Bexar County), with western extension into Medina County (Castroville) and the Westover Hills / Texas Research Park concentration.

Key nodes: ***Microsoft Castroville*** cluster (SAT 80, SAT 81, SAT 82 — roughly 734,000 square feet across the campus, with SAT 82 a 244,676-square-foot $482.6 million facility breaking ground April 2026 for early 2028 completion);

Energy Ranch is an oil-and-gas-capital pivot into AI-grade BTM compute at a scale (1–3 GW) that could justify supernode classification on its own. The structural signature parallels Waco InfraKey, CloudBurst, and Fermi America — energy-aligned capital using gas infrastructure expertise and behind-the-meter generation to serve hyperscale AI workloads.

Stream San Antonio III (up to 200 MW across five 40 MW AI-Ready buildings with an on-site 334 MW CPS substation, West Military Drive and Loop 1604);

CyrusOne, Vantage, Equinix, CloudHQ, and Amazon operations across the metropolitan area; the broader CPS Energy data center customer base of 21 operating campuses with 59 future projects in various stages of planning.

Corridor edges: I-10 as the east-west backbone connecting San Antonio to Houston and Brookshire; I-35 as the north-south spine connecting to San Marcos–Uhland and Austin–Taylor; I-37 as the southward corridor to the Gulf and to the emerging Atascosa County anchor; the CPS Energy transmission and substation infrastructure as the metropolitan grid backbone.

Supernode density: The cluster is dense in aggregate capacity — roughly 900 MW operating as of mid-2025 — but the integration quality is currently more aggregation than coherence. Operators are co-located by market access, CPS Energy reliability, and the regional workforce rather than by an integrated industrial stack. The pattern parallels the DFW Data Center Belt structurally — enterprise cloud and hyperscale concentration without a frontier-AI training anchor. CPS Energy's queue now reflects roughly 2 GW of AI-driven additions, with the broader future-project pipeline projecting energy needs in the tens of thousands of megawatts. Whether the AI workload migration pulls the corridor toward operational coherence, or leaves it as a parallel enterprise-cloud capital with regional AI presence but no frontier-training role, is the open structural question.

Comparable: DFW Data Center Belt for the enterprise-cloud-with-emerging-AI structural pattern, though San Antonio's southward integration with Atascosa County and its position on the I-35 spine give it a different connectivity geometry.

Tesla Brookshire Energy Campus

Type: Vertically integrated energy manufacturing — grid-scale battery storage, terrestrial solar, integrated grid nodes, and autonomous dispatch software produced as a single industrial stack.

Geographic anchor: Brookshire (Waller County), 30 miles west of downtown Houston along the I-10 corridor.

Key nodes: ***Tesla Megapack Megafactory*** (Megapack 3 and Megablock production beginning H2 2026, 50 GWh of annual storage capacity at full ramp); ***Brookshire Solar Manufacturing*** Complex (vertically integrated terrestrial solar fab — ingot through wafer through cell through panel within one site); Tesla distribution operations supporting the western Houston metropolitan service area; the broader Empire West Business Park as the master-planned industrial platform.

Corridor edges: I-10 as the east-west spine between Houston and San Antonio, and as the workforce and logistics corridor; the Houston metropolitan industrial workforce base as the labor substrate, drawing from a century of refining, petrochemical, and offshore expertise now transitioning into energy manufacturing; the Texas natural gas pipeline network supplying utility load to the campuses; the ERCOT transmission infrastructure connecting the manufacturing operations to the grid they produce hardware for.

Supernode density: The only geography in the United States, and likely in the world, where a single operator manufactures grid-scale battery storage, integrated grid nodes, terrestrial solar generation hardware, and the autonomous dispatch software that orchestrates the deployed fleet — all at one anchor, within one vertical-integration architecture. The supernode quality is vertical integration across the full grid-scale energy stack at one site. The structural signature is the closure of the manufacturing loop for an entire industrial layer at a single geography.

Comparable: No precise analog. The vertical integration across the full grid-scale energy stack at one site is structurally new.

Potential Supernodes

Major announced facilities whose surrounding cluster has not yet formed, but whose scale and strategic role suggest a supernode trajectory.

Fermi America AI Cluster (Project Matador)

Type: Distributed multi-anchor AI concentration around the Amarillo metropolitan region — one centralized generational-scale energy-and-compute campus pursuing a four-phase architecture across natural gas, conventional nuclear, and small modular reactors alongside three distributed hyperscaler sites under power-first co-location, integrated by the Texas Panhandle workforce shed, the Panhandle wind generation base, and adjacency to the Pantex nuclear weapons facility.

Geographic anchor: Carson, Armstrong, Roberts, and Gray Counties forming a ring around the Amarillo metro in the Texas Panhandle. Carson County hosts Fermi America's Project Matador adjacent to the Pantex nuclear weapons facility; Armstrong, Roberts, and Gray Counties host Google's distributed Panhandle footprint announced as part of Google's $40 billion Texas commitment in November 2025.

Key nodes: **Fermi America Project Matador** (Carson County near Pantex; Fermi America as developer in partnership with the Texas Tech University System on 5,800 leased acres; four-phase master plan totaling approximately 20 GW — Project Matador gas-fired phase at 6 GW operational 2027, Project Matador East gas-fired phase at 5 GW, timeline TBD, Fermi Nuclear at 4.4 GW conventional via four Westinghouse AP1000 reactors targeted 2032, Fermi SMR Expansion at 4.6 GW targeted 2036; 18 million square feet of planned data center floor area; $746 million raised in October 2025 IPO; Doosan Enerbility contracted for reactor pressure vessels and steam generators, Hyundai Engineering & Construction contracted for nuclear FEED, Siemens letter of intent for up to 1.1 GW of gas turbines, Energy Transfer pipeline interconnect targeted Q1 2026);

Google Gray County / Meitner Energy Center (Pampa, $3 billion campus with Google as anchor tenant and Intersect Power as energy partner, 840 MW initial data center capacity backed by over 1 GW of on-site wind, solar, and battery storage, power-first co-location model integrating data center and generation as a single procurement, multi-building campus);

Google Armstrong County (announced November 2025 as part of the $40 billion Texas commitment, capacity and timeline not publicly disclosed at site-specific level, AES as energy partner on the broader commitment);

Google Roberts County (announced November 2025 as part of the same commitment, capacity and timeline not publicly disclosed at site-specific level).

Corridor edges: I-40 east-west as the transcontinental backbone connecting Amarillo to Oklahoma City east and Albuquerque west; US 287 as the north-south spine to DFW; US 60 as the secondary east-west route through Pampa; the Texas Panhandle wind generation base as the distinctive renewable energy substrate edge — the region produces a substantial share of ERCOT's wind generation and has existing transmission infrastructure connecting Panhandle generation to ERCOT load centers; the natural gas pipeline network supplying the Fermi gas phases as the fossil-fuel substrate edge; the Texas Tech University System partnership with Fermi as the institutional substrate edge; the Pantex nuclear weapons facility in Carson County as the classified-substrate concentration edge, the only US final-assembly and disassembly site for nuclear warheads and the structural rationale for Fermi's site selection adjacent; Amarillo metro workforce shed.

Supernode density: Anticipated rather than realized, at the largest combined planned capacity in the appendix graph. Fermi's 20 GW four-phase architecture plus Google's distributed Panhandle footprint at an estimated 3–5 GW combined would establish total planned capacity of 23–25 GW, exceeding both Permian and Borderplex. The supernode quality parallels Permian in the multi-anchor pattern but with three distinctive structural elements unique to this cluster: the first explicit nuclear-AI supernode planned for Texas, with Fermi's 4.4 GW conventional nuclear by 2032 and 4.6 GW SMR by 2036 representing the longest-horizon energy commitments in the appendix graph; the centralized mega-anchor plus distributed hyperscaler topology, with Fermi as the single largest planned campus in the United States by capacity if executed and Google as a four-county distributed presence; and the renewable-heavy substrate, with the Panhandle wind generation base supplying both Fermi's renewable component and Google's power-first Intersect Power co-location. Fermi's four-phase architecture is the analytical signature — gas in the near term for speed-to-power, conventional nuclear and SMR in the longer term for clean-baseload AI compute. The construction timeline across the four phases is the question; the Panhandle's emerging-supernode status converts to operating only as the Fermi sequence materializes.

Comparable: Permian Basin AI Corridor for the multi-anchor gigawatt-scale pattern with non-Triangle geography; El Paso–Borderplex for the cross-jurisdictional and defense-adjacent character — Pantex parallels Fort Bliss as the defense-substrate edge, though Pantex is nuclear-weapons-stewardship rather than DoD-tenant. Distinguishing characteristics versus both: the four-phase nuclear-inclusive Fermi anchor at 20 GW, the renewable-heavy Panhandle wind substrate, and the largest combined planned capacity of any emerging supernode in the appendix graph.

SpaceX Terafab Complex (Gibbons Creek)

Type: Single facility in development at generational capital scale — leading-edge semiconductor fabrication on a brownfield-inherited industrial platform.

Geographic anchor: Grimes County, central Texas, adjacent to the Gibbons Creek reservoir and the retired Gibbons Creek coal facility.

Key nodes: ***Terafab Production Facility*** – the second node of the Terafab semiconductor architecture, following the Terafab Research Fab co-located at Giga Texas.

Corridor edges: Corridor to College Station and ***Texas A&M***; SH 90 connecting to the Bryan–College Station metropolitan workforce; the Gibbons Creek reservoir water rights and retired coal infrastructure as the brownfield inheritance platform.

Supernode density: Anticipated. Supernode formation depends on whether the Terafab anchor draws a supplier ring and adjacent operations into Grimes County over the next decade. The Gibbons Creek brownfield inheritance is the platform. The CHIPS Act funding base and the Tesla–Samsung supply relationship provide the demand pull. Whether a full supernode forms around the anchor is the structural question for the next ten years.

Waco-McGregor AI-Space Corridor

Type: Emerging single-anchor data center concentration with adjacent aerospace neighbor – behind-the-meter natural gas compute at the I-35 midpoint between Austin and DFW.

Geographic anchor: Waco (McLennan County) and the surrounding I-35 corridor, with McGregor (30 miles southwest) as a structural neighbor.

Key nodes: ***InfraKey Capital*** Waco (1,200 MW behind-the-meter natural gas data center, $10 billion committed capital); ***Cipher Digital*** (Project McLennan) data center 10 miles east of Waco up to 500 MW; ***McGregor SpaceX*** engine test facility as a related but distinct cluster; the regional Texas State Technical College workforce base.

Corridor edges: I-35 as the workforce and logistics backbone, with the Waco anchor at the midpoint between the Austin–Taylor supernode and the DFW Data Center Belt; the regional natural gas pipeline network supplying the behind-the-meter generation.

Supernode density: Emerging single-anchor with future supernode potential. The construction milestones over the next 24 to 36 months will determine whether a supporting cluster forms around the InfraKey anchor. The McGregor neighbor may eventually anchor a related aerospace cluster, distinct from but adjacent to the data center buildout.

Nexus Hubbard Gas-AI Campus

Type: Single facility in construction at generational capital scale — behind-the-meter natural-gas-powered AI hyperscale campus with sovereign-grade infrastructure positioning.

Geographic anchor: Hubbard, Hill County, along State Highway 171 approximately 30 miles north of Waco and 80 miles south of DFW on the I-35 corridor.

Key nodes: ***Nexus Hubbard Campus*** (Nexus Data Centers as developer, headquartered in The Woodlands; Transition Equity Partners as principal capital backer; 2,000-acre site with approximately 800 acres for the data center footprint; continuous generating capacity of approximately 7,230 MW at full buildout across 31 combined-cycle and simple-cycle turbines plus 24 reciprocating engines; behind-the-meter natural gas; Phase 1 of 500 MW operational target H2 2026; partnership with AirJoule Technologies deploying Metal-Organic Framework waste-heat-to-water harvesting; construction commenced late 2025).

Corridor edges: State Highway 171 as the local access corridor; I-35 as the workforce and logistics backbone connecting Hubbard to Waco southward and DFW northward; the regional natural gas pipeline network supplying utility load to the campus.

Supernode density: Anticipated. At roughly 7.2 GW of continuous generating capacity, Nexus Hubbard is one of the largest single-site power-generation projects in Texas. The structural signature is sovereign-grade positioning — Nexus Core Systems is explicitly engineering for operation outside Patriot Act and Cloud Act jurisdictional exposure, supporting national data residency and emerging AI governance frameworks. This is the first appendix entry with sovereign-AI infrastructure as the explicit market positioning rather than a downstream consequence.

Data City Hydrogen-AI Campus

Type: Single proposed campus at generational capital scale — off-grid behind-the-meter data center platform with planned hydrogen-transition energy strategy with salt cavern located near Corpus Christi.

Geographic anchor: Webb County near Laredo, southwest Texas on the US–Mexico border, at the southern terminus of I-35.

Key nodes: ***Data City*** (Energy Abundance Development Corporation as developer, Houston-based, formerly Green Hydrogen International; 50,000 acres; 5 GW master plan; over 15 million square feet of leasable space at full buildout; behind-the-meter and fully islanded from ERCOT; initially natural-gas-powered, transitioning to green hydrogen sourced from the adjacent

Hydrogen City project; first 300 MW and 1 million square feet phase scheduled 2026); Hydrogen City (companion project, 2 TWh hydrogen storage in salt cavern infrastructure, first phase 2028).

Corridor edges: I-35 as the southern terminus connection to San Antonio (approximately 150 miles north) and to the broader Texas Triangle; the regional natural gas pipeline network as the initial energy substrate; the planned Hydrogen City salt-cavern storage infrastructure as the future renewables substrate edge.

Supernode density: Anticipated. Data City is structurally novel — the only entry with an explicit hydrogen-transition energy strategy and salt-cavern energy storage paired with the data center buildout. The 50,000-acre footprint is the largest single-site data center master plan in Texas by acreage. The Webb County position extends the I-35 spine to its southern endpoint at the US –Mexico border. Supernode formation depends on execution of both Data City and Hydrogen City over the next decade.

West Texas Energy-AI Corridor

Type: Emerging AI-energy corridor anchored by multiple gigawatt-scale AI campuses that leverage West Texas wind, solar, battery storage, natural gas, and ERCOT transmission infrastructure. Unlike the Permian Basin, which is centered on behind-the-meter natural gas generation, this corridor combines diverse energy strategies with hyperscale AI development along the Interstate 20 spine.

Geographic anchor: The Sweetwater–Borden County–Big Spring–Midland–Odessa corridor of West Texas, following Interstate 20 and the CREZ (Competitive Renewable Energy Zone) transmission network. The corridor is centered on a growing concentration of hyperscale AI campuses supported by abundant renewable generation, strategic transmission capacity, utility-scale battery storage, and regional fiber infrastructure.

Key campuses: *IREN Sweetwater* (2 GW renewable-powered AI campus and one of the largest operational AI developments in North America); ***Galaxy Helios*** (planned 1.6 GW hyperscale AI campus reinforcing Sweetwater as a major AI hub); ***SB Energy Borden County*** (1+ GW AI campus integrated with the 418 MW Juno Solar project and the planned 500 MW Lyra battery energy storage system); ***Texas Critical Data Centers (TCDC)*** Odessa (planned gigawatt-class AI campus utilizing behind-the-meter natural gas generation).

Corridor edges: Sweetwater forms the northern AI anchor; Borden County extends the renewable-energy platform southward; Big Spring provides industrial land, transmission, and logistics opportunities; Midland–Odessa anchors the southern end with energy-integrated AI development. Interstate 20, regional fiber routes, and the CREZ transmission network provide the corridor backbone.

Supernode density: Four independent AI campuses totaling more than 5 GW of planned AI capacity distributed along approximately 170–200 miles of the Interstate 20 corridor. Unlike a compact supernode such as Abilene, this is a linear AI-energy corridor in which multiple campuses are linked by shared transmission infrastructure, renewable generation, battery storage, natural gas resources, fiber, and transportation networks.

Comparable: Comparable to the Permian Basin AI Supernode in overall strategic importance, but differentiated by its hybrid energy model. Whereas the Permian Basin is built primarily around behind-the-meter natural gas generation, the West Texas AI-Energy Corridor combines utility-scale wind, solar, battery storage, grid-connected hyperscale campuses, and selective behind-the-meter generation. Its defining characteristic is not a single campus, but a geographically distributed concentration of AI infrastructure built on one of North America's richest electricity-producing regions.

West Texas Energy Triangle

Type: Not a facility-based supernode but the substrate-focused West Texas analogue to the Texas Triangle in the east. State-scale energy backbone connecting the largest concentrations of wind, solar, battery storage, natural gas production, transmission infrastructure, and future AI development sites in Texas.

Geographic anchor: Extends from the Texas Panhandle through the South Plains and into the Permian Basin, linking Amarillo, Abilene, Lubbock, Sweetwater, Big Spring, Midland/Odessa, and Pecos into a continuous energy and infrastructure corridor.

Key nodes: Amarillo (Panhandle wind anchor); Lubbock (transmission and logistics gateway); Sweetwater and Big Spring (renewable-energy core); Midland/Odessa (Permian industrial anchor); Pecos (AI-campus and energy-development frontier). The West Texas Energy-AI Corridor lies inside.

Corridor edges: ***CREZ transmission network***; ERCOT backbone infrastructure; I-20 and I-40 freight corridors; regional fiber routes; natural-gas pipeline systems; future high-capacity transmission projects; battery-storage and renewable-energy interconnections.

Supernode density: Few regions in North America combine such high concentrations of wind, solar, battery storage, natural gas production, long-haul transmission, pipeline infrastructure, and available industrial land within a single connected geography. The corridor functions as an energy substrate.

Comparable: A hybrid of the Permian Basin and the Tennessee Valley Authority corridor model. Distinguishing characteristics include simultaneous concentration of renewable and hydrocarbon energy resources, extensive transmission infrastructure, and suitability for future gigawatt-scale AI and industrial campuses.

Closing the Supernode set

At its center sits the Austin Supercluster: five supernodes within ~35 miles — integrated through shared workforce, shared corridor infrastructure, and the academic, design, and institutional substrate that no other US region has assembled.

The Inner Triangle of Giga Texas, the SpaceX Bastrop Compound, and Samsung Taylor region sits inside the supercluster as its densest concentration.

Behind-the-meter natural gas has become the dominant energy strategy for absorbing grid-interconnection delay across most of the new buildout.

West Texas/Permian Basin, the Panhandle, and the El Paso Borderplex are emerging as a parallel AI-industrial geography to the Triangle, anchored by abundant gas and renewable generation.

A small set of structural firsts — cross-border supernodes, defense-anchored AI buildout, nuclear-AI substrate, hydrogen-transition energy strategies, sovereign-grade infrastructure positioning — that may or may not become dominant patterns but are visible at gigawatt scale today.

Appendix B — AI-Industrial Nodes

This appendix documents the 75-odd primary facilities that constitute the Texas AI-industrial graph as of mid-2026. Appendix A describes the supernodes — the geographic clusters and corridors within which these facilities sit. This one documents the facilities (nodes) themselves. Together the two appendices supply the empirical substrate the body of the book describes analytically.

Entries are organized by category —Data Centers, Semiconductor Fabs, Gigafactories, Aerospace, Energy Generation, Critical Materials — so the reader can move directly to a category of interest or browse across them. The structure rewards browsing more than sequential reading.

Criterion A — Scale or distinctiveness. Landmark capital ($10B+ committed), generational-scale anchor capital, or structural first-of-kind feature: novel energy substrate, unprecedented capacity ceiling, defining supplier ring, or category-establishing role.

Criterion B — Convergence participation. Demonstrable role in the AI-industrial convergence thesis — AI data center, fab, gigafactory, aerospace and defense, energy, or critical materials — as a primary rather than incidental workload. Facilities whose primary workload is Bitcoin mining, with AI capacity only as a secondary or convertible role, are documented as convergence-adjacent and called out explicitly.

The capital scale tiers below classify committed capital at the entry level. Status and all figures reflect mid-2026 conditions.

Capital Scale Tiers

- Foundational: under $1B committed.
- Significant: $1B to $10B committed.
- Landmark: $10B to $40B committed.
- Generational: $40B and above committed.
- Not Disclosed: no public capital figure available.

***Note**: Appendix BA documents the state of the buildout as of manuscript completion in June, 2026. Announced capacities, capital commitments, construction timelines, and operational status are dynamic across every facility documented. Capacities scale, funding tranches evolve, timelines shift, and phase transitions occur continuously across the buildout. Readers checking specific facility details against later reporting should treat the appendix as the structural baseline.*

Data Centers

The Texas dataset holds 484 data centers as of September 2025 (297 operational, 109 under construction, 124 planned) with forecasted total capacity surpassing 70 GW by 2030 per the UT Austin COMPASS program. Northern Virginia currently leads the data center market; industry practitioners project Texas as the largest data center market globally by end of decade.

Fermi America Hypergrid (Project Matador)

Carson County

AI compute coupled with on-site energy generation under an architecture spanning natural gas, conventional nuclear, and small modular reactors. No primary grid dependency.

Category: Data Center; Federated Supernode

Status: Pre-Construction; four-phase master plan totaling approximately 20 GW: Project Matador (gas, 6 GW, expected 2027); Project Matador East (gas, 5 GW, TBD); Fermi Nuclear (conventional, 4.4 GW, expected 2032); Fermi SMR Expansion (4.6 GW, expected 2036)

Energy: Behind-the-meter; four-phase architecture combining natural gas turbines for near-term speed-to-power, conventional nuclear reactors for longer-horizon clean baseload, and small modular reactor units for multi-decade clean baseload; approximately 20 GW combined campus design target at full buildout across multiple reactor and turbine units; Panhandle wind generation base as the renewable substrate.

Water: Not disclosed; significant cooling requirements for nuclear and data center load.

Capital: Not disclosed; novel coupling of nuclear and data center build; capital framework not publicly filed.

Supplier Ring: Anchor facility; gas turbine and combined-cycle equipment supply chain, SMR construction and licensing, nuclear fuel supply, conventional nuclear reactor equipment, data center compute supply chain.

Note: First explicit nuclear-AI supernode planned for Texas, with nuclear and SMR phases representing the longest-horizon energy commitments in the dataset. Gas phases provide near-term speed-to-power for AI workloads; nuclear baseload phases (2032 and 2036) extend the energy substrate into multi-decade horizons. The energy architecture — combining the speed advantages of gas with the durability and cleanness of nuclear under one campus program — distinguishes Fermi from all other dataset entries.

Crusoe Lancium Abilene Campus (Stargate One)

Abilene, Taylor County

AI training and inference compute. OpenAI anchor program.

Category: Data Center; Multi-campus

Status: Under Construction

Energy: Grid-connected plus onsite natural gas; 1,200 MW Phase 1

Water: City of Abilene municipal plus Lake Fort Phantom Hill

Capital: Generational; $100B minimum committed; $500B full national program announced

Vantage Frontier Abilene Campus (Stargate)

Shackelford County, near Abilene

Frontier-scale AI training mega-campus. Ultra-high-density GPU workloads.

Category: Data Center; Multi-campus

Status: Under Construction; first delivery H2 2026

Energy: Grid-connected; 1,400 MW full buildout; combined air-cooling and liquid-to-liquid cooling; 250 kW ultra-high-density racks

Water: Not disclosed

Capital: Generational; $25B committed; largest single-campus data center capital commitment in Texas dataset

Note: 10 single-story data centers, 3.7 million sq ft, 1,200 acres. 250 kW rack density is the highest specification in the dataset, purpose-built for frontier AI training.

Crusoe Microsoft Abilene Campus

Abilene, Taylor County

Microsoft-dedicated frontier AI training campus. Largest single AI compute deployment in Texas.

Category: Data Center; Multi-campus

Status: Operational plus Expanding; Phase 1 (1.2 GW) coming online through 2026; Phase 2 (900 MW) first building mid-2027

Energy: Grid-connected; 2,100 MW combined capacity at full buildout

Water: City of Abilene municipal plus closed-loop cooling

Capital: Not Disclosed; Microsoft anchor tenant; combined 2.1 GW program scale

Note: Part of the Abilene Supernode alongside Stargate and Vantage Frontier. Combined cluster capacity exceeds 4 GW.

SB Energy Borden County Campus

Gail, Borden County; West Texas

SB Energy multi-resource AI data center campus. Part of SB Energy's multi-gigawatt portfolio supporting Stargate.

Category: Data Center; Cluster

Status: Pre-construction; tax abatement under active discussion with Borden County early 2026

Energy: Multi-resource onsite energy substrate; natural gas, wind, and solar generation combined; SB Energy previously developed the 418 MW Juno Solar Project CdTe solar farm in western Borden County, establishing prior energy infrastructure presence; 1 GW master plan

Water: Rainwater collection system included in the abatement; closed-loop cooling; Texas Water Development Board review underway

Capital: Not Disclosed; SB Energy as developer; part of SB Energy's multi-gigawatt national portfolio backed by the $1B OpenAI–SoftBank investment and $800M Ares preferred equity; scale comparable to Landmark tier

Note: Borden County population 631; Gail population 249. The 1 GW capacity in a 631-person county represents one of the most extreme infrastructure-to-population ratios in the dataset. Located between Midland/Odessa to the south and Lubbock to the north, with US 180 as the local corridor. Approximately 100 miles northwest of Stargate Abilene with no direct highway corridor; structurally distinct from the Abilene Supernode. SB Energy plans a Gail job fair and subcontractor registration process; the necessity of a "man camp" to house construction workers in the sparsely populated county is under evaluation. Borden County Data Center may also be referenced as Borden County Data Center 1 or Data Center 5 in SB Energy's internal numbering.

Pacifico Energy GW Ranch Campus

North of Fort Stockton, Pecos County

Off-grid Permian Basin private-grid energy and compute campus. Purpose-built for hyperscale AI workloads at the gas source.

Category: Data Center

Status: TCEQ-permitted February 2026; first 1 GW power delivery targeted 2028; 5 GW by 2030; full buildout 2031

Energy: Off-grid private grid; 7.65 GW TCEQ air permit (largest single power-plant permit issued in the United States); high-efficiency natural gas turbines combined with solar generation and 1.8 GW battery storage

Water: Pecos County groundwater; specifics not disclosed

Capital: Not Disclosed; Pacifico Energy Group as developer; scale comparable to Generational tier

Note: 8,000-plus acres approximately 17 miles north of Fort Stockton along SH 18. Air permit exceeds every operating Texas power plant including W.A. Parish; approximately three times Comanche Peak Nuclear's combined capacity. Structural signature: co-located at the gas source rather than transporting gas to compute. First Texas AI anchor sited inside the Permian Basin proper. Off-grid architecture bypasses the ERCOT interconnection queue.

Nexus Hubbard Campus

Hubbard, Hill County; State Highway 171

Behind-the-meter natural gas AI hyperscale campus. Sovereign-grade infrastructure positioning.

Category: Data Center

Status: Under Construction; commenced late 2025; Phase 1 (500 MW) operational target H2 2026; full buildout phased over multi-year horizon

Energy: Behind-the-meter natural gas; approximately 7,230 MW continuous generating capacity at full buildout across 31 combined-cycle and simple-cycle turbines plus 24 reciprocating engines, with 6 emergency diesel generators

Water: AirJoule Technologies waste-heat-to-water harvesting using Metal-Organic Framework technology (2025 Nobel-recognized); closed-loop cooling from waste-heat capture

Capital: Not Disclosed; Nexus Data Centers (The Woodlands) as developer; Transition Equity Partners as principal capital backer; scale comparable to Generational tier

Note: 2,000-acre site with approximately 800 acres for data center footprint. Nexus Core Systems positions explicitly for operation outside Patriot Act and Cloud Act jurisdictional exposure, supporting national data residency and emerging AI governance frameworks (GDPR-aligned, sovereign-grade). First dataset entry with sovereign-AI infrastructure as explicit market positioning rather than downstream consequence. Located on the I-35 spine approximately 30 miles north of Waco and 80 miles south of DFW.

Energy Abundance Data City Campus

Webb County near Laredo; US–Mexico border region

Off-grid behind-the-meter AI hyperscale campus. Hydrogen-transition energy substrate paired with salt-cavern energy storage.

Category: Data Center

Status: Phase 1 in Development; 300 MW and 1 million sq ft scheduled 2026; full 5 GW buildout phased over decade

Energy: Behind-the-meter, fully islanded from ERCOT; initially natural gas, transitioning to green hydrogen from adjacent Hydrogen City; combined wind, solar, batteries, and dual-fuel gas turbines

Water: Closed-loop cooling; Webb County base; specifics not disclosed

Capital: Not Disclosed; Energy Abundance Development Corporation (Houston, formerly Green Hydrogen International) as developer; scale comparable to Generational tier at full buildout

Note: 50,000-acre footprint — the largest single-site data center master plan in Texas by acreage. Over 15 million sq ft of leasable space at full buildout. Webb County position extends the I-35 spine to its southern endpoint at the US–Mexico border. Only dataset entry with explicit hydrogen-transition energy strategy paired with the data center buildout. Management team previously developed, built, and financed over 3 GW of power projects globally.

HC Energy Ranch Campus (San Antonio adjacent)

Atascosa County, approximately 58 miles south of San Antonio

Hyperscale AI/cloud campus. Oil-and-gas capital pivot into behind-the-meter AI compute.

Category: Data Center

Status: Pre-construction; first 250 MW phase early 2028; scaling in phases to 1–3 GW total capacity

Energy: Behind-the-meter natural gas; modular natural-gas reciprocating generation; planned 765-kV high-voltage transmission corridor expected to reach the region later this decade as future grid edge; integrated proximity to multiple natural gas pipelines

Water: Closed-loop cooling; Atascosa County base; specifics not disclosed

Capital: Significant; $2B–$4B first-phase capital announced; HC Capital Partners and Herrmann Family Companies as principals; Wrangler Infrastructure as development and operating platform

Note: 1,500-plus acres on Herrmann-owned land. Structural signature: oil-and-gas infrastructure expertise applied to AI hyperscale buildout, with on-site generation, secure fuel supply, thermal cooling, and water infrastructure integrated under one campus architecture. Anchors the southern extension of the I-35 / I-37 corridor below San Antonio.

Microsoft Castroville Supernode (San Antonio adjacent)

Castroville, Medina County; 18844 Farm-to-Market Road

Microsoft regional cloud and AI capacity. Multi-building hyperscale campus.

Category: Data Center; Multi-campus cluster

Status: Operational plus Expanding; first two SAT buildings delivering through 2026; SAT 82 breaking ground April 2026 for early 2028 completion

Energy: Grid-connected via CPS Energy infrastructure; aggregate campus draw not disclosed; new transmission and substation planning underway in coordination with CPS Energy AI-driven queue

Water: Medina County municipal and groundwater base

Capital: Significant; SAT 82 alone at $482.6 million; aggregate campus capital across SAT 80, SAT 81, SAT 82 estimated $1.5 billion-plus; Microsoft as operator

Note: Approximately 734,000 sq ft across SAT 80, SAT 81, and SAT 82 buildings combined. Anchors the western edge of the San Antonio Compute Corridor (Appendix A). Microsoft's broader San Antonio metropolitan footprint also includes the existing 15000 Lambda Drive facility with 79,385-sq-ft expansion. Cloud and AI regional capacity rather than frontier-training scale. Microsoft additionally leases approximately 700 MW from Crusoe in Abilene adjacent to the Stargate campus — a separate Abilene-supernode capacity arrangement.

IREN Sweetwater Campus

Nolan County (Sweetwater) with Sweetwater 2 in adjacent Jones County; West Texas approximately 60 miles west of Abilene

2 GW grid-connected hub across two contiguous campuses — Sweetwater 1 at 1.4 GW and Sweetwater 2 at 600 MW — linked by direct fiber loop. Designed for direct-to-chip liquid cooling at 200 kW rack densities for next-generation Nvidia GPUs.

Category: Data Center

Status: Sweetwater 1 energized April 2026 at 1.4 GW grid-connected capacity (1,400 MW bulk substation operational at 345 kV/138 kV with primary substations at 138 kV/34.5 kV); Sweetwater 2 (~28 miles from Sweetwater 1) targeted for late 2027 at 600 MW; full 2 GW hub at completion; design capacity supports up to ~700,000 liquid-cooled GPUs; commercial contracts pending across the hub

Energy: ERCOT grid-connected, leveraging CREZ-era West Texas transmission infrastructure originally built for wind energy delivery; 345 kV/138 kV bulk substation owned and operated by IREN at Sweetwater 1; primary 138 kV/34.5 kV substations on-site; access to regional wind and solar generation through ERCOT spot procurement; no behind-the-meter generation disclosed

Water: Closed-loop direct-to-chip liquid cooling for AI/HPC workloads; arid West Texas water context implies minimized water consumption versus evaporative cooling; specifics not disclosed

Capital: IREN Limited, formerly Iris Energy, rebranded November 2024) as developer and operator; ~1,800–2,000 acres assembled across Nolan and adjacent counties; freehold land model with company-owned substations and transmission infrastructure; Sweetwater positioned as IREN's largest infrastructure project, building on the operational template established at the company's 750 MW Childress campus

Note: IREN's vertically integrated model — owning land, substations, buildings, and grid interconnect rather than leasing — gives the company time-to-power advantages that newer Permian and Panhandle entrants lack. The Sweetwater Hub sits inside the same West Texas CREZ-era transmission corridor that Helios (Dickens County), Crusoe's Abilene campus, Stargate Abilene, and Vantage Frontier (Shackelford) draw from, making the I-20 / US 84 corridor between Abilene and Lubbock the densest grid-connected AI concentration outside the Permian Basin's behind-the-meter cluster. Crypto-mining-to-AI conversion is the structural signature, paralleling the Helios trajectory: the same low-cost ERCOT power and high-density rack engineering that worked for Bitcoin mining now converts directly to AI training and inference workloads with liquid cooling retrofits.

Galaxy Helios Coreweave Campus

Dickens County (Afton); approximately one hour east of Lubbock, four miles northwest of Afton off County Road 112

Crypto-mining-to-AI conversion campus pivoted from Bitcoin mining to hyperscale AI/HPC under a 15-year, 800 MW CoreWeave anchor lease. Phase 2 buildout (Helios 2/3/4) targeting 1.6 GW approved capacity with long-term potential to 3.5 GW.

Category: Data Center

Status: Phase 1 (800 MW) under construction with first power to CoreWeave beginning early 2026; ERCOT approved an additional 830 MW in January 2026 bringing total approved capacity to 1.6 GW; Phase 2 permits filed in early 2026 for Helios 2 ($900M, 199,124 sq ft), Helios 3 ($1.3B, ~350,000 sq ft), and Helios 4 ($1.3B, ~350,000 sq ft), combined $3.5B; long-term capacity potential of 3.5 GW with an additional 2.7 GW under various load study stages; site expanded from original 160 acres to over 1,500 acres in three years

Energy: ERCOT grid-connected via the adjacent Cottonwood switching station, one of the largest electrical switching facilities in Texas; service agreements with AEP Texas and transmission coordination with Wind Energy Transmission Texas (WETT); access to CREZ-era West Texas wind transmission; hundreds of megawatts of private backup generation infrastructure disclosed

Water: ~10 million gallon fresh-water pond on site supporting closed-loop liquid cooling for next-generation GPU densities; 130–140 kW rack densities standard at Phase 1

Capital: Galaxy Digital Holdings as developer and operator; acquired from Argo Blockchain December 2022 for $65 million; $1.4 billion debt facility closed May 2026 (Galaxy contributed $350 million equity, 36-month term debt secured by Phase 1 assets); CoreWeave (Nvidia-backed AI cloud, as anchor tenant with full 800 MW Phase 1 capacity committed under 15-year lease projected to generate $1 billion average annual revenue and approximately $15 billion total contract value over the term

Note: The Argo-to-Galaxy-to-CoreWeave trajectory makes Helios the canonical crypto-mining-to-AI conversion in the Texas dataset. Argo built the original 125,000 sq ft mining facility at 180 MW in May 2022; Galaxy operated mining there through November 2024, then cleared all Bitcoin mining infrastructure and pivoted to AI/HPC hosting. The Cottonwood switching station adjacency is the structural anchor — one of the largest electrical switches in Texas, providing reliable bulk power without the multi-year ERCOT interconnect queue. Multi-terabit fiber connectivity provides under-15 ms round-trip latency to the DFW metro at ~250 miles, making Helios viable for AI inference workloads as well as training. The CREZ transmission corridor positions Helios alongside IREN Sweetwater, Crusoe Abilene, Vantage Frontier (Shackelford), and Stargate Abilene as anchors of the I-20 / US 84 grid-connected AI concentration; collectively these projects represent the densest CREZ-era utilization in Texas.

Texas Critical Data Centers (TCDC)

Ector County (Odessa); Permian Basin, adjacent to existing Vistra and Calpine gas-fired generation assets

1+ GW behind-the-meter natural-gas-powered AI/HPC campus on a 438-acre site, leveraging Permian Basin gas and helium production for net-zero energy with carbon capture integration potential.

Category: Data Center

Status: Phase 1 (100 MW) targeting first power December 2026 with full 250 MW operational by mid-2027; multi-phase development scaling toward 1+ GW total capacity; 438 acres owned with a 54-acre corridor pending; Phase Two engineering commenced October 2025 with site clearing underway; non-binding LOI signed April 2026 with Stream Data Centers and a third-party institutional investor to form a joint venture for development and financing

Energy: Behind-the-meter natural gas generation; 250 MW power supply MoU signed with PowerForward Energy in June 2025 (responsible for manufacturing, installation, and operation of power generation assets); separate behind-the-meter gas LOI signed September 2025 with Mawgan Capital; natural gas supply agreement with New Era Helium for fuel; carbon capture integration planned via the helium-and-gas production substrate; campus adjacent to existing Vistra and Calpine merchant generation providing grid-backup optionality

Water: Permian arid context implies closed-loop liquid cooling for AI/HPC GPU densities; specifics not disclosed

Capital: New Era Energy & Digital, Inc., formerly New Era Helium) as sole developer after acquiring Sharon AI's 50 percent stake December 2025 for $70 million in cash, equity, and convertible notes; site secured through GROW Odessa partnership; Stream Data Centers (a Tier-1 US data center operator) joining as joint-venture partner April 2026; institutional investor providing equity capital and arranging debt financing; CEO E. Will Gray II

Note: TCDC is the first hyperscale data center campus integrated with helium and natural gas production at the wellhead level, with New Era's parent operation supplying both the gas feedstock for power generation and the helium byproduct streams. The carbon-capture-with-helium-coproduction model is structurally distinctive in the Texas dataset and provides a sustainability narrative most behind-the-meter gas projects lack. Adjacency to Vistra and Calpine merchant gas plants gives TCDC the option of grid-backup plus behind-the-meter generation in a hybrid architecture, reducing single-point-of-failure risk. Located in the Permian Basin.

AWS Project Spectrum Comanche Peak Campus

Hood County (Granbury); approximately 50 miles southwest of Fort Worth, adjacent to the Comanche Peak Nuclear Power Plant in neighboring Somervell County

$5 billion AWS hyperscale campus on 1,265 acres anchored by a 20-year nuclear PPA with Vistra subsidiary Luminant. First nuclear-anchored hyperscaler campus in Texas.

Category: Data Center

Status: 1,265 acres assembled at 2300 Coates Rd in Granbury; initial 600 MW phase submitted to the ERCOT interconnection queue February 2026; first stage planned for up to 18 two-story buildings across approximately 435 acres; construction breaking ground early 2027 with initial energization late 2027; full PPA capacity of up to 1,200 MW scaling over multi-year deployment with option to extend an additional 20 years beyond the initial 20-year term

Energy: 20-year Power Purchase Agreement with Luminant Generation (Vistra Corp. subsidiary, for up to 1,200 MW of carbon-free nuclear electricity from the 2.4 GW Comanche Peak Nuclear Power Plant; NRC license extended through 2053 by the July 2024 approval, providing 28 years of remaining licensed operation; campus also included in Vistra's broader industry-leading ~3,800 MW nuclear PPA portfolio with AWS spanning multiple Vistra nuclear sites disclosed in Vistra's FY2026 guidance

Water: Closed-loop liquid cooling architecture implied for the 1.2 GW deployment; Comanche Peak's existing reservoir provides cooling water for the nuclear plant itself; specifics for the data center campus not disclosed

Capital: Amazon Web Services as anchor tenant ($5 billion campus investment reported); Vistra Corp. as power supplier via Luminant subsidiary; Vistra has not officially confirmed Amazon as the counterparty in SEC filings (listed as undisclosed customer), though Amazon-as-buyer has been reported by Hood County News, Data Center Dynamics, and WFAA

Note: Will be the first operational nuclear-anchored hyperscaler campus in Texas, completing the AWS multi-site nuclear strategy that began with the Talen Energy / Susquehanna co-location in Pennsylvania. The structural parallel with Sailfish Comanche Circle is striking: both projects are gigawatt-class plans in Hood County leveraging the same Comanche Peak Nuclear Plant for clean-baseload economics. Combined with Pacifico Energy's planned behind-the-meter gas project also in Hood County and Mara's existing cryptomine, Hood County is consolidating as a clean-baseload-plus-gas hybrid AI cluster anchored by Comanche Peak. Project Spectrum's nuclear-PPA model differs structurally from the Permian Basin gas model (Chevron–Microsoft, GW Ranch, Alpha Digital, TCDC) by trading off speed-to-power for clean-energy certification and the long-term operating cost advantage of nuclear baseload — the same trade Fermi America is making in the Panhandle but with existing rather than newbuild nuclear capacity.

Sailfish Comanche Circle Supernode

Tolar, Hood County; adjacent to Glen Rose

Nuclear co-located hyperscale AI campus. Adjacent to Comanche Peak Nuclear plant.

Category: Data Center; Multi-campus Supernode

Status: Proposal Stage; early discussions; community and environmental review underway

Energy: Grid-connected; 5,000 MW proposed capacity co-located with the Vistra Comanche Peak 2,400 MW nuclear plant

Water: Paluxy River watershed; subject of active community opposition over water usage

Capital: Not Disclosed; 2,600 acres; proposal stage

Note: If realized, would be the largest single data center campus by power capacity in the dataset. Community opposition mirrors Hill County moratorium movement.

Meta Temple Campus

Temple, Bell County

AI training compute. Meta foundation models.

Category: Data Center

Status: Under Construction

Energy: Grid-connected; 2,000 MW design capacity

Water: City of Temple municipal

Capital: Landmark; $10B-plus operator announced

Supplier Ring: Anchor facility; hyperscale MEP construction, power infrastructure, and cooling systems; dataset at AustinIO.com

Rowan Digital Project Temple Cluster

Temple, Bell County

300 MW hyperscale campus serving a single top-tier global technology tenant.

Category: Data center; Multi-campus

Status: Under Construction (groundbreaking March 2026); Operations expected 2027.

Energy: Oncor grid-connected; 300 MW full buildout across approximately 700 acres. Leverages existing power infrastructure from the adjacent Moriah site to avoid new transmission and substation construction

Water: Closed-loop cooling; utility service agreement with City of Temple

Capital: $700 million construction cost; $3 billion green debt financing (Rowan's largest)

Note: Adjacent to Rowan's earlier Moriah campus. Quinbrook has telegraphed up to 1.6 GW combined Temple capacity across approved and future Rowan phases through 2030, indicating supernode-class potential at Temple alongside Meta and forward Stargate-adjacent projects in Bell-Milam corridor.

SB Energy Stargate Milam Campus

Burlington, Milam County

OpenAI Stargate AI training and inference campus. SB Energy developed and operated; OpenAI tenant.

Category: Data Center

Status: Phase 1 Under Construction; commenced October 2025; Phase 1 (548,950 sq ft, four data halls) completion targeted October 2026; full 1.2 GW buildout subsequent phases

Energy: SB Energy is building new on-site power generation to support the campus and minimize Texas grid impact; precise architecture not fully disclosed; 1.2 GW design capacity

Water: SB Energy-designed to minimize water usage; closed-loop architecture; specifics not disclosed

Capital: Landmark; Phase 1 estimated $470 million per TDLR filing; full project $3 billion estimated; broader capital includes $1B OpenAI–SoftBank joint investment in SB Energy January 2026 and $800M Ares preferred equity

Note: 595-acre site between Austin and Waco on County Road 134. OpenAI's third confirmed Stargate site alongside Stargate Abilene and Stargate Shackelford County (Vantage Frontier). Part of the OpenAI–SoftBank–Oracle $500 billion Stargate initiative. SB Energy acquired Studio 151 in January 2026 to deepen execution capabilities. Project head Zoheir Taheri (SB Energy) came from Studio 151. Burlington is approximately 15 miles northeast of Rockdale along US 79, placing this campus within the same Milam County US 79 corridor as the Sandow Lakes / Xebec advanced manufacturing site.

Skybox PowerCampus Hutto

Hutto, Williamson County

Hyperscale data center campus; six to eight buildings.

Category: Data Center

Status: Phase 1 Under Construction

Energy: Grid-connected; 600 MW Phase 1

Water: City of Hutto municipal plus closed-loop cooling

Capital: Landmark; $10B minimum, $20B full buildout per operator announcement

Iron Mountain Hutto Campus

Hutto, Williamson County

Seven-building hyperscale campus is directly south of Samsung Austin Semiconductor Taylor. Iron Mountain's first major Texas footprint and largest single addition to its global capacity base.

Category: Data Center

Status: Pre-Construction; First grid-connected buildings targeted approximately 2030; behind-the-meter natural gas option under evaluation for accelerated phase delivery

Energy: Grid plus optional behind-the-meter natural gas; 7 buildings across 500 acres with on-site substation; MW capacity not publicly disclosed

Water: Closed-loop cooling; sanitary sewer via Hutto agreement (~ 30,200 gal/day projected)

Capital: Not Disclosed

Note: Samsung Taylor adjacency places this facility on the fab-data center convergence axis. Capacity scale at 7 buildings extrapolates to a 300 to 700 MW class campus once disclosed.

Colovore Project Raptor Hutto

Hutto, Williamson County

First dedicated liquid-cooled high-density AI/HPC colocation facility in the Austin metro. Tenant class includes Cerebras, Lambda Cloud, and Cirrascale.

Category: Data Center

Status: Approved (December 2024); Construction targeted early 2026

Energy: Grid-connected; MW capacity not publicly disclosed; 180,000 sq ft at 250 kW/rack design class indicates 40 to 80 MW range

Water: Closed-loop liquid cooling; minimal evaporative water use by design

Capital: $500 million; Colovore (Santa Clara) on 30 acres within the Ironwood Tract industrial park

Note: Category-establishing role as Austin Supercluster's first liquid-cooled AI colo; convergence-fit on workload despite sub-100 MW scale.

KDC Project Comal Campus

Taylor, Williamson County

360 MW hyperscale campus immediately adjacent to Samsung Austin Semiconductor's Taylor fab.

Category: Data Center

Status: Approved (Taylor City Council unanimous, March 26, 2026); 3.5-year construction phase

Energy: Oncor grid-connected; 360 MW full buildout across 6 data center buildings and 2 substations on 220 acres

Water: Closed-loop cooling; municipal water through Taylor

Capital: $2.5 billion estimated; KDC Real Estate Development & Investments (Dallas)

Note: First KDC entry into data center development. Fab-adjacent positioning makes this a structurally distinct convergence node within the Samsung Taylor supplier ring.

InfraKey Capital Waco Data Park

North of Lacy Lakeview, McLennan County

Hyperscale data center campus with behind-the-meter natural gas generation.

Category: Data Center

Status: Permitting; petition filed with City of Waco for ETJ release

Energy: Behind-the-meter; 1,200 MW onsite natural gas generation plant; largest BTM gas data center commitment in dataset

Water: Not disclosed

Capital: Landmark; $10B committed

Note: 500-plus acres on I-35 corridor midpoint between Austin and DFW. 30 miles from SpaceX McGregor engine test facility.

Cipher Digital Project McLennan Campus

Riesel, McLennan County (approximately 10 miles east of Waco)

500 MW grid-connected hyperscale data center on 300 acres adjacent to the Sandy Creek coal-fired power plant. One of ten sites in Cipher Digital's 4.2 GW Texas portfolio, marking the company's pivot from Bitcoin mining to HPC/AI hosting.

Category: Data Center

Status: Project code "McLennan"; option agreement signed October 2024, option exercised February 2026, site purchase in process as of mid-2026; energization targeted 2028; up to 500 MW total capacity; tenant commitments not yet disclosed

Energy: ERCOT grid-connected via application to the state power grid; proximity to high-capacity transmission lines on the Sandy Creek interconnect (the 932 MW Sandy Creek coal-fired generating station is adjacent, though Sandy Creek itself is offline through 2027 due to equipment failure); no on-site behind-the-meter generation disclosed

Water: Cipher has ruled out using City of Riesel water for cooling; brackish or recycled water treatment solutions consistent with the company's other West Texas sites; specifics not finalized pending tenant determination

Capital: Cipher Digital Inc., formerly Cipher Mining, rebranded February 2026) as developer led by CEO Tyler Page; ~300 acres acquired off Rattlesnake Road through a series of land deals beginning October 2024 with additional purchases completing February 2026; site is one node in Cipher's broader 4.2 GW portfolio across ten sites; total contracted backlog approximately $9.3 billion anchored by three hyperscale leases (15-year 300 MW AWS lease, Fluidstack, and a third unnamed hyperscaler); $810 million senior secured notes issued for the 100 MW Stingray West Texas project provides financing template

Note: Cipher Digital's mining-to-HPC pivot at scale places McLennan in the same crypto-mining-to-AI conversion category as Helios (Galaxy/CoreWeave) and IREN Sweetwater (formerly Iris Energy), though Cipher's approach differs by sourcing greenfield sites for hyperscale buildout rather than converting existing mining facilities in place. McLennan's adjacency to the Sandy Creek coal plant is structurally distinctive — the 932 MW coal-fired plant provides existing high-capacity transmission infrastructure that newbuild greenfield sites must construct from zero, similar to the brownfield-revival logic at Sandow Lakes (Alcoa heritage) and Amp Z Lufkin (paper mill heritage). Located on the I-35 corridor approximately midway between Austin and DFW, McLennan extends the corridor data center concentration anchored by CloudBurst (San Marcos), InfraKey (Waco), and the Lacy Lakeview-area projects.

Switch The Rock Campus (Dell)

Round Rock, Williamson County

Tier 5 hyperscale campus on land acquired from Dell Technologies adjacent to Dell's global headquarters. Switch's fifth Prime campus and flagship Texas asset.

Category: Data Center

Status: Operational with phased expansion; Initial 1.5 million sq ft

Energy: Oncor grid-connected; 185 MW initial Texas design class (Rock plus former Data Foundry footprint); multi-hundred MW class at full Rock buildout

Water: Closed-loop cooling

Capital: Not Disclosed; DigitalBridge / IFM Investors

Note: Dell HQ adjacency anchors the Round Rock corridor as a hybrid enterprise-and-hyperscale cluster distinct from Hutto-Taylor's pure-hyperscale concentration.

Sabey Round Rock Campus (TACC Horizon)

Round Rock, Williamson County

NSF LCCF academic supercomputing. Horizon system.

Category: Data Center

Status: Pre-Production

Energy: Grid-connected; 85 MW

Water: City of Round Rock municipal plus closed-loop cooling

Capital: Significant; $457M minimum, $647M full buildout per NSF disclosure

Texas Advanced Computing Center (TACC)

Austin, Travis County

Academic supercomputing. Frontera, Stampede3, Vista systems.

Category: Data Center

Status: Operational plus Expanding

Energy: Grid-connected; precise campus draw not publicly disclosed

Water: Austin Water municipal

Capital: Significant; $200M-plus estimated across multiple NSF system generations

AWS Creekside Campus

Cedar Creek, Bastrop County

Amazon Data Services 1,300-acre land acquisition. One of the largest single Central Texas land grabs tied to hyperscale data center development to date.

Category: Data Center

Status: Land Acquired; No construction announcement filed publicly

Energy: Capacity not yet disclosed; 1,300-acre site scale supports a 1 GW class campus at typical AWS density

Water: Not specified; closed-loop cooling assumed per AWS Texas standard

Capital: Not Disclosed; land previously held by CTX Capital Partners (Austin) for the abandoned Creekside Master Planned Community

Note: Co-located with the Bastrop Complex and the EdgeConneX Cedar Creek campus.

EdgeConneX Cedar Creek Campus

Cedar Creek, Bastrop County

Four-building hyperscale campus on 140 acres. AI and cloud computing workload designation

Category: Data Center

Status: Under Construction; AUS02 (Phase 1, 578,000 sq ft) targeted mid-2026 completion; remaining 3 buildings phased

Energy: Grid-connected; approximately 96 MW per building × 4 buildings ≈ 384 MW campus full buildout

Water: Closed-loop cooling

Capital: $1.44 billion (initial four-building campus) to $4.4 billion

Note: Tax abatement structure and four-phase commitment signal long-term hyperscale anchor for the Bastrop Complex. Adjacent recent acquisition (180-acre parcel) creates expansion runway.

CloudBurst San Marcos Data Center

San Marcos, Hays and Guadalupe Counties

Behind-the-meter natural gas AI training and inference campus. Three-phase, 400 MW each.

Category: Data Center

Status: Phase 1 Under Construction

Energy: Behind-the-meter natural gas; Energy Transfer Oasis Pipeline 10-year supply agreement; 1,200 MW full buildout

Water: Closed-loop cooling; Hays and Guadalupe County base

Capital: Landmark; $14.5B total project value per April 2026 Guadalupe County filing

Tract Caldwell Valley Technology Park

Uhland, Caldwell County

Master-planned hyperscale data center campus.

Category: Data Center

Status: Pre-Construction

Energy: Grid-connected plus onsite natural gas; 4,000 MW full buildout design

Water: Bluebonnet Electric Cooperative service territory plus Caldwell County underlying base

Capital: Not Disclosed; Tract Capital Management multi-gigawatt national portfolio framework

PowerHouse Data Centers Uhland

Caldwell County

700 MW hyperscale campus within or adjacent to Tract Caldwell Valley Technology Park

Category: Data Center

Status: Application Filed; Regulatory contestation underway over jurisdiction; County-level approval process initiated

Energy: Grid plus likely on-site augmentation per gigawatt-class regional pattern; 700 MW proposed capacity on approximately 500 acres

Water: Closed-loop cooling; jurisdiction-dependent utility servicing

Capital: Not Disclosed

Note: Embedded within the Caldwell Valley supernode (Tract 4 GW master plan) creating within-campus density layering. ETJ regulatory outcome is the gating risk on capacity delivery timeline. Adds to a Caldwell County stack now totaling 5 GW-plus across PowerHouse, Tract, Prime, and Edged Energy proposals.

Sabey SDC San Marcos Campus

San Marcos, Hays County

786-acre campus. Clusters with CloudBurst GigaCenter and the Prime Maxwell campus.

Category: Data Center

Status: Pre-Construction; San Marcos City Council negotiation phase opened June 2025

Energy: Grid-connected via Pedernales Electric Cooperative (PEC) substation on-site; MW capacity not yet publicly disclosed.

Water: Closed-loop cooling

Capital: Not Disclosed; Carson Select Investments LP land partnership selected Sabey mid-2024

Note: Sabey's Texas portfolio anchor alongside the existing 84 MW SDC Austin in Round Rock (which hosts TACC's supercomputer infrastructure). Forms a Hays-Caldwell southern counterpart to the Williamson County hyperscale concentration; co-locates with CloudBurst's 1.2 GW BTM gas

GigaCenter in what is emerging as a distinct San Marcos supernode anchored by Hays Energy Power Plant proximity.

Borderplex Project Jupiter Campus (Stargate)

Santa Teresa, Doña Ana County, New Mexico; cross-border integration with El Paso, Texas

AI training compute. OpenAI Stargate program site. OpenAI and Oracle as principal tenants.

Category: Data Center

Status: Doña Ana County industrial revenue bonds approved September 2025; pre-construction; phased development under 30-year incentive agreement

Energy: Behind-the-meter natural gas; isolated generation system; will not connect to El Paso Electric; commitment to carbon-neutral operation by 2045

Water: Closed-loop cooling system; approximately 10 million gallons one-time fill across four data centers; 7.2 million gallons annual ongoing consumption

Capital: Generational; $50B committed in first five years; $165B total over 30 years; OpenAI and Oracle as principal tenants; BorderPlex Digital Assets (Austin) as initial developer; Stack Infrastructure (Colorado) as development partner

Note: Located across the Texas–New Mexico state line in Doña Ana County. Included in the dataset as the integrating cross-border node of the El Paso–Borderplex AI Corridor (Appendix A) — the supernode's structural coherence requires Jupiter alongside the Texas-side anchors at Meta Seafox and Fort Bliss. Four data centers designed for AI training. One of announced Stargate sites alongside Shackelford County, Texas (adjacent to Stargate Abilene) and an undisclosed Midwest site.

Carlyle Group Fort Bliss Data Center

Fort Bliss, far East El Paso, El Paso County

DoD-anchored AI compute campus. First major Department of Defense AI infrastructure buildout in the Texas dataset.

Category: Data Center

Status: Conditional agreement between Fort Bliss and Carlyle Group announced March 2026; community engagement underway April 2026; initial operating capacity approximately 100 MW targeted 2027; full 3 GW by 2029

Energy: Grid-connected via El Paso Electric (115 kV onsite, 345 kV at East Bliss range); behind-the-meter generation provisioned; El Paso Natural Gas

(Kinder Morgan) interstate pipeline and ONEOK intrastate pipeline available onsite

Water: El Paso Water Utilities municipal; water use and conservation under active community review

Capital: Not Disclosed; US Army as anchor tenant; Carlyle Group as development and operating partner under ground lease; scale comparable to Generational tier at full 3 GW buildout

Note: 1,384 acres of unused Fort Bliss land. Multiple I-10 fiber providers available onsite including Cogent, Lumen, TransTelco, and Windstream. First dataset entry with DoD as anchor tenant. Part of broader DoD private-capital data center initiative including Fort Bragg, Fort Hood, and Dugway Proving Ground, Utah (CyrusOne / KKR). Supports both battlefield AI and DoD day-to-day operations.

Meta Project Seafox

Northeast El Paso, El Paso County

AI-optimized hyperscale data center campus. Meta foundation model infrastructure.

Category: Data Center

Status: Under Construction

Energy: Grid-connected; 1,000 MW by 2028

Water: El Paso Water Utilities municipal

Capital: Landmark; $10B committed

Note: Approximately 1,000 acres. Geographically distinct from the Austin corridor: El Paso is approximately six hours west of Giga Texas. Evidence that the convergence's pull extends to Texas's geographic limits.

Chevron–Engine No. 1 Microsoft Campus (Project Kilby)

Pecos County (Fort Stockton corridor); Permian Basin

Behind-the-meter gas-fired hyperscale campus co-located with Chevron natural gas operations. First named-hyperscaler anchor for a Permian Basin AI campus.

Category: Data Center

Status: FID targeted early 2026; first power 2027; 2.5 GW initial capacity, expandable to 5 GW for a single hyperscale customer

Energy: Behind-the-meter natural gas; 7 GE Vernova combined-cycle gas turbines on order; Chevron-supplied gas from in-basin Permian operations via existing gathering and processing infrastructure

Water: Permian arid context implies closed-loop or air-cooled architecture; specifics not disclosed

Capital: ~$7 billion estimated for the initial phase; partnership between Chevron Corporation, Microsoft Corporation, and energy investment firm Engine No. 1; Microsoft entered exclusivity April 2026

Note: First Chevron "power foundry" pilot of a strategy the company intends to replicate across the Southeast, Midwest, and West. Structurally significant because it vertically integrates Permian natural gas production into hyperscale electron delivery, bypassing ERCOT entirely. Microsoft's named involvement is the first time an Anthropic/Google/Meta/Microsoft-class hyperscaler has anchored a Permian campus rather than a Central Texas or DFW site. CFO Eimear Bonner's framing of the strategy at Chevron's November 2025 investor day was deliberately blunt: "We've got the gas." Future supernode candidate at the Permian cluster level rather than as a standalone site.

MSB Matrix Campus

Hopkins County (Sulphur Springs); I-30 corridor northeast of Dallas

Phased 3 GW campus combining solid-oxide fuel-cell, geothermal, and hybrid microgrid architectures. Dual-utility grid access via Oncor and TXU.

Category: Data Center

Status: Phase I (1.5 GW) targeting Q1 2026 first power; Phase II (1.2 GW geothermal) under planning on 780 adjacent acres; Phase III (1 GW) targeted 2029; total 3.7 GW at buildout

Energy: Phase I powered entirely by on-site Bloom Energy solid-oxide fuel-cell microgrid; Phase II via geothermal microgrid as part of a stated net-zero strategy; dual-utility grid access from Oncor and TXU; developer-claimed monthly rollout cadence of 100 MW

Water: Geothermal architecture implies closed-loop or hybrid cooling; specifics not publicly disclosed

Capital: MSB as developer; financing structure not publicly disclosed; comparable to Generational tier at full buildout, though execution risk is materially higher than peer-tier sites

Note: Lowest public visibility of any 3 GW-class Texas project. The Bloom fuel-cell microgrid for Phase I is technically distinctive — most behind-the-meter campuses use gas turbines or reciprocating engines rather than solid-oxide fuel cells. The geothermal Phase II further differentiates the energy substrate from the gas-dominant Texas data center model and is one of the few large-scale geothermal-anchored DC plans in the US. Sulphur Springs lacks the existing cluster gravity of Permian or Panhandle sites; current trajectory is single-anchor. Warrants verification of permit filings, financing milestones, and Bloom Energy supply commitments before committing to a supernode-class projection.

Amp Z Project Lufkin Campus

Angelina County (Lufkin); East Texas Piney Woods

2.1 GW master-planned campus on the former Southland Paper Mill site. First major data center investment in East Texas, leveraging legacy industrial-scale interconnection.

Category: Data Center

Status: Announced 2025; 55 MW grid interconnect operational today; 175 MW grid scale-up targeted late 2026; 1.1 GW grid expansion by 2028; 1 GW behind-the-meter self-generation by 2029; total potential 2.1 GW; site expandable from 1,041 acres to ~4,000 acres

Energy: Hybrid utility + on-site self-generation model; behind-the-meter gas-fired generation planned for the 2029 self-gen tier; specific generation technology not yet disclosed

Water: Leverages former paper-mill water infrastructure (legacy 20 MGD water rights from mill operations); developer-stated facility use ~500 GPD; closed-loop cooling design implied

Capital: Approximately $1 billion project; developer Amp Z (Amp Energy); Verdant Industries as ground-lease counterparty (acquired the mill site at auction for $11 million in 2010); positioned by Amp Z as an "AI-era digital-infrastructure platform"

Note: First major data center investment in the East Texas Piney Woods. The brownfield play is structurally distinctive — the Southland Paper Mill operated from the late 1930s until 2007 and left behind grid interconnect, water infrastructure, and rail access that greenfield sites must construct from zero. Amp Z's broader pipeline (three sites at 3.8 GW combined, plus diligence on four additional sites representing 4.7 GW) suggests Lufkin is one node in a portfolio strategy rather than a standalone bet. Future supernode candidate at the East Texas regional level only if Amp Z's pipeline materializes in the same geographic corridor.

LandBridge–PowerBridge Alpha Digital Campus

Reeves County (near Pecos, Texas); Permian Basin near the Waha natural gas hub

2 GW behind-the-meter initial-phase campus on LandBridge surface acreage, co-located with the Waha hub for direct gas supply.

Category: Data Center

Status: Option-to-lease agreement signed April 2026 for up to ~3,400 acres; first power targeted 2027; Generation Interconnection Request (GINR) filed; long-lead-time power equipment ordered; 2 GW initial co-located generation capacity

Energy: Behind-the-meter gas-fired generation via private use network architecture; direct Waha hub gas supply via LandBridge surface rights; PowerBridge power partners and specific turbine OEMs not yet disclosed

Water: Permian arid context implies closed-loop cooling; specifics not disclosed

Capital: PowerBridge LLC as developer (Five Point Infrastructure portfolio company, backed by $1 billion equity commitment); LandBridge Company LLC as surface-rights lessor; PowerBridge CEO Alex Hernandez was previously founder and CEO of Cumulus Data and CEO of Talen Energy Corp

Note: LandBridge's surface-rights business model — royalty income from Permian oil and gas tenants, now extending into data center ground leases — is itself structurally distinctive. Each gigawatt of data center capacity on LandBridge acreage generates $10–20 million annually in recurring royalty income for LandBridge. Hernandez's Cumulus Data was the first multi-gigawatt data center campus co-located with wholesale power generation in the US, giving PowerBridge the most credible operational pedigree of any Permian newcomer. Combined with adjacent Chevron–Microsoft–Engine No. 1 (Pecos County) and GW Ranch (Pecos County), Alpha Digital anchors the Reeves-Pecos border as a candidate Permian supernode at the 12–15 GW combined-capacity range.

Provident–PowerHouse Grand Prairie Campus

Dallas County (Grand Prairie); DFW Corridor adjacent to Google Midlothian

2 GW campus densifying the established DFW data center cluster. One of the most aggressive first-power timelines published for a 2 GW-class US site.

Category: Data Center

Status: 768-acre campus; first power delivery targeted May 2026; phased buildout to 2 GW

Energy: Grid-anchored with on-site supplemental generation planned; leverages Google Midlothian transmission corridor adjacency; specifics not publicly disclosed

Water: Closed-loop cooling architecture implied; specifics not publicly disclosed

Capital: Joint development by Provident Realty Advisors and PowerHouse Data Centers; full capital stack not publicly disclosed

Note: Adjacency to Google's Midlothian–Red Oak Ellis County complex is the structural signal. The DFW Corridor already functions as a generational supernode — Google operational + Meta hyperscale interest + multiple colocation operators including Compass, Aligned, and Digital Realty.

Provident–PowerHouse adds 2 GW directly into that gravity well rather than emerging as a separate cluster. The May 2026 first-power target is one of the fastest published timelines for a 2 GW-class campus in the US, which signals either an aggressive disclosure posture or a substantially de-risked permitting and interconnect position inherited from the Midlothian-adjacent transmission infrastructure. Densifies an existing supernode rather than creating a new one.

Intersect Google Meitner Energy Center Campus

Gray County (Pampa); Texas Panhandle

$3 billion multi-building Google campus co-located with the Meitner Energy Center. 840 MW initial data center capacity backed by over 1 GW of on-site wind, solar, and battery storage developed by Intersect Power.

Category: Data Center

Status: Announced November 2025 as part of Google's $40 billion Texas commitment; 840 MW initial capacity; multi-building campus; commissioning timeline not publicly disclosed

Energy: 1 GW+ on-site wind, solar, and battery storage developed and operated by Intersect Power under a "power-first" co-location model that ensures the data center comes online alongside generation infrastructure rather than waiting on grid expansion; co-located generation reduces transmission requirements and grid burden

Water: Air-cooling architecture consistent with Google's published Panhandle water-scarcity strategy; water consumption limited to critical campus operations (kitchens, sanitary)

Capital: $3 billion campus; Google as anchor tenant; Intersect Power as energy partner; structure mirrors Google's broader Texas commitment and parallels its Wilbarger County, Armstrong County, Haskell County, and Roberts County announcements

Note: Pampa position places this in the Panhandle Triangle alongside Fermi Project Matador (Carson County, 11 GW planned) and Google's Armstrong, Haskell, and Roberts County sites. The Panhandle is consolidating into an emerging supernode anchored by Google's distributed footprint and Fermi's centralized one — a different cluster topology from Permian (single-county density) or DFW (legacy-hub densification). Power-first co-location with Intersect Power is the architectural template Google is using across its Texas Panhandle expansion: data center and generation as a single integrated investment rather than separate procurements. Future supernode candidate at the Panhandle cluster level rather than the single-site level.

Hut 8 Beacon Point Campus

Nueces County, near Corpus Christi

AI infrastructure. Hyperscale AI campus designed for large-scale GPU deployments.

Category: Data Center

Status: Under development; announced May 2026

Energy: Grid-connected; planned 1 GW campus; utility-scale AI infrastructure with liquid-cooling support for high-density AI workloads

Water: Not publicly disclosed; cooling system details have not yet been released

Capital: Approximately $10 billion long-term lease value

Note: Beacon Point is Hut 8's flagship AI campus, distinct from the operational Vega Bitcoin mining campus. The project is backed by a long-term AI infrastructure lease and represents Hut 8's strategic transition from cryptocurrency mining toward hyperscale AI infrastructure. It forms part of the emerging Corpus Christi–South Texas AI corridor, complementing nearby energy-first developments such as Energy Ranch and Data City..

Soluna Curtailment Platform

The four Soluna projects below represent the convertible-compute pattern: facilities currently operating primarily as Bitcoin mining infrastructure with AI workload capacity provisioned as secondary or future capability. They are included as convergence-adjacent rather than convergence-primary — their structural value to the dataset is the conversion pathway and the curtailment-driven batchable-compute architecture they pioneer (excess solar or stranded wind generation converted into computable revenue). Workload mix is noted explicitly for each entry. Conversion to AI-primary workloads is a watch-worthy pattern across this category.

Project Fei (Soluna)

Northern Texas, precise location not disclosed

Solar co-located batchable compute. AI and Bitcoin mining.

Category: Data Center

Status: Announced August 2025

Energy: Behind-the-meter solar; 240 MW solar farm co-location; 100 MW compute capacity

Water: Minimal; solar BTM, no evaporative cooling disclosed

Capital: Not Disclosed

Note: Primary workload is Bitcoin mining; AI workloads secondary. Curtailment model converts excess solar generation into batchable compute revenue.

Project Gladys (Soluna)

Nueces County

Wind co-located batchable compute. AI and HPC workloads.

Category: Data Center

Status: Planned

Energy: Behind-the-meter wind; 150 MW compute capacity

Water: Minimal; wind BTM, no evaporative cooling disclosed

Capital: Not Disclosed

Note: Primary workload is Bitcoin mining; AI and HPC secondary. Part of Soluna's 350 MW Texas expansion.

Project Rosa (Soluna)

Rio Grande Valley, precise location not disclosed

Wind co-located batchable compute. Stranded wind energy monetization.

Category: Data Center

Status: In Development

Energy: Behind-the-meter wind; 187 MW compute capacity from stranded wind

Water: Minimal; wind BTM

Capital: Not Disclosed

Note: Converts curtailed wind energy that cannot be economically transmitted to load centers into compute revenue.

Project Annie (Soluna)

Northeast Texas, precise location not disclosed

Solar co-located batchable compute.

Category: Data Center

Status: In Development

Energy: Behind-the-meter solar; 114 MW solar farm co-location; 75 MW compute capacity

Water: Minimal; solar BTM

Capital: Not Disclosed

Note: Primary workload is Bitcoin mining; AI secondary.

Semiconductor Fabs

The fab category represents the silicon layer of the book's argument. Disclosed capital exceeds $140 billion at minimum committed figures. The two-node Terafab architecture comprises the Research Fab co-located inside the Giga Texas campus (Travis County) and the Production Facility at Gibbons Creek (Grimes County). The Bastrop semiconductor advanced packaging facility is a separate SpaceX program for Starlink kit hardware, categorized under Aerospace alongside Project Echo Solar and Starlink Factory Bastrop.

Samsung Taylor

Taylor, Williamson County

Advanced node logic semiconductors. First-gen 4nm to 2nm process at Fab 1; SF2P at Fab 2.

Category: Fab

Status: Under Construction; limited operations underway early 2026

Energy: Grid-connected; precise draw not disclosed; leading-edge fab load estimated at several hundred MW sustained

Water: Carrizo-Wilcox Aquifer, Milam County; confined aquifer selected for semiconductor-grade purity; sourcing decision separates fab water demand from Williamson County BRA municipal systems

Capital: Generational; $44B committed; largest foreign direct investment in Texas history

Supplier Ring: Anchor facility; Linde Taylor onsite air separation, Tekscend Round Rock photomask, Valex Round Rock ultra-high-purity components, plus broader WFE and chemical supply chain.

Samsung Austin

Austin, Travis County (Parmer Lane semiconductor corridor in north Austin)

Foundry logic and memory semiconductors ranging from 65nm and 28nm planar transistor processes through 14nm 3D FinFET. Samsung's first US manufacturing operation and the largest single fab in the United States by floor area until Samsung Taylor surpasses it.

Category: Fab

Status: Operational; First Fab (200mm/8-inch) opened 1996–1997 originally for DRAM, continuing in a mix of memory production; Main Fab (300mm/12-inch) opened 2007 originally for NAND flash at 50nm, transitioned to foundry operation in 2017 producing 65nm to 14nm logic and embedded processes; both fabs in current production

Energy: Grid-connected via Austin Energy; precise draw not disclosed; combined 200mm and 300mm fab load estimated at several hundred MW sustained across the two-fab campus

Water: Austin Water municipal supply; the Austin fabs are among Austin Water's largest single customers; water sourcing pressure is the structural reason Samsung selected the Carrizo-Wilcox Aquifer in Milam County for the Taylor expansion rather than expanding Austin draw

Capital: Generational; $18+ billion total cumulative investment since 1996, originally the largest foreign direct investment in Texas history (now surpassed by the $44 billion Taylor commitment); 600+ acres in land holdings; 2.45 million square feet of floor space; 4,500+ Samsung Austin Semiconductor employees across the combined Austin-Taylor campus footprint; combined Austin and Taylor operations contributed $19.8 billion in 2024 economic impact to Central Texas

Supplier Ring: Established 30-year supplier ecosystem anchoring the semiconductor corridor that Samsung Taylor now inherits and extends; foundry customer history includes Qualcomm, NVIDIA, and Tesla (Tesla's first Autopilot computer was fabricated at Samsung Austin); Linde, Air Liquide, Praxair industrial gases; photomask, ultra-high-purity components, WFE servicing, and chemical supply chain infrastructure; the Austin supplier base provides the template and trained labor pool transitioning to Samsung Taylor as the new fab ramps.

TI Sherman (SM1)

Sherman, Grayson County

300mm analog and mixed-signal semiconductors.

Category: Fab

Status: Phased Operational

Energy: Grid-connected; precise draw not disclosed; 300mm analog fab load substantial

Water: City of Sherman municipal plus groundwater

Capital: Generational; $30B full program across SM1 through SM4 phases disclosed

TI Richardson (RFAB1 + RFAB2)

Richardson, Dallas County (north Dallas Telecom Corridor / US 75 spine)

300mm analog and embedded processing semiconductors. RFAB1 opened 2009 as the world's first 300mm analog wafer fab; RFAB2 connected to RFAB1, started production September 2022 and is 30 percent larger.

Category: Fab

Status: Phased Operational; RFAB1 in production since 2009; RFAB2 in production since September 2022 and continuing to ramp; combined ~630,000 sq ft of cleanroom space across the two connected fabs; 15 miles of automated overhead delivery systems linking RFAB1 and RFAB2; combined design capacity exceeds 100 million analog chips per day at full production

Energy: Grid-connected via Oncor; precise draw not disclosed; combined 300mm analog fab load substantial across the two connected facilities; LEED Gold-certified at both RFAB1 (the world's first LEED Gold-certified analog semiconductor manufacturing facility) and RFAB2

Water: City of Richardson municipal supply; sustainability commitments via dual LEED Gold certification include water reclaim and conservation; specifics not disclosed

Capital: Generational; part of TI's $60+ billion total US manufacturing investment program; TI's North Texas analog and embedded processing manufacturing anchor, complemented by DMOS6 (Dallas) and the four-fab Sherman buildout (SM1 through SM4, $30 billion); customers include Apple, Ford, Medtronic, Nvidia, and SpaceX

Substrate Inc. / Project Factory One

Bryan, Brazos County

Commercial semiconductor manufacturing. First-of-its-kind US independent fab.

Category: Fab

Status: Approved plus Pre-Construction

Energy: Grid-connected; power requirements not disclosed

Water: Brazos County water groundwork plus groundwater

Capital: Generational; $10B minimum committed, $108B full buildout announced

Note: Operator semiconductor manufacturing track record is limited; full buildout figure should be treated as announced rather than verified.

NXP Austin

Austin, Travis County (Oak Hill headquarters at 6501 William Cannon Blvd in southwest Austin; Ed Bluestein manufacturing campus in north Austin)

Legacy 200mm automotive microcontroller, RF, sensor, and power-management semiconductors. The largest legacy automotive-analog fab operation in Texas, on a multi-year wind-down trajectory as NXP transitions globally to 300mm production at lower-cost geographies.

Category: Fab

Status: Operational with declining trajectory; two-campus footprint producing automotive MCUs, microprocessors, power-management ICs, RF transceivers, and sensors on 200mm wafers; ~2,700 employees across both campuses; Oak Hill headquarters listed for sale December 2025; Ed Bluestein manufacturing campus continues operations on rumored 2028 shutdown timeline (industry chatter, unconfirmed by NXP); headquarters relocating to 225,000 sq ft at Champion Office Park in Northwest Austin (lease signed February 2026, full interior renovation beginning June 2026, move-in by March 2027) — approximately 15 percent of the Oak Hill campus footprint; new CEO Rafael Sotomayor (San Jose-based) took the helm October 28, 2025 succeeding Kurt Sievers

Energy: Grid-connected via Austin Energy across both campuses; precise draw not disclosed; 200mm legacy fab load substantially lower than 300mm advanced-node loads at Samsung Austin or TI Richardson on a per-wafer-square-inch basis

Water: Austin Water municipal supply across both campuses; ultrapure water consumption tracking the declining production trajectory; specifics not disclosed

Capital: NXP Semiconductors N.V., Eindhoven-headquartered; Austin operations inherited from the Motorola Semiconductor / Freescale Semiconductor lineage that anchored Austin's original 1970s-1980s semiconductor presence (Motorola Semiconductor spun out as Freescale Semiconductor in 2004; Freescale acquired by NXP in 2015); $290.8 million; no confirmed CHIPS Act award for Austin operations; NXP now pursuing smaller-scale Texas Semiconductor Innovation Fund (TSIF) funding instead; broader corporate program announced to close four 200mm fabs globally as part of a 10-year transition to 300mm production

Note: NXP Austin is the legacy counterpart to TI Sherman's $30 billion four-fab expansion — the same automotive-analog and embedded-processing product category, one contracting (NXP, 200mm, Austin) and one expanding (TI, 300mm, Sherman), in the same state on overlapping timelines. NXP's withdrawal from the $290.8M Chapter 380 deal and decision to forgo CHIPS Act pursuit signal a corporate-level decision to consolidate long-term automotive-analog manufacturing capacity at lower-cost geographies (industry chatter cites Malaysia product-line transfers in 2026) and at NXP's existing 300mm operations in Europe, rather than reinvest in Austin's 200mm legacy facilities. The Ed Bluestein campus carries the "ghost of Motorola" lineage that built Austin's original semiconductor industry — the same Motorola facility that trained the engineering workforce now staffing Samsung Austin, Samsung Taylor, Apple, AMD, and Tesla's chip-design operations. The structural contrast with TI Sherman is the analytical signal: regional automotive-analog capacity is shifting from

NXP's legacy Austin presence to TI's newbuild Sherman complex, with Samsung Taylor's foundry capacity providing the advanced-node complement to that shift.

TIE NGMM Center

Austin, Travis County

Advanced packaging and heterogeneous integration prototyping. UT Austin Texas Institute for Electronics.

Category: Fab

Status: Phase 1 of development

Energy: Grid-connected; research facility load profile

Water: Austin Water municipal

Capital: Significant; $440M minimum, $1.4B full buildout disclosed; DARPA plus Texas state funded

Texas A&M Semiconductor Institute (TSI)

Bryan, Brazos County; RELLIS Campus

Semiconductor R&D and workforce development. 300mm cleanroom.

Category: Fab

Status: Under Construction

Energy: Grid-connected; research facility load profile

Water: Brazos County water base

Capital: Significant; $431M disclosed; Texas CHIPS Act funded

Terafab Production Facility (under development)

Anderson area, Gibbons Creek, Grimes County

Vertically integrated 2nm / Intel 14A semiconductor production. SpaceX / Tesla / Intel joint program. Second node of the two-node Terafab architecture. First node is Terafab Research Facility on the Tesla Giga Texas complex.

Category: Fab

Status: Pre-Permitting

Energy: Grid-connected plus brownfield legacy generation foundation; former TMPA coal plant infrastructure

Water: Gibbons Creek Reservoir; legacy water rights from former Texas Municipal Power Agency coal plant; 2,400-acre reservoir sized for 470 MW power plant cooling. Separates fab water demand from nearest municipal systems.

Capital: Generational; $55B minimum committed per May 2026 Grimes County tax abatement filing; $119B full buildout announced

Supplier Ring: Anchor facility; Intel 18A process partnership, WFE queue access across full equipment stack, Cyclotron Institute at Texas A&M for D3 radiation qualification.

Gigafactories

The gigafactory category spans electric vehicle and robotics manufacturing, battery energy storage manufacturing, solar cell manufacturing, orbital compute satellite manufacturing, and industrial megasite development. Tesla Giga Texas in Travis County is the gravitational center of the convergence; its four-program campus runs vehicle production, Cortex AI training, Optimus factory site preparation, and the Terafab Research Fab. Giga Texas annual water consumption rose 68 percent over two years to 556 million gallons, making it Austin Water's third-largest customer; the Terafab Research Fab's coming UPW demand will compound this significantly. T1 Energy G2_Austin at the Sandow Lakes / Xebec campus is the US's first large-scale domestic silicon solar cell manufacturing facility; Phase 1 at 2.1 GW exceeds the entire prior US solar cell manufacturing capacity.

Giga Texas Complex

Austin, Travis County

Four-program campus: vehicle production, Cortex AI training, Optimus factory, Terafab Research Fab.

Category: Gigafactory / Anchor Cluster

Status: Operational plus Expanding; Optimus factory and Terafab Research Fab in site preparation

Energy: Grid-connected plus onsite solar; total campus draw not disclosed.

Water: Austin Water municipal; third-largest Austin Water customer as of 2025; Terafab Research Fab will add UPW demand of 1 to 2 million gallons per day minimum; 556 million gallons annual water consumption rising 68 percent in two years.

Capital: Landmark; $10B-plus operator announced; ongoing expansion capital not separately disclosed

Supplier Ring: Anchor facility; 80-plus documented onsite and adjacent suppliers across four concurrent programs spanning vehicle, compute, robotics, and silicon supply chains.

Sandow Lakes / Xebec Manufacturing Campus

Rockdale, Milam and Lee Counties

33,000-acre former Alcoa industrial campus. Master-planned advanced manufacturing megasite.

Category: Gigafactory / Anchor Cluster

Status: Operational plus Expanding; multiple concurrent tenants

Energy: Grid-connected plus onsite natural gas; 3,900 MW full buildout; reconstructing former 1,200 MW Sandow Power Plant on brownfield footprint

Water: 15 onsite lakes plus groundwater; Alcoa legacy water rights dating to 1954 operations

Capital: Significant; $240M Xebec acquisition base; aggregate tenant capital includes T1 Energy G2_Austin $850M full buildout; additional tenant capital not consolidated.

Tesla Lithium Refinery

Robstown, Nueces County

First large-scale domestic lithium hydroxide refinery. Transitional feedstock from Canada; target domestic Smackover sourcing as EnergyX scales.

Category: Gigafactory

Status: Under Construction

Energy: Grid-connected; draw not disclosed

Water: Not disclosed

Capital: Significant; $375M minimum, $1B full buildout operator announced

Supplier Ring: Anchor facility; lithium feedstock inputs, Port of Corpus Christi logistics, chemical processing infrastructure.

T1 Energy G2_Austin Solar Factory

Rockdale, Milam County; Xebec Advanced Manufacturing Campus

Largest US silicon solar cell manufacturing facility. First fully domestic utility-scale solar supply chain.

Category: Gigafactory

Status: Under Construction; Phase 1 production targeted end 2026

Energy: Grid-connected; Alcoa legacy power infrastructure repurposed

Water: Alcoa legacy water rights; Sandow Lakes campus

Capital: Significant; $350M Phase 1 committed; $850M full buildout announced

Note: Phase 1 at 2.1 GW TOPCon cells exceeds entire existing US silicon solar cell manufacturing capacity. Supply chain: Hemlock/Corning Michigan polysilicon and wafers → G2_Austin cells → G1_Dallas modules (Wilmer, TX); first fully domestic utility-scale solar supply chain in the US.

Tesla Brookshire Megapack Factory

Brookshire, Waller County; Empire West Business Park

Grid-scale battery storage manufacturing. Megapack 3 and Megablock architecture. The energy storage half of the co-located Tesla Brookshire complex.

Category: Gigafactory

Status: Operational; Megapack 3 and Megablock production starting H2 2026 at 50 GWh annual capacity. Currently the only Megablock production facility in operation globally

Energy: Grid-connected; manufacturing facility draw not disclosed

Water: Not disclosed

Capital: Foundational; $200M estimated

Note: Co-located with the Tesla Brookshire Solar Manufacturing Complex on the same campus. The two facilities together constitute Tesla Energy's manufacturing footprint.

Tesla Brookshire Solar Complex

Brookshire, Waller County; Empire West Business Park, ~35 miles west of Houston

Fully vertically integrated terrestrial solar manufacturing. Ingot growth, wafer slicing, photovoltaic cell production, and finished panel assembly under one operation. Anchor of Tesla's 100 GW domestic solar manufacturing target.

Category: Gigafactory

Status: Under Construction; multiple simultaneous facility builds underway. Buildings 9 and 10 leased (1.65 million sq ft combined), with an additional 600,000 sq ft planned

Energy: Grid-connected; cleanroom-grade manufacturing draw not disclosed

Water: Not disclosed; solar cell manufacturing carries significant ultrapure water requirements

Capital: Significant; $250M-plus initial capital expenditure across simultaneous facility builds, separate from the $2.9B solar manufacturing equipment program reported in procurement

Supplier Ring: Anchor facility; ingot growth and crystal feedstock, wafer slicing equipment, photovoltaic cell production lines, panel assembly, and the solar manufacturing equipment supply chain.

Note: Building 9 (approximately 1 million sq ft) houses the vertically integrated manufacturing floor from ingot growth through panel assembly; Building 10 (approximately 600,000 sq ft) is warehousing and logistics. The ingot-growth stage distinguishes this facility from most US solar manufacturing, which begins from purchased wafers. Distinct from SpaceX Project Echo Solar at Bastrop, which produces radiation-hardened solar cells for orbital deployment at a different scale and for a different purpose. Distinct also from the T1 Energy G2_Austin facility, an unaffiliated domestic solar cell operation.

Aerospace

The aerospace category represents the space layer of the book's argument. The Bastrop Complex is a multi-program SpaceX industrial compound whose pull is producing a secondary data center cluster formation (Amazon 1,300 acres, EdgeConneX $1.4B campus). Blue Origin's Project Blue Hub potential siting in Hutto would place the second major commercial orbital launch operator inside the Inner Triangle's northern vertex. CesiumAstro's 270,000 sq ft west Austin manufacturing campus is the founder flywheel operating at the frontier aerospace technology layer. Defense-related facilities are omitted from this listing.

SpaceX Bastrop Compound

Bastrop, Bastrop County

Multi-program SpaceX industrial compound. Starlink terminals, Project Echo solar cells, semiconductor advanced packaging for Starlink kit hardware, The Boring Company operations, Snailbrook company town.

Category: Aerospace and Defense / Anchor Cluster

Status: Operational plus Expanding; multiple concurrent construction programs

Energy: Grid-connected; aggregate campus draw not disclosed

Water: City of Bastrop municipal plus industrial bedrock

Capital: Significant; $280M TSIF grant confirmed; $200M-plus Starlink facility estimated; total campus capital not consolidated

Supplier Ring: Anchor facility; terminal manufacturing supply chain, aerospace-grade solar cell materials, advanced packaging consumables.

Note: Multiple distinct industrial functions under one operator: Starlink terminal manufacturing, orbital-grade solar cell production, aerospace semiconductor advanced packaging, tunneling automation development, and worker housing. Densest single-operator captive vertical integration in the dataset outside Giga Texas itself.

SpaceX Starbase

Boca Chica, Cameron County

Starship development, production, and launch. Starlink satellite manufacturing.

Category: Aerospace and Defense

Status: Operational plus Expanding

Energy: Grid-connected plus onsite generation; draw not disclosed; deluge water system requires industrial-scale water management per launch

Water: Brownsville PUB municipal plus Rio Grande adjacency

Capital: Significant; $3B-plus estimated; precise capital not publicly filed

Supplier Ring: Anchor facility; 50-plus documented suppliers spanning propellant supply, ground support equipment, satellite manufacturing components, and maquiladora border manufacturing relationships.

SpaceX Gigasat Factory

Bastrop, Bastrop County; adjacent to Project Echo Solar Factory

Orbital compute satellite manufacturing.

Category: Aerospace and Defense

Status: Beginning Construction; production targeted end 2027

Energy: Grid-connected; draw not disclosed

Water: City of Bastrop municipal

Capital: Not Disclosed

Note: 1,000 acre, 11 million sq ft two-building facility. Radiation-tolerant orbital compute data center satellites (AI1) qualify against the same radiation environment as D3 chips from Terafab; qualification testing at Cyclotron Institute, Texas A&M alongside D3.

SpaceX Project Echo Solar Factory

Bastrop, Bastrop County; adjacent to Starlink Factory

Aerospace-grade solar cell manufacturing for orbital compute infrastructure. Radiation-tolerant PV panels for orbital deployment.

Category: Aerospace and Defense

Status: Under Construction; equipment installation underway; production targeted end 2026

Energy: Grid-connected; draw not disclosed

Water: City of Bastrop municipal

Capital: Not Disclosed; $2B-plus equipment acquisition announced; SpaceX describes as one of the largest solar cell manufacturing facilities in the world

Note: 1.1 million sq ft two-story facility. Radiation-tolerant solar cells qualify against the same radiation environment as D3 chips from Terafab; qualification testing at Cyclotron Institute, Texas A&M alongside D3.

Starlink Factory Bastrop

Bastrop, Bastrop County

Starlink V2 user terminals. Largest printed circuit board manufacturing facility in the United States.

Category: Aerospace and Defense

Status: Operational plus Expanding

Energy: Grid-connected; draw not disclosed

Water: City of Bastrop municipal

Capital: Significant; $200M-plus estimated

Note: Produces 15,000 terminals per day; 4.68 million annually. Raw plastic and aluminum arrive; complete Starlink kits ship. Satellites manufactured in Redmond, WA; Bastrop produces ground terminals only.

SpaceX Semiconductor Advanced Packaging Facility

Bastrop, Bastrop County

Advanced packaging for Starlink kit hardware. SpaceX captive advanced packaging program.

Category: Aerospace and Defense

Status: Under Construction

Energy: Grid-connected; draw not disclosed

Water: City of Bastrop municipal

Capital: Significant; $280M TSIF grant confirmed; $17.3M TSIF allocation disclosed

Note: Largest printed circuit board and panel-level packaging facility in North America. Co-located with Project Echo Solar. Distinct from the Terafab semiconductor program; serves Starlink kit hardware and SpaceX in-house advanced packaging applications rather than orbital AI compute silicon.

Firefly Aerospace

Cedar Park, Williamson County

Alpha rocket, Eclipse rocket, Blue Ghost lunar lander, Elytra orbital vehicle. First successful US commercial lunar landing 2025.

Category: Aerospace and Defense

Status: Operational plus Expanding

Energy: Grid-connected; facility draw not disclosed

Water: City of Cedar Park municipal

Capital: Significant; $500M-plus invested; $93.3M NASA CLPS contract.

CTSDC Spaceport and Defense Innovation Campus

Cedar Park, Williamson County

Aerospace and defense innovation cluster. Firefly, Wright One, AUMOVIO, and prospective operators.

Category: Aerospace and Defense

Status: Cluster Buildout

Energy: Grid-connected; cluster draw not disclosed

Water: City of Cedar Park municipal

Capital: Foundational; $128M-plus committed across cluster operators to date

Blue Origin Project Blue Hub (under evaluation)

Hutto, Williamson County

Aerospace manufacturing, R&D, warehouse, and logistics campus. Second major commercial orbital launch operator in Williamson County.

Category: Aerospace and Defense

Status: Site Selection Finalizing; Hutto EDC public hearing May 2026

Energy: Grid-connected; draw not disclosed

Water: City of Hutto municipal

Capital: Landmark; $650M to $1B estimated; 2,000 jobs projected

Note: 1.3 million sq ft. Blue Origin also operates Launch Site One at Corn Ranch, Van Horn (165,000 acres). Project Sunrise FCC filing March 2026: 50,000-satellite orbital compute constellation. TeraWave: 5,000-satellite connectivity network.

Energy Generation

The energy category represents the underlying base layer of the book's argument. Comanche Peak Nuclear is the anchor of the Glen Rose nuclear co-location cluster whose formation validates the energy sovereignty architecture argument at the most extreme scale in the dataset. The

Permian Basin's 8 million barrels per day of produced formation water is an emerging lithium feedstock source; Element3 Fort Worth opened the first US lithium carbonate plant from oilfield produced water in half a century. The $33 billion Strategic Transmission Expansion Plan (765-kV program) is the energy foundation's most consequential current infrastructure development, targeted for 2030 completion, spanning three lines totaling approximately 700 miles through 40-plus Texas counties.

Comanche Peak Nuclear

Glen Rose, Somervell County

Nuclear electricity generation; Units 1 and 2. Key asset in hyperscaler AI-PPA strategies. Anchor of Glen Rose nuclear co-location cluster.

Category: Energy

Status: Operational

Energy: Grid-connected nuclear generation; 2,400 MW combined capacity

Water: Squaw Creek Reservoir; purpose-built closed-loop cooling reservoir

Capital: Landmark; $11B-plus estimated original construction; current replacement value substantially higher

Note: Glen Rose nuclear co-location cluster assembling: Frasier Cole 190-acre campus, AWS multi-building campus, Comanche Circle 5 GW proposal by Sailfish Development on 2,600 acres adjacent to plant. Nuclear adjacency driving data center siting in Somervell and Hood Counties.

Hydrogen City

Duval County (Piedras Pintas Salt Dome); approximately 90 miles west of Corpus Christi; pipeline-connected to Data City in Webb County and to the Port of Corpus Christi

Salt-cavern green hydrogen production and storage hub at full-buildout target of 60 GW production capacity. Multi-customer pipeline distribution network serving Data City, Corpus Christi ammonia, Brownsville aviation fuel, and statewide hydrogen power-plant off-takers.

Category: Energy

Status: Phase 1 operational 2026 with 2 GW production capacity and two salt caverns; intermediate-capacity 2 TWh storage tier targeted later this decade, supplying Data City's gas-to-hydrogen transition; full buildout to 60 GW production, 50-plus storage caverns, and 6 TWh of total energy storage capacity as a multi-decade program

Energy: Salt-cavern hydrogen storage architecture in the Piedras Pintas Salt Dome — approximately 24,000 tons (~0.8 TWh) at current operator-disclosed capacity, scaling to 6 TWh (~180,000 tons) at the 50-cavern full buildout; green hydrogen production at full buildout of 2.5 million tons annually via electrolysis powered by 60 GW of behind-the-meter solar and wind generation, with additional renewable energy procured from ERCOT during low-price periods; salt-cavern storage buffers intermittent renewable generation against continuous customer offtake, performing for hydrogen the role that Henry Hub performs for natural gas

Water: Salt cavern brine displacement during cavern creation; electrolysis water input continuous during production; specifics not disclosed

Capital: $12 billion total project capital reported; up to $1.2 billion federal funding allocated through the HyVelocity Hub Department of Energy initiative; Energy Abundance Development Corporation (formerly Green Hydrogen International) as developer led by founder Brian Maxwell; INPEX Corporation (Japanese oil-and-gas major) as joint study partner via October 2023 agreement; Energy Estate (Australian renewable energy developer) as development assistance

Note: Targeting the largest planned green hydrogen production and storage hub in the world by completed capacity. The multi-customer pipeline distribution network is the structural signature distinguishing Hydrogen City from a captive-supply hydrogen project: an initial 75-mile pipeline to Corpus Christi supplies a 1-million-ton-per-annum green ammonia production facility and local Port-of-Corpus-Christi off-takers; planned pipeline extension westward supplies Data City in Webb County for behind-the-meter feedstock through Data City's gas-to-hydrogen transition; additional pipelines planned to the Port of Brownsville for sustainable aviation fuel production, with potential green methane rocket-fuel supply to SpaceX Starbase via hydrogen combined with CO2; multiple statewide hydrogen power-plant fuel-switching customers under negotiation.

Samson Solar Energy Center

Multi-county Northeast Texas

Solar electricity generation. Among the largest single solar installations in the United States.

Category: Energy

Status: Operational

Energy: Grid-connected solar generation; 1,310 MW capacity

Water: Minimal; utility-scale solar requires no process water

Capital: Significant; $1.6B operator announced

Roscoe / King Ranch Wind Cluster

Multi-county West and South Texas

Wind electricity generation. RWE and multiple operators.

Category: Energy

Status: Operational

Energy: Grid-connected wind generation; 782 MW combined capacity

Water: Minimal; wind generation requires no process water

Capital: Significant; $1B-plus estimated across cluster operators

Rodeo Ranch Energy Storage

Pecos, Reeves County

Standalone battery energy storage. Lithium-ion two-hour duration.

Category: Energy

Status: Operational

Energy: Grid-connected; 300 MW discharge capacity

Water: Closed-loop battery thermal management; minimal consumption

Capital: Significant; $488M disclosed

Critical Materials

The critical materials category represents the supply chain sovereignty layer. Texas now has the most complete domestic critical materials processing and refining stack of any US state: lithium brine extraction (EnergyX Hooks), lithium hydroxide refining (Tesla Robstown), virgin REE sintered magnets (MP Materials Fort Worth), and recycled REE sintered magnets (Noveon Magnetics San Marcos). MP Materials and Noveon both claim to be among the first US commercial-scale NdFeB magnet manufacturers; their approaches are distinct. MP Materials uses virgin feedstock from Mountain Pass mine; Noveon uses recycled end-of-life magnet material via EcoFlux technology. Together they address both the virgin supply and circular recovery dimensions of the rare-earth magnet sovereignty problem.

MP Materials Fort Worth Magnet Plant

Fort Worth, Tarrant County

Sintered NdFeB rare-earth magnets from virgin REE feedstock. DoD majority shareholder.

Category: Critical Materials

Status: Operational plus Expanding

Energy: Grid-connected; draw not disclosed

Water: City of Fort Worth municipal

Capital: Significant; $700M minimum, $10B full buildout operator announced

Note: Among first US commercial-scale sintered NdFeB magnet manufacturers; virgin rare-earth feedstock from MP Materials' Mountain Pass mine, California. DoD majority shareholder reflects national security significance.

Round Top Mountain Project

Sierra Blanca, Hudspeth County

Heavy rare-earth concentrates; Dy, Tb, plus broader REE basket. Virtually all 17 rare-earth elements.

Category: Critical Materials

Status: Pre-Production

Energy: Grid-connected; draw not disclosed

Water: Hudspeth County groundwater plus Pecos Valley aquifer

Capital: Significant; $350M minimum, $1.5B full buildout

Note: Texas General Land Office ownership; leased to USA Rare Earth Inc. Eleven of 35 DoD-designated critical minerals in one deposit. DoD research funding via Defense Logistics Agency.

EnergyX Project Lonestar

Hooks, Bowie County; TexAmericas Center industrial park

First US industrial-scale DLE lithium extraction from Smackover Formation brine. Demonstration plant operational March 2026.

Category: Critical Materials

Status: Demonstration Operational; commercial Phase 1 (12,500 tons/year) targeted 2028; 50,000 tons/year by 2030

Energy: Grid-connected; TexAmericas Center industrial park base

Water: Smackover Formation brine; closed-loop DLE process; dedicated rail line for lithium transport

Capital: Significant; $30M demonstration invested; $1B-plus commercial phase committed. Backed by GM Ventures, POSCO, Eni Next

Note: EnergyX controls 47,500 acres of Smackover Formation in Texas and Arkansas. GET-Lit DLE technology cuts production time from 18 months (evaporation ponds) to days. At commercial scale, could supply significant fraction of Tesla Robstown refinery feedstock from domestic sources.

Noveon Magnetics

San Marcos, Hays County

Sintered NdFeB rare-earth magnets from recycled feedstock. EcoFlux closed-loop manufacturing.

Category: Critical Materials

Status: Operational plus Expanding; scaling beyond 2,000 tons per year

Energy: Grid-connected; 132,000 sq ft manufacturing center

Water: City of San Marcos municipal

Capital: Significant; $215M Series C closed January 2026; $200M led by One Investment Management

Note: Addresses recycling sovereignty gap: recovers Nd, Pr, Dy, Tb from end-of-life EV motors, wind turbines, and defense systems. Distinct from MP Materials' virgin feedstock approach. Together the two Texas operators constitute the most complete domestic NdFeB magnet sovereignty architecture in the US.

Closing the Node Set

Reading across the entries, the patterns are visible.

A small relatively set of operators holds most of the gigawatt-scale facilities

Behind-the-meter natural gas has become the dominant energy strategy for new gigawatt-scale buildout, deployed for speed-to-power against the multi-year ERCOT interconnection queue.

The Permian gas substrate, the Panhandle wind base, and the West Texas solar geography together form a multi-resource energy region no other US territory can match. Within that pattern, a small set of facilities test alternative substrates at gigawatt scale — Fermi America's nuclear-plus-SMR architecture, hydrogen-transition strategies in the Borderplex, cross-border supernode arrangements spanning US–Mexico jurisdictions. These may become dominant patterns or remain category-of-one experiments. They are visible at gigawatt scale today.

Category interlock is the structural signature. Gigafactories within the supplier rings of fabs. Fabs within driving distance of AI training infrastructure. Launch facilities operating from the same metropolitan radius as humanoid robotics assembly. Energy substrate co-located with compute load. The facilities are not independent industries that happen to colocate in Texas — they are integrated as a single industrial program executing across multiple operators.

Geographic distribution is asymmetric. Most facilities cluster in the Triangle, with extreme concentration in the Austin Supercluster. West Texas, the Panhandle, and the Borderplex are emerging as a parallel AI-industrial geography anchored by abundant gas and renewable generation rather than by metropolitan workforce concentration. The two geographies do different work in the same program.

Appendix C — Inherited Brownfield Facilities

One of the recurring patterns this book documents is brownfield inheritance: the AI-industrial buildout repeatedly siting itself on industrial land built for a prior era's industry, and inheriting that land's physical platform. The inheritance is not only the acreage. It is the heavy-industrial zoning, the rail access, the road infrastructure, the grid interconnection, the on-site or adjacent power generation, and, most consequentially, the water rights — all of which were established for an industry that has since declined or departed. The industrial platform outlasts the industry it was built for, and the next industry finds it ready.

The pattern matters because it is the market's discovered solution to several of the binding constraints the book documents. A buildout that requires new freshwater allocations, new grid interconnection, and new heavy-industrial permitting on greenfield land moves at the speed of those permitting and construction processes. A buildout that inherits a brownfield site moves faster, because the constraints were resolved decades earlier for a different occupant. This appendix lists the principal inheritances in the Texas AI-industrial buildout, with each site's prior industrial life, the bedrock it carried forward, and its current use. The pattern is one of the clearest expressions of the book's central argument that the stage was built before the builders arrived.

Sandow Lakes Ranch

Prior industrial life: Alcoa aluminum smelter and associated coal-fired power generation. The smelter broke ground in November 1951 to supply aluminum for the Korean War effort, sited in Milam County for the region's lignite coal reserves. It was fully curtailed in 2008 and permanently closed in 2017. The roughly 31,000-acre site was sold in October 2021.

Substrate inherited: Heavy-industrial zoning and permitting; rail access; grid interconnection; the footprint of the former Sandow power plant; decades of established water rights dating to the 1950s aluminum operations; approximately 31,000 acres of contiguous industrial and pasture land with on-site lakes used originally for smelter and power-plant cooling.

Current AI-industrial use: Master-planned advanced manufacturing megasite under Xebec, with multiple concurrent tenants. The former Sandow power plant footprint is being redeveloped for behind-the-meter generation through Sandow Lakes Energy, a planned 1,200 MW natural gas facility sized to anchor the broader campus load. T1 Energy's G2_Austin solar cell factory is sited on the campus on a 100-acre lease and purchase option, representing approximately $850 million of investment and 5 GW of annual solar cell manufacturing capacity at full ramp — the aluminum smelter that closed because US electricity prices made aluminum smelting uneconomical is now becoming a solar manufacturing site. Bitdeer occupies the former Alcoa smelter buildings themselves, football-field-length structures from the 1950s retrofitted with concrete to fill the smelting-pot floor concaves, operating as one of the largest Bitcoin mining facilities in North America. Riot Platforms' Whinstone facility, on an adjacent 100-acre parcel within the broader Sandow property, was built previously undeveloped land but inherits the 700 MW Alcoa-era grid connection, dedicated substations, water supply, and fiber routes; Riot announced in 2025 its first large-scale AI data center lease with AMD at the Rockdale site, beginning the conversion of that inherited power envelope from Bitcoin to AI workload.

Amp Z / Southland Paper Mills

Prior industrial life: Southland Paper Mills, the first commercial newsprint plant in the world to use southern pine as raw fiber. Incorporated in 1938 and beginning production in 1940, the mill was sited east of Lufkin in Angelina County for the region's pine timber, water, fuel, and rail access. It closed in 2007 after roughly 67 years of continuous operation. A failed 2009 sale was followed in 2010 by an auction acquisition by Verdant Industries for $11 million, which has held the property since.

Substrate inherited: Heavy-industrial zoning and eight decades of established environmental and industrial permitting precedent; siting outside Lufkin city limits, which places land use and permitting under county rather than municipal jurisdiction and puts the site inside a designated Opportunity Zone with Foreign Trade Zone eligibility; rail access via the Angelina & Neches River Railroad line originally built to serve the mill; a 55 MW grid interconnection with dedicated on-site substation carried over from the paper-machine load, wired to the Oncor transmission network; roughly 1,041 acres under purchase option to Amp Z, expandable to approximately 4,000 acres of contiguous industrial land; location on the Neches River drainage with the industrial water infrastructure and water rights developed to support one of the most water-intensive industrial processes in the region — a dedicated water supply system built and still controlled by the City of Lufkin specifically to serve heavy industry at the site, and available for sale to a successor tenant without displacing the city's residential and commercial water base; existing fiber routes and heavy-haul road access accumulated across the mill's operating life; and a large-scale industrial building footprint whose obsolete structures are slated for removal to make way for warehouse-style data center facilities on the same pad.

Current AI-industrial use: Master-planned gigawatt-scale AI data center campus under Amp Z Champion Data Center Holdings LLC, a Denver-based subsidiary of the Toronto-headquartered Amp Energy platform capitalized in part by a $374 million Carlyle Group investment. The Project Lufkin buildout is phased against the inherited electrical envelope: the existing 55 MW paper-mill grid connection is planned to expand to 175 MW by late 2026 and 1.1 GW by 2028 through Oncor infrastructure agreements, with an additional 1 GW of behind-the-meter self-generation planned through the proposed Longleaf Power Generation Plant — a natural-gas facility sited on the property under a ten-year Chapter 381 tax abatement granted by Angelina County — bringing total campus capacity to 2.1 GW by 2029. Initial capital investment is estimated at $1.05 billion at first phase, with total project value cited by county officials at approximately $1.5 billion once on-site generation is included. The historical inversion at the site is exact: the mill was built to end Texas newspapers' dependence on Canadian newsprint through a domestic supply-chain substitution enabled by a chemistry breakthrough, and it closed as the digital transition eroded newsprint demand entirely — the site is now being rebuilt as physical infrastructure for the AI compute layer that displaced newsprint in the first place, on the same footprint, drawing on the same water rights, and stepping up from the same 55 MW grid interconnection that once ran a paper machine.

Terafab Gibbons Creek (Grimes County)

Prior industrial life: Coal-fired power plant operated by the Texas Municipal Power Agency, with an associated lignite mine and a 2,400-acre cooling reservoir sized for the plant's generation load. The plant operated from 1982 through 2018, when it was retired. Charah Solutions acquired the roughly 6,000-acre property in 2021 for demolition and remediation and subsequently sold it off in parcels.

Substrate inherited: The Gibbons Creek Reservoir and its legacy water rights, originally established for power-plant cooling, which separate the site's industrial water demand from the nearest municipal systems; power-generation interconnection infrastructure; heavy-industrial land. The coal ash at the adjacent former mine has been noted as a potential rare-earth-element feedstock.

Current AI-industrial use: The Terafab Production Facility, the second node of the Terafab semiconductor architecture and the high-volume manufacturing complement to the Research Fab pilot line at Giga Texas. SpaceX's filing with Grimes County envisions a Phase 1 capital commitment of approximately $55 billion, with the program scaling to as much as $119 billion across subsequent phases and a power envelope at full buildout exceeding 10 GW. The legacy reservoir water rights are a primary reason the site can support a semiconductor fab's water demand without competing with municipal supply.

Tesla Gigafactory Texas

Prior industrial life: A 2,100-acre Martin Marietta sand and gravel quarry along the Colorado River at the eastern edge of Austin's industrial belt. Described in the public record at the time of Tesla's acquisition as roughly 2,000 acres of extractive craters, the site had served as the source of construction aggregate for Austin's long building boom. The land had also been swampland before the quarry, requiring extensive soil stabilization in the redevelopment.

Substrate inherited: Heavy-industrial zoning established for extractive use; Colorado River water rights; proximity to the SH 130 transmission corridor; rail access along the existing freight network; industrial-grade road infrastructure. Tesla acquired the site for approximately $5 million in 2020 — a fraction of the cost of comparable greenfield agricultural land — because the prior extractive use had absorbed the permitting and entitlement costs that conventional industrial siting would have required.

Current AI-industrial use: Tesla corporate headquarters; Model Y, Cybertruck, and Roadster manufacturing; Cortex AI training cluster; 4680 battery cell production; the announced Optimus humanoid robot factory; and the Terafab Research Fab pilot line for chip-and-fab co-design experiments. The Research Fab at Giga Texas is the upstream half of the Terafab program that lands its production fab at Gibbons Creek, making the two sites a closed-loop research-to-production pair, both sited on brownfield substrate.

Comanche Peak

Prior industrial life: Not a brownfield site in the decommissioned sense, but an inheritance of a different kind: the Comanche Peak nuclear plant, an operating two-unit station, and the industrial and grid infrastructure built around it.

Substrate inherited: Proximity to 2,400 MW of operating zero-carbon baseload nuclear generation; the transmission infrastructure built to move that generation; the regulatory and industrial familiarity of a county that has hosted a nuclear plant for decades.

Current AI-industrial use: A cluster of nuclear-co-located data center campuses is taking shape, drawn to the site specifically by the baseload generation adjacency. The inheritance here is the generation asset and its grid infrastructure rather than a decommissioned industrial platform.

Permian Basin

Prior industrial life: Decades of oil and gas production infrastructure: produced-water handling systems, pipeline networks, gas-processing capacity, and the associated grid and road infrastructure.

Substrate inherited: The Permian Basin's produced-water infrastructure handles industrial cooling demand without drawing on the freshwater allocation system; the natural gas production and pipeline network is the dispatchable fuel supply for behind-the-meter generation; the oil-and-gas workforce supplies transferable skills to the energy and construction trades.

Current AI-industrial use: Behind-the-meter natural gas data center campuses; produced-water-cooled industrial siting; lithium extraction from oilfield produced water. The hydrocarbon industry's infrastructure is an underlying base the AI-industrial buildout draws on without displacing the hydrocarbon industry itself.

Crypto-era substrate inheritance

A structurally distinct class of brownfield inheritance has emerged from the Bitcoin mining buildout of 2018 through 2023. Cryptocurrency miners, drawn to Texas for stranded wind generation and aggressive interconnection siting, built out substations, transmission interconnections, behind-the-meter power agreements, and high-density industrial cooling at scales that conventional data center developers would have required a decade to permit. The Bitcoin price cycle and rising mining difficulty rendered much of that infrastructure economically marginal for its original use by 2024 — exactly when frontier AI training crossed the gigawatt threshold and began searching for industrial-power substrate it could acquire at the speed of an asset purchase rather than the speed of grid interconnection. The crypto industry, in effect, built part of the brownfield platform that AI is now inheriting. The dirt is usually not classical brownfield in the contaminated-land sense, but the electrical and operational substrate qualifies under the broader pattern this appendix documents, and the aggregate megawatt count of crypto-to-AI inheritance in Texas is large.

Galaxy Helios

Prior industrial life: Argo Blockchain Bitcoin mining facility, built out 2021 through 2022 as one of the largest crypto operations in Texas, sited specifically to interconnect with stranded wind generation in the Texas Panhandle and South Plains.

Substrate inherited: Substation infrastructure, transmission interconnection, behind-the-meter wind power agreements, and the operational shell of a working high-density compute facility. Galaxy Digital acquired the site from Argo for $65 million in 2022 after the Bitcoin price cycle made the original economics untenable.

Current AI-industrial use: Conversion to AI data center under a 15-year, approximately $15 billion contract with CoreWeave, with 800 MW initial committed capacity and 3.5 GW at full buildout. Galaxy has raised approximately $1.4 billion in dedicated debt financing plus an additional $350 million in equity to fund the conversion. Helios is the largest single illustration of crypto-to-AI substrate inheritance in North America.

Core Scientific Pecos

Prior industrial life: Bitcoin mining operation with approximately 300 MW of installed capacity, sited in the Permian Basin to leverage stranded power and proximity to gas-fired generation.

Substrate inherited: Substation, transmission interconnection, gas pipeline access, and the 300 MW of existing industrial-power envelope being repurposed for AI workloads. Core Scientific has acquired more than 200 additional acres locally for expansion, funded through approximately $3.3 billion in senior secured notes and an existing $1 billion credit facility from Morgan Stanley.

Current AI-industrial use: 1.5 GW conversion to high-density AI colocation, of which approximately 1 GW is being made available for external lease. First data hall foundation complete; operational target early 2027.

Core Scientific Denton

Prior industrial life: Core Scientific's original Bitcoin mining facility, established 2018 as one of the first commercial Bitcoin mines in Texas, sited adjacent to the Denton Energy Center at 8171 Jim Christal Road.

Substrate inherited: Substation and grid interconnection at hyperscale-relevant capacity, proximity to the Denton Energy Center's dispatchable generation, and the operational footprint of an active high-density compute facility.

Current AI-industrial use: A $6.1 billion conversion of the original Bitcoin mining operation into an AI data center supporting GPU colocation and high-performance computing workloads.

Hut 8 Vega (Texas Panhandle)

Prior industrial life: Acquired by Hut 8 as a Panhandle high-density compute site, with the wind-power interconnection and behind-the-meter substrate characteristic of the crypto-era Panhandle buildout.

Substrate inherited: Behind-the-meter wind farm interconnection plus front-of-the-meter ERCOT grid connection; designed-in direct-to-chip liquid cooling capable of 180 kW per rack — sufficient for both Bitcoin ASICs and current-generation AI GPU racks.

Current AI-industrial use: 205 MW facility at full energization, built around a proprietary direct-to-chip liquid cooling system designed specifically for the convertible Bitcoin-or-AI workload model that defines the inheritance class.

Coleto Creek

Prior industrial life: Coal-fired power plant originally commissioned 1980 with 600 MW of nameplate capacity, operating for more than four decades on coal under successive ownership from Central Power & Light, Topaz Power, Dynegy, and finally Vistra. Scheduled retirement in 2027 after sustained pressure from emissions compliance costs and natural gas economics.

Substrate inherited: The cooling water rights and reservoir; the switchyard and transmission interconnection at 600 MW; heavy-industrial zoning; the site's existing dispatchable-generation interconnection envelope. Vistra has announced a natural gas repowering at up to 600 MW to preserve the interconnection capacity through the transition.

Current AI-industrial use: The natural gas repowering, scheduled to be operational by 2027, positions the site as a candidate for data center colocation under the behind-the-meter generation pattern the Texas AI buildout has converged on. No tenant has been announced; the substrate is actively in developer evaluation.

Soluna Project Kati

Prior industrial life: Rio Grande Valley site engineered from inception around an EDF Renewables wind interconnection and a behind-the-meter agreement designed for flexible Bitcoin and high-performance computing loads. The substrate is renewable-power infrastructure rather than classical retired-industry land, but the inheritance pattern is structurally identical: the interconnection envelope and power-purchase architecture were built for an earlier workload and are now being inherited by AI.

Substrate inherited: Renewable-power interconnection envelope from the EDF Renewables, Masdar, and Spring Lane Capital partnership; design intent for dual-purpose Bitcoin and AI workload from inception. Galaxy Digital's relocation of 48 MW of proprietary Bitcoin mining from its Helios campus to Kati makes the site the anchor of a flexible-load colocation model.

Current AI-industrial use: 166 MW dual-purpose facility supporting both Bitcoin hosting and AI/HPC workloads, demonstrating the convertible-substrate pattern that underlies the broader crypto-to-AI inheritance class.

Early-stage substrate inheritance

Two corridors of decommissioned coal-plant substrate are at earlier stages of the inheritance cycle than the sites documented above. Each has the platform — the cooling water rights, the switchyard, the transmission interconnection, the heavy-industrial zoning — but the tenant has not yet landed. Both corridors are tracked here because the structural logic that delivered Sandow Lakes and Gibbons Creek to the AI-industrial buildout is identical to what will likely deliver these corridors over the next several years. The book's central argument predicts these inheritances; the appendix documents them in advance of the tenant announcements that will confirm the pattern.

Northeast Texas Coal Corridor

Prior industrial life: A cluster of large-scale lignite coal generation in East and Northeast Texas, built in the 1970s for baseload service to the rapidly growing Texas grid. Big Brown (Freestone County, 1,150 MW, retired 2018) operated for 47 years on lignite from the adjacent Turlington Mine, with Fairfield Lake as its cooling reservoir. Monticello (Titus County, three units, retired 2018) and Pirkey (Hallsville, Harrison County, SWEPCO, retired 2023) followed similar economic trajectories — built for lignite-fueled baseload, displaced by natural gas and renewables, retired ahead of original design life.

Substrate inherited: At each site, the cooling reservoir and associated water rights; the switchyard and 138 kV or 345 kV transmission interconnection; heavy-industrial zoning; remediation pathways being executed by current owners. Commercial Liability Partners (CLP) acquired both Big Brown and Monticello from Luminant in 2021 with explicit redevelopment intent and is working through demolition and site preparation.

Current AI-industrial use: No tenants have been announced as of mid-2026. The sites are the largest unallocated coal-plant brownfield substrates in Texas and structurally analogous to Gibbons Creek before SpaceX filed for Terafab. The Northeast Texas corridor is a watch-list region: tenants have not yet landed, but the inheritance logic that delivered the Milam and Grimes County corridors will, with high probability, deliver this corridor to AI-industrial use over the next several years.

Texas Panhandle Coal-to-Gas Corridor

Prior industrial life: Xcel Energy's Panhandle coal fleet. Harrington Station (Potter County, three units) supplied baseload to the Panhandle grid for decades on Powder River Basin coal. Tolk Station (Lamb County, two units) operated alongside Harrington in the same fleet, sited near the West Texas wind corridor that has dominated the region's generation mix in recent years.

Substrate inherited: Switchyards and transmission interconnection at multi-hundred MW scale; cooling water rights; proximity to the Texas Panhandle wind generation envelope; heavy-industrial zoning. Xcel has converted Harrington to natural gas and announced the same conversion for Tolk by 2028, preserving the interconnection envelope through the transition.

Current AI-industrial use: Xcel's chairman has publicly stated that the Panhandle gas conversions are intended to support both regional electrification load and the transformation of crypto mining sites into a data center complex. The corridor is structurally aligned with the Galaxy Helios and Hut 8 Vega inheritance pattern — Panhandle wind substrate plus repowered coal interconnection plus convertible crypto-to-AI demand. Specifics on AI tenancy are still surfacing; the inheritance trajectory is underway.

Summary

The brownfield inheritances above share a structural logic. In each case, an earlier industry — aluminum smelting, coal generation, hydrocarbon production, aggregate extraction, crypto mining — established the heavy-industrial platform: the zoning, the water rights, the grid interconnection, the rail, the workforce. In each case, that industry declined, departed, or shifted, leaving the platform substantially intact. And in each case the AI-industrial buildout found the platform ready and occupied it, moving faster than a greenfield siting could because the binding constraints had been resolved decades earlier for a different purpose.

Appendix D — Tesla Silicon Roadmap

Tesla silicon process nodes, foundries, deployment timelines, and geographic distribution for the chip families documented throughout the book.

HW3 Family

HW3 (2019–2023 production, still in legacy fleet) — Samsung 14nm FinFET. Fabricated at Samsung Hwaseong, South Korea. 260 mm^2 die, ~6 billion transistors. Twelve Arm Cortex-A72 CPUs across three quad-core clusters at 2.2 GHz; two custom NPUs delivering 36 TOPS each, 144 TOPS per dual-chip system. Tesla's first custom-designed FSD silicon; replaced Nvidia DRIVE PX 2 (HW2). Single-source Korea.

AI4/HW4 Family (Legacy Baseline, 2023–2027)

AI4/HW4 (2023–present) — Samsung 7nm EUV (7LPP). Fabricated at Samsung Hwaseong, South Korea. Maintains Samsung Exynos IP architecture. Twenty Arm CPU cores across five quad-core clusters at 2.35 GHz; three enhanced NPUs at ~50 TOPS each. Ten cameras in the AI4 sensor suite. Currently the volume node for every production Tesla including Cybertruck and Semi. Single-source Korea.

AI4.5 (2026 Model Y, Fremont) — Samsung 7nm-class, three-chip configuration. Never formally announced; identified through teardown of 2026 Model Y vehicles produced at Fremont. Same Hwaseong production site as AI4.

AI4+ / HW4.1 (announced for 2027 production) — Samsung 7nm with facility modifications pending. 32 GB RAM per SoC, 64 GB total system memory (double AI4 baseline). Approximately 10 percent compute and memory-bandwidth uplift over AI4. Two AI4+ chips per HW4 Plus system. Production site likely remains Hwaseong.

Geographic note: All AI4-family chips — AI4, AI4.5, AI4+ — have been or will be fabricated in Korea. Samsung Austin S2 (Parmer Lane) is a mature-node fab for image sensors, displays, and RF, and is not in the Tesla autonomy supply chain at any point.

AI5 Generation (US Production, 2027–2030)

AI5 (codename Helios) — Taped out April 15, 2026. Triple-sourced. TSMC Arizona Fab 21 on N3P (3nm) for the launch production stream; Samsung Taylor on SF2 (2nm GAA) for the second stream; Terafab Gibbons Creek (Intel 14A licensed) as the third production leg once Intel 14A reaches qualified production. Tesla reportedly the largest customer at the TSMC AZ node when high-volume production ramps. Mass production late 2026 to H1 2027.

Performance — ~2,000–2,500 TOPS, roughly 5x AI4 raw throughput, ~50x AI4 on certain inference workloads per Tesla's Q1 2026 disclosure. Power envelope ~250W after FSD-specialization.

Deployment ordering — Optimus Gen 3 first (2026, ~5–10K units initial run, ramping to ~250K in 2027). Cortex 2 training cluster second (2026–2027). Consumer vehicles third, in H1 2028. Cybercab pilot production launches on AI4, not AI5.

Geographic significance — AI5 is the first Tesla autonomy chip in history fabricated on US soil. The Samsung Taylor equipment move-in ceremony took place April 24, 2026, with qualified 2nm risk production targeted for end of Q2 2027.

AI6/AI6.5 Generation (Triple-Source, 2028–2030)

AI6 — Tape-out target December 2026 (with risk of slip). Mass production late 2027 (slipped six months from original mid-2027 target due to Samsung 2nm yield issues). Triple-sourced Samsung Taylor (2nm GAA), TSMC Arizona Fab 21 Phase 3 for AI6.5 (N2 brought to Arizona), and Terafab Gibbons Creek once online. $17B Samsung deal anchor; original 16,000 wafer starts per month, Tesla negotiating toward 40,000 wpm. Performance ~2x AI5, ~100x AI4 cumulative.

AI6.5 — Within-generation refresh. Same triple-source split, with unconfirmed reporting suggesting partial TSMC allocation may shift to Intel under sourcing-policy pressure.

Deployment — Optimus Gen 4 (2028), Cybercab refresh (2028–2029), Model 3/Y / Cybertruck / Semi refresh cohorts (2029), Cortex 3 training infrastructure (2028–2029). Broad fleet penetration 2030.

Co-design character — First Tesla silicon for which the foundries adapted node-level process parameters to Tesla's power-performance requirements rather than the chip being designed to standard PDKs. Closer to captive co-design than AI5 but still inside merchant-foundry relationship.

D3 Generation (Terafab, 2028–2032)

D3 — Sole-source Terafab Gibbons Creek on Intel 14A. Tesla operates the research and pilot line at the Giga Texas campus ($3B, several thousand wafers per month). The pilot feeds into the high-volume Gibbons Creek production facility, which SpaceX operates under Intel 14A licensing. Intel EMIB (Embedded Multi-die Interconnect Bridge) for module assembly and test. Tesla becomes Intel's first major 14A customer, anchoring the survival of Intel's foundry business.

Deployment estimate — 2028–2029 for first orbital deployment via SpaceX Starlink and successor satellite platforms. Cortex 3 terrestrial training cluster receives D3 silicon in the same window.

Sole-source rationale — D3 is the only chip in the Tesla silicon roadmap for which Terafab is the sole-source foundry. The radiation-hardened process modifications are uneconomical for any merchant foundry to develop. AI5 and AI6 are also produced at Terafab as additional capacity, but their existence at Samsung Taylor and TSMC Arizona means they are not what makes Terafab necessary. D3 is what makes Terafab necessary.

D3 Radiation hardening process modifications

Total ionizing dose tolerance — Thinner gate oxides, optimized dopant profiles, radiation-hardened shallow trench isolation. Charged particles accumulate trapped charge in transistor active layers, shifting threshold voltages and degrading oxide integrity over time. Standard commercial process nodes are not designed to survive these effects at fluence levels encountered in low Earth orbit.

Single-event effect mitigation — Hardened cell libraries with transistor sizing and layout techniques that reduce single-event upset cross-section; deep n-well guard rings isolating sensitive analog and mixed-signal circuits from ground noise and latch-up current paths; suppression of destructive single-event burnout in power devices.

Architectural redundancy – Error-correcting code on every memory array without exception; triple modular redundancy on safety-critical control paths where a single bit flip cannot be tolerated.

Process engineering investment – These modifications require changes to the foundry's baseline process that affect yield and design rules. No commercial foundry will make this investment for a single customer's orbital compute program. Captive fabrication is the only economic structure that supports the engineering work.

Geographic Distribution

Three geographies produce Tesla autonomy silicon, and a fourth is often assumed to but does not.

Korea – the legacy production base. Samsung's Hwaseong fab campus in South Korea has fabricated every Tesla autonomy chip from HW3 in 2019 through the AI4 family, and will continue producing AI4-family silicon through approximately 2030 as Tesla's legacy fleet is refreshed. HW3, AI4, AI4.5, and AI4+ are all single-source Korea. No AI4-family chip is or will be produced anywhere else. When AI4+ production ends, Korean fabrication of Tesla autonomy silicon ends with it.

The United States – the multi-node stack, first activated 2027. Three US fabs come online in sequence to produce the AI5 and AI6 generations, with each facility running a different process node and each generation triple-sourced across all three.

Samsung Taylor, Williamson County, Texas – SF2 (2nm GAA) production. Qualified 2nm risk production targets end of Q2 2027. Produces AI5, AI6, and AI6.5.

TSMC Arizona Fab 21, Phoenix – N3P (3nm) at Phase 1/2 for the AI5 launch stream, transitioning to N2 as Phase 3 brings 2nm capacity to Arizona. Produces AI5, AI6, and AI6.5. Tesla is reportedly the largest customer at the AZ node when high-volume production ramps.

Terafab Gibbons Creek, Grimes County, Texas – Intel 14A under license. Produces AI5 and AI6 as supplemental capacity, and D3 as sole-source.

A fourth US site, the Terafab research facility at Giga Texsa in Travis County, operates the pilot line at several thousand wafers per month across mixed nodes, feeding qualified designs into high-volume production at Gibbons Creek. It is a research and pilot facility, not a production-volume node in the roadmap.

D3 is Terafab only. AI5 and AI6 are produced across the three US fabs because their process modifications are within what merchant foundries will accommodate at Tesla scale. D3 is not. Its radiation-hardened Intel 14A process is fabricated at Terafab Gibbons Creek exclusively, with the Giga Texas pilot line supplying qualified designs upstream. No other fab on Earth produces D3, and none is expected to.

Samsung Austin S2 (Parmer Lane) — not in the Tesla autonomy supply chain. Austin's S2 fab is a mature-node facility producing image sensors, displays, RF, and legacy automotive silicon. It is significant to Austin's semiconductor ecosystem accounting and to the region's total wafer starts, but it fabricates no Tesla autonomy silicon at any node in the roadmap, and it is not scheduled to.

Structural Arc

The Tesla silicon roadmap moves along two axes. Geographically, single-source Korean production through the AI4 family gives way to a US-domiciled multi-node stack beginning with AI5 in 2027. Structurally, single-source procurement of a merchant foundry's off-the-shelf process gives way to triple-sourced production with foundry co-design at AI6, and finally to captive sole-source fabrication at D3.

D3 is the fulcrum. AI5 and AI6 process modifications sit within what merchant foundries will accommodate at Tesla scale. D3's radiation-hardening modifications do not — they change yield curves and design rules across every wafer on the line, and no merchant foundry will absorb that engineering cost for one customer's orbital compute program. Captive fabrication is the only remaining path, and D3 is what makes captive fabrication necessary.

Terafab exists in Grimes County because D3 exists, and D3 exists because Starlink's evolution into an orbital compute platform requires silicon that survives low Earth orbit. What Appendix B catalogs as facility clusters, Appendix D catalogs as chip families. They are the same phenomenon at different resolutions.

Appendix E — The Musk Empire

This appendix focuses on the Musk-controlled entities because they constitute the only structurally integrated multi-layer operator in the Texas AI-industrial buildout, not because they are the only operators of significance. Other operators — Samsung, TSMC, Apple, Google, Amazon, the data center operators — participate in the buildout as single-layer entities and are covered in the body chapters at the analytical depth their role requires.

The entities documented here are frequently analyzed in isolation: Tesla as an automotive company, SpaceX as a launch services provider, The Boring Company as a tunneling contractor. Held simultaneously and mapped against the requirements of the colonization program the book's outward arc describes, they resolve differently. Each entity produces a capability the program requires — mobility, labor, intelligence, silicon, energy, transportation, connectivity, communications, subsurface construction — and the union of those capabilities across the entity set is structurally complete for the program.

Whether that completeness reflects coordination architected across two decades or emergent alignment produced by capability accretion whose match to the program became visible only in retrospect is a historical question this appendix does not attempt to resolve. Both readings are live. What can be documented — and what this appendix documents — is the capability set itself. The coordination is at the capability-set level. The intentionality is left to the historian.

Tesla

Tesla is a single publicly traded company that reports its operations in two principal segments: automotive (which encompasses vehicles, autonomy, and the robotics and silicon programs Tesla has not yet broken out as separate reportable segments) and energy generation and storage. This entry treats those two segments as sub-entries, in the same structural format as the SpaceX entry, with one distinction noted for accuracy: SpaceX's Starlink and SpaceXAI are genuine internal divisions of a privately held parent, while Tesla's automotive and energy operations are reportable segments of a single public company. The structural breakout below reflects the analytical reality that the two segments serve different layers of the Mars colonization program, not a claim that Tesla Energy is a separately incorporated entity.

Primary terrestrial function: Vertically integrated automotive, autonomy, robotics, and energy company. The automotive segment produces electric vehicles, the FSD autonomy stack, the Optimus humanoid robot, the Cortex AI training infrastructure, and the Terafab semiconductor fabrication program. The energy segment produces grid-scale battery energy storage, residential storage, solar generation hardware, and the energy management software that operates them. The two segments together make Tesla the only entity in the ecosystem with simultaneous operational programs across vehicle manufacturing, AI training infrastructure, humanoid robotics, semiconductor fabrication, and grid-scale energy storage.

Tesla Automotive, Autonomy, Robotics

Primary terrestrial function: Electric vehicle manufacturing, the FSD autonomy platform, the Optimus humanoid robot program, the Cortex AI training infrastructure, and the Terafab semiconductor fabrication program. The mobility, labor, intelligence, and silicon layers of the Tesla stack.

Mars colonization layer: Surface mobility: pressurized electric rover architecture derived from the Cybertruck platform (both human-driven and autonomous variants). Physical labor: Optimus units establishing lunar and Mars base infrastructure before human arrival, the pre-human construction workforce whose absence of life-support requirement makes the colonization mass budget workable. Intelligence: Cortex AI training infrastructure produces the models that FSD's Mars surface adaptation and Optimus's construction capability run on. Silicon: Terafab's chip roadmap produces the radiation-tolerant AI inference silicon that lunar and Mars surface compute requires across successive process generations beyond D3. Manufacturing: the machine that builds the machine, whose architecture produces the Optimus units and the vehicles the program depends on.

Texas footprint: Giga Texas campus (Austin, Travis County): vehicle production, Cortex AI training, Optimus factory site preparation, Terafab Research Fab. Terafab Production Facility (Gibbons Creek, Grimes County): site preparation. Samsung Taylor supply agreement: $16.5 billion through 2033.

Loop participation: Loop 1 anchor at Cortex. Loop 2 anchor at Giga Texas vehicle production and the FSD fleet. Loop 3 anchor at the Optimus factory. Loop 4 anchor at the D3 Terafab program. The automotive segment participates in four of the five loops directly.

Current status (early 2026): Vehicle production operational at scale. Cortex 1 operational, Cortex 2 expanding. Optimus factory in site preparation; Optimus units in active deployment at Giga Texas measured in hundreds. Terafab Research Fab and Production Facility in site preparation. Samsung Taylor supply agreement operational from second half 2027. Cybercab launch underway in Austin.

Unique capability gap filled: The mobility, labor, intelligence, and silicon layers of the colonization program under one operator, co-developed at one campus, with the manufacturing architecture that produces both the machines and the machines that build them.

Tesla Energy (segment)

Primary terrestrial function: Grid-scale and facility-scale battery energy storage, residential energy storage, solar generation hardware, and the energy management software layer that operates them. Tesla Energy is the most vertically integrated energy operation in the electrification ecosystem, spanning generation, storage, power conversion, grid-interface hardware, and dispatch software.

Mars colonization layer: Energy storage: the Megapack 3 and Megablock architecture, adapted for lunar and Mars surface microgrid management under thermal cycling and dust storm variability. The Megablock's 20 MWh per-unit capacity, -40°C to 60°C operating range, 10,000-plus cycle longevity, and 25-year lifespan are the specifications lunar South Pole and Mars surface microgrids require. Power management: Autobidder and Opticaster, the autonomous energy trading, forecasting, and dispatch software, are the layer that converts an installed storage fleet into a managed grid resource; the same dispatch intelligence that optimizes Megapack fleets against terrestrial grid conditions is the architecture that manages a Mars surface microgrid against dust-storm generation variability. Tesla Energy pairs with the SpaceX Project Echo radiation-hardened photovoltaic generation to form the complete lunar and Mars surface power stack: generation from SpaceX, storage and dispatch from Tesla Energy.

Texas footprint: Tesla Brookshire complex (Brookshire, Waller County, Empire West Business Park): two co-located Tesla Energy manufacturing facilities. The Megapack Megafactory produces Megapack 3 and Megablock from second half 2026 at 50 GWh annual capacity, currently the only Megablock production facility in operation globally. The Brookshire Solar Manufacturing Complex is a fully vertically integrated terrestrial solar fab, ingot growth through finished panel, anchoring Tesla's 100 GW domestic solar target; it is distinct from the SpaceX Project Echo orbital solar program at Bastrop. Giga Texas: behind-the-meter solar and storage operating the campus microgrid. Energy management software developed across the Austin engineering footprint.

Loop participation: Loop 5 primary expression: the grid-scale energy storage and power management layer is the energy infrastructure component of Tesla's vertical integration. Loop 1 supporting role: Tesla Energy's storage and dispatch infrastructure provides the behind-the-meter resilience layer that the compute-intensive training operations of Loop 1 depend on for power cost certainty and ride-through.

Current status (early 2026): Megapack in production. Megapack 3 and Megablock production beginning at the Brookshire Megapack Megafactory second half 2026. Autobidder and Opticaster operational across multiple grid markets. Solar Roof and the broader solar generation hardware line in production. The 100 GW terrestrial solar manufacturing program is anchored at the Brookshire Solar Manufacturing Complex, with construction underway across multiple simultaneous facility builds and the supporting equipment program in procurement.

Unique capability gap filled: The only operator producing grid-scale battery storage, the integrated grid node that replaces field-built substations, the solar generation hardware, and the autonomous dispatch software as one vertically integrated stack. Tesla Energy is the storage and dispatch half of the lunar and Mars surface power architecture and the grid-hardware backbone of the terrestrial data center buildout.

SpaceX

The current corporate structure consolidates several previously separate Musk-controlled entities. The progression: xAI was founded in March 2023 and operated as a standalone entity through early 2026. In March 2025, xAI acquired X.com (the former Twitter, which Musk had personally acquired in October 2022 and renamed in July 2023) in an all-stock transaction that valued xAI at $80 billion and X at $33 billion. In February 2026, SpaceX acquired xAI in a separate all-stock transaction valuing the combined entity at $1.25 trillion. Shortly after, xAI as a standalone was dissolved and its programs became the SpaceXAI division within SpaceX. The result, as of mid-2026: SpaceX is the parent entity, with Starlink and SpaceXAI as internal divisions alongside the core Starship and launch operation that is the parent's namesake function.

Primary terrestrial function: Vertically integrated orbital infrastructure, artificial intelligence, and connectivity company combining launch services (Falcon 9, Starship), satellite internet connectivity (Starlink), satellite manufacturing, in-orbit compute development, and frontier large language model development (SpaceXAI) under unified corporate structure. The first organization in the world to combine orbital launch capability, global satellite connectivity, in-orbit compute development, and frontier AI model development under unified ownership and capital allocation.

Mars colonization layer: Transportation: Starship as the Earth-to-Moon-to-Mars vehicle. Communications: Starlink constellation extended through Mars relay satellites in areosynchronous orbit. Orbital compute: Starlink connectivity substrate plus the proposed AI1 satellite in-orbit compute layer. Power generation: Project Echo Solar Warehouse (Bastrop) producing aerospace-grade radiation-hardened solar cells whose process variants serve orbital compute satellites, lunar South Pole rim sites, and Mars surface power generation. The PV generation pairs with Tesla Megablock storage to produce the lunar and Mars surface microgrid architecture. Intelligence: Grok frontier model architecture as the cognitive substrate for lunar and Mars colony AI management systems. Colossus AI training infrastructure as the compute that produces the intelligence layer's models. The integration accelerates the convergence between orbital infrastructure and AI capability that the book's Loop 4 analysis identified as the most consequential compounding interaction in the system.

Texas footprint: Starbase (Boca Chica): Starship production, launch operations, Starlink satellite manufacturing. McGregor: Raptor engine production and testing. Bastrop Complex: Starlink terminal manufacturing, Project Echo Solar Warehouse (aerospace-grade solar cells for orbital compute and surface power applications), SpaceX advanced semiconductor research and chip packaging for Starlink kit hardware. Austin engineering presence: substantial, spanning core SpaceX engineering teams and former xAI engineering teams now integrated under SpaceXAI division. Terafab Production Facility (Gibbons Creek, Grimes County): SpaceXAI operator of the facility producing D3 orbital compute silicon.

Loop participation: Loop 1: Colossus as AI training infrastructure (under SpaceXAI division). Loop 4: primary anchor across all stages: Starlink connectivity, orbital compute, OTA update infrastructure, D3 silicon. Loop 5: capital allocation toward vertical integration in launch, satellite manufacturing, orbital infrastructure, and frontier AI development now under unified direction.

Current status (early-to-mid 2026): Merger completed February 2026; xAI dissolved as standalone and integrated as SpaceXAI division shortly after. Falcon 9 operational at high cadence. Starlink at 10,500-plus satellites. Grok models operational. Colossus operational in Memphis with pull toward Texas strengthened by organizational consolidation. Starship development ongoing. Terafab Production Facility in site preparation. Organizational integration of legacy SpaceX and xAI teams underway.

Unique capability gap filled: The consolidated entity is the first organization in the ecosystem, and in the world, to combine orbital launch capability, global satellite connectivity, in-orbit compute development, and frontier AI model development under unified ownership and capital allocation. The merger closes the organizational gap between the intelligence layer and the orbital infrastructure layer that the book's Loop 4 analysis identified as the system's most consequential compounding interaction. Tesla's eventual consolidation into the same structure would close the remaining organizational gap between the terrestrial manufacturing and energy layers and the orbital and intelligence layers.

Starlink (SpaceX division)

Primary terrestrial function: Global satellite internet connectivity service providing broadband access to users in locations where terrestrial telecommunications infrastructure is absent or inadequate. The world's largest satellite constellation by unit count.

Mars colonization layer: Planetary communications infrastructure. Extended through Mars relay satellites in areosynchronous orbit, Starlink provides the communications layer that connects Earth-based mission control to Mars surface operations across the 3-to-22-minute one-way light travel time that Mars's orbital position imposes. The communications delay makes real-time Earth-Mars teleoperation impossible and requires Mars surface systems to operate with substantially higher on-device autonomy than Earth operations need. Starlink's architecture provides the data link that training updates, mission planning, and emergency consultation require while the autonomous systems handle the operational decisions that the communication delay cannot support in real time. Lunar deployment is more bandwidth-tractable given the 1.3-second one-way delay but still requires the relay infrastructure that direct Earth-line-of-sight cannot guarantee at the lunar South Pole.

Texas footprint: Bastrop satellite manufacturing facility: largest US satellite manufacturing operation by unit count. Boca Chica launch operations: primary launch substrate for constellation deployment and replenishment. Austin engineering presence: design and systems engineering teams integrated with the broader SpaceX Austin footprint.

Loop participation: Loop 4 primary expression: the connectivity layer that enables OTA updates to globally deployed Tesla vehicles, Optimus units, Megapack grid installations, and eventually D3 orbital compute satellites. Without Starlink's global coverage, Loop 4's OTA update architecture is geographically constrained to locations with reliable terrestrial connectivity. With it, the loops operate globally and eventually planetarily.

Current status (mid-2026): 10,500-plus satellites operational. Global coverage active. FCC filing for million-satellite constellation in review. Starshield defense variant operational for national security customers. Direct-to-cell capability expanding.

Unique capability gap filled: The global and eventually interplanetary connectivity layer that makes every deployed autonomous system — Tesla vehicles, Optimus units, lunar and Mars surface robots — reachable by the training infrastructure that improves them.

SpaceXAI (SpaceX division)

Primary terrestrial function: Frontier artificial intelligence research and product development division. Inherits the Grok large language model series, the Colossus AI training infrastructure, and the X.com social platform from the prior xAI entity that operated independently from 2023 through May 2026 and was integrated into SpaceX after the SpaceX-xAI merger that same month.

Mars colonization layer: Voice and conversational interface for human-AI collaboration at the colony. The lunar base and Mars colony operational model has three AI layers stacked vertically: D3-and-successor silicon as the hardware substrate (Terafab program), Optimus on-device perception and motion models as the autonomous behavior layer (Tesla program), and Grok as the conversational layer through which human colonists communicate with both the base AI management systems and the Optimus units performing physical tasks. The colonist physician talking to an Optimus medical assistant, the colonist farmer coordinating with Optimus units in the hydroponics facility, the colonist habitat operator querying the AI management system about life support status, all use Grok as the natural-language interface. Local deployment is mandatory: a Grok instance running on D3-successor silicon physically present at the colony bypasses the 1.3-second lunar and 3-to-22-minute Mars communication delays that would otherwise make Earth-hosted query routing operationally unworkable. The on-device intelligence requirement scales with the communication delay, and Grok's deployment at the colony rather than from Earth is what makes natural-language human-AI collaboration tractable at colony scale.

Texas footprint: Austin engineering presence: substantial, with legacy xAI teams now integrated alongside core SpaceX engineering teams. Colossus AI training infrastructure: currently in Memphis, with pull toward Texas strengthened by the merger's organizational consolidation and the broader Texas Triangle AI compute geography.

Loop participation: Loop 1: Colossus as one of the major AI training infrastructure anchors in the broader AI-industrial complex. Loop 4: the model intelligence layer that runs on the D3 silicon substrate, closing the loop between orbital infrastructure and AI capability under unified SpaceX capital allocation.

Current status (early-to-mid 2026): xAI standalone dissolved following the Feb 2026 SpaceX-xAI merger; programs operating under SpaceXAI division within SpaceX. Grok frontier model series operational. Colossus AI training infrastructure operational in Memphis. X.com platform operational. Organizational integration with core SpaceX engineering ongoing.

Unique capability gap filled: The frontier conversational AI model that serves as the natural-language interface layer between human colonists and the autonomous systems managing every critical lunar and Mars base function. Grok is the voice through which the colony talks to its AI.

Neuralink

Primary terrestrial function: Investigational medical device company developing an implantable brain-computer interface. As of 2026 the device is in early human clinical trials and is implanted in a small number of participants with specific motor and neurological impairments, principally severe paralysis from spinal cord injury or conditions such as ALS. The clinical purpose is the restoration of digital control, allowing a person who cannot use their hands to operate a computer through neural signals. Neuralink is a medical device program at the clinical-trial stage. It is not, in 2026, a general-purpose human-AI interface, and it is the most speculative entity in this appendix by a wide margin.

Mars colonization layer (speculative): This field is explicitly speculative, and more so than any other entry in this appendix. Neuralink has no colonization-relevant program, no announced deep-space roadmap, and no demonstrated capability beyond the restoration of digital control for clinically impaired participants. What can honestly be said is a long-horizon conditional: if brain-computer interface technology matures over the multi-decade timeframe that crewed deep-space operations occupy, a high-bandwidth neural interface could in principle become relevant to operators whose roles demand faster human-machine information exchange than a conversational interface allows. That is a possibility, not a program. It rests on capability that does not exist in 2026 and has no stated path to existing. Neuralink appears in this appendix because it is a Musk-controlled entity that is frequently grouped with the others, not because it has a defined role in the colonization architecture the other entries document.

Texas footprint: Austin headquarters: relocated from the Bay Area. Clinical trial operations and engineering development based in Austin. The proximity to Tesla's Giga Texas campus and the broader Austin technology cluster places Neuralink's development within the same labor market and institutional ecosystem as the other entities in the convergence.

Loop participation: None direct. Neuralink does not participate in the five loops in any current operational sense. Its inclusion in this appendix is as a Musk-controlled entity grouped with the others, not as a component of the loop architecture.

Current status (early 2026): Early human clinical trials underway. A small number of participants with severe motor impairment have been implanted with the device. Reported results demonstrate that implanted participants can control a computer cursor and basic digital interfaces through neural signals. The program remains at the investigational medical-device stage. There is no consumer product, no non-medical application, and no announced timeline for either.

Unique capability gap filled: In 2026, within its actual domain: the restoration of digital control for people with severe motor impairment, a genuine and demonstrated medical capability. In the colonization architecture this appendix documents: none currently. Any future role is conditional on technology maturation that has not occurred.

The Boring Company

Primary terrestrial function: Tunneling and underground infrastructure company developing lower-cost tunnel boring technology and deploying it for urban transit, utility infrastructure, and private transportation networks.

Mars colonization layer: Subsurface habitation infrastructure. Underground pressurized tunnels on Mars are not an engineering convenience. They are the radiation shielding solution that makes multi-year human surface residence medically acceptable. The Martian surface's thin atmosphere and absent global magnetic field expose surface inhabitants to galactic cosmic ray flux and solar energetic particle events at doses that exceed acceptable long-term exposure limits for above-surface habitation. Underground construction using Martian regolith as radiation shielding mass reduces surface dose rates to acceptable levels. The Boring Company's tunneling technology, adapted for Martian subsurface geology, is the construction tool whose Mars application was implicit in its development from the beginning. The same logic applies to the lunar South Pole where underground tunneling into crater walls or subsurface lava tubes provides radiation shielding and thermal stability that surface habitat structures cannot match.

Texas footprint: Austin: proposed tunnel projects in various stages of planning and permitting. Fort Worth: reported tunnel project discussions. The Texas regulatory environment's permissiveness toward infrastructure development makes Texas a natural expansion market. The Las Vegas Convention Center Loop, outside Texas, remains the primary current revenue generator.

Loop participation: Loop 3 indirect: the infrastructure construction capability that Optimus units will eventually operate in lunar and Mars surface tunneling operations, creating a manufacturing-to-robotics-to-manufacturing loop at the colony base construction level. Loop 4 indirect: the underground infrastructure that protects compute and communications equipment operating in the lunar and Martian subsurface from the radiation environment that above-surface deployment cannot survive long-term.

Current status (early 2026): Las Vegas Convention Center Loop operational. Multiple city projects in various planning and permitting stages. Prufrock tunnel boring machine development ongoing, targeting order-of-magnitude cost reduction in tunnel boring relative to conventional TBM operations.

Unique capability gap filled: The underground construction capability that makes both lunar and Martian surface habitation medically viable for multi-year human residence. No other entity in the ecosystem produces the physical infrastructure that radiation shielding requires at the scale that a permanent surface colony demands.

Glossary

Terms are listed alphabetically. Coined terms are concepts developed in this book that do not appear in standard references.

A

ADAS *autonomy*

Driver assistance features — lane keeping, adaptive cruise control, automatic emergency braking — that support a human driver without replacing them. Distinct from autonomous driving systems that operate without continuous human supervision.

Advanced packaging *semiconductor*

Techniques for integrating multiple chips into a single package — chiplets, 3D stacking, heterogeneous integration — improving performance and power efficiency beyond what a single monolithic chip can achieve.

AI7 *semiconductor / space*

Alternate naming for D3, see D3.

AMR (Autonomous Mobile Robot) *autonomy*

A warehouse or logistics robot that navigates its environment autonomously using sensors and mapping, without fixed tracks or infrastructure. The most commercially mature segment of the broader robotics deployment.

Anchor node *geographic*

A strategically distinctive facility whose capital scale, supply chain relationships, or loop participation makes it a center for supplier and talent concentration in its surrounding geography.

Artemis *space*

NASA's program to return humans to the lunar surface.

Autobidder *energy*

Tesla's AI-powered grid optimization software that manages Megapack and grid-scale battery storage dispatch, optimizing for cost, grid stability, and revenue generation in real-time electricity markets.

B

Behind-the-meter (BTM) *energy*

Generation and storage assets located on the customer's side of the utility meter, providing power without flowing through the grid. BTM operators eliminate wholesale price exposure, transmission charges, and grid reliability dependency for their covered loads.

BESS (Battery Energy Storage System) *energy*

A system combining battery cells, power electronics, and management software to store and dispatch electrical energy. Grid-scale BESS systems like Tesla Megapack provide frequency regulation, peak shaving, and renewable integration services.

BMI (Brain-Machine Interface) *autonomy*

A direct communication pathway between the brain and an external device. Neuralink's high-resolution cortical BMI aims to replace the keyboard-and-screen interface with a direct neural connection whose bandwidth is orders of magnitude higher.

BRA (Brazos River Authority) *geographic*

The Texas state agency managing water resources in the Brazos River basin, including Lake Georgetown, Lake Stillhouse Hollow, and Lake Belton — primary water sources for Georgetown and the northern Williamson County corridor.

Brownfield substrate inheritance *coined*

The pattern by which AI-industrial operators acquire pre-existing industrial infrastructure — water rights, power systems, permitting frameworks — from prior-generation industrial users, compressing development timelines relative to greenfield sites. Gibbons Creek Reservoir and Sandow Lakes are the clearest Texas examples.

C

Captive fab *semiconductor*

A semiconductor fabrication facility that produces primarily or exclusively for a single customer or parent company, enabling process customization not available through merchant foundry relationships.

Carrizo-Wilcox Aquifer *energy / water*

A confined deep aquifer system in Milam County providing Samsung Taylor's process water supply. Selected for its capacity, isolation from surface contamination, and consistency of supply at semiconductor-grade purity requirements.

CDP (Census-Designated Place) *geographic*

A concentration of population identified by the US Census Bureau for statistical purposes, without incorporated municipal status.

Chip-and-fab co-design *coined*

A development approach in which the chip architecture and the fabrication process are developed jointly rather than the chip designer accepting a fixed process from an external foundry. Enables performance characteristics unavailable within standard merchant foundry process design kits.

CHIPS Act *semiconductor*

The 2022 US federal legislation providing approximately $52 billion in funding for domestic semiconductor manufacturing, research, and workforce development, with the goal of reducing US dependence on foreign-operated fabs.

Cognitive amplification *coined*

The layer of the AI-industrial convergence in which LLM and AI tooling amplifies individual human productive capability — enabling solo-builders to achieve institutional-scale research, content production, and software development output.

Colossus *AI / compute*

SpaceX's AI training supercomputer, now operating as a SpaceXAI division asset following the Feb 2026 integration. Currently located in Memphis with pull toward Texas as the broader ecosystem consolidates.

Convergence, the *coined*

The process by which multiple compounding systems — silicon, compute, energy, autonomy, intelligence — interact in the same geography simultaneously, producing outcomes that none of the systems would generate in isolation. The specific phenomenon the Texas Triangle is producing. Multiplicative rather than additive effect.

Corridor *geographic*

An emergent linear concentration of industrial or residential development following infrastructure threads — power transmission, fiber, water, roads — rather than a planned designation.

Cortex *AI / compute*

Tesla's captive AI training infrastructure at Giga Texas — a hyperscaler-class compute cluster whose output is FSD capability, Optimus autonomous operation, and the AI stack running every Tesla-deployed product.

Cybercab *autonomy*

Tesla's purpose-built robotaxi — a vehicle designed from inception for autonomous operation without a steering wheel or pedals, targeting the transportation-as-a-service market at unit economics competitive with personal vehicle ownership.

Cyclotron Institute *space / semiconductor*

The radiation physics research facility at Texas A&M University in College Station, providing particle beam testing and radiation qualification for D3 and other space electronics before orbital deployment.

D

D3 *semiconductor / space*

The radiation-tolerant chip being developed in Tesla's Terafab program for SpaceX orbital compute applications. Requires process modifications — hardened cell libraries, ECC on all memory, TMR on safety-critical paths — that no commercial merchant foundry will develop for a single customer.

Deregulated electricity market *energy*

A market structure in which electricity generation and retail are open to competition rather than controlled by a regulated utility monopoly. ERCOT operates a deregulated wholesale market whose design produces structurally lower industrial electricity rates than regulated state markets.

E

ECC (Error-Correcting Code) *semiconductor*

Memory protection technology that detects and corrects bit errors caused by radiation or electrical noise. Required in radiation-hardened chips for space applications.

Edge compute / Edge inference *AI / compute*

AI inference performed locally on a device rather than in a centralized data center. Reduces latency, eliminates dependence on network connectivity, and enables autonomous operation where cloud round-trips are impractical or impossible.

Edwards Aquifer *energy / water*

A karst limestone aquifer underlying San Antonio and extending north, managed by the Edwards Aquifer Authority. Primary water source for the Austin-San Antonio megaregion's southern end.

Energy sovereignty *coined*

The condition in which an operator, household, or facility generates and stores its own power without dependence on the shared grid for primary supply. Distinguished from simple renewable energy adoption by the intent to eliminate grid dependency rather than reduce it.

ERCOT (Electric Reliability Council of Texas) *geographic / energy*

The grid operator managing approximately 90 percent of Texas's electric load. Operates in deliberate isolation from the Eastern and Western Interconnections, giving Texas control over its own power market and rates.

EUV (Extreme Ultraviolet Lithography) *semiconductor*
The lithography technology used to pattern features at leading-edge process nodes below 7nm. EUV tools are produced by ASML at approximately 50 to 60 units per year globally, making them a binding constraint in advanced fab buildouts.

F

FED (Fleet Energy Depot) *coined*
An integrated site providing fleet charging, energy storage, and often microgrid and compute capabilities for EV fleets.

Formation event *coined*
The observable, ongoing process by which a geographic concentration of industrial activity reaches self-sustaining momentum — when the system's own gravity becomes the primary determinant of its trajectory rather than external inputs.

Founder flywheel *coined*
The self-compounding mechanism by which anchor employer alumni become founders of adjacent companies, whose success recycles talent, capital, and expertise back into the ecosystem, generating the next wave of anchor employers and founders.

FSD (Full Self-Driving) *autonomy*
Tesla's autonomous driving software stack, running on the vehicle's onboard compute hardware. As of early 2026 the fleet has accumulated more than 11 billion real-world driving miles.

G

GAA (Gate-All-Around) *semiconductor*
A transistor architecture used at 3nm and below in which the gate material surrounds the channel on all four sides, improving electrostatic control and enabling continued scaling beyond FinFET limits.

GCR (Galactic Cosmic Rays) *space*
High-energy particles originating outside the solar system that penetrate spacecraft and planetary surfaces. A primary radiation hazard for

electronics and humans in space, more severe beyond Earth's magnetic field.

Gibbons Creek Reservoir *energy / water*

A 2,400-acre reservoir in Grimes County originally constructed for coal plant cooling. Its water rights flow with the property to Terafab's production facility —brownfield inheritance applied to water infrastructure.

Golden Dome *space / defense*

A US missile defense architecture targeting hypersonic threat intercept, whose compute requirements are a direct driver of the Trusted Foundry program.

H

Humanoid robot *autonomy*

A general-purpose robot with a human-like form factor designed to operate in environments built for humans. Optimus is Tesla's humanoid robot program, with the factory in site preparation as of early 2026.

Hyperscale *AI / compute*

Data center facilities operating at very large scale — typically 100 MW or more of power capacity — serving the compute requirements of major cloud and AI platform operators.

I

IDM (Integrated Device Manufacturer) *semiconductor*

A semiconductor company that both designs and manufactures its own chips. Texas Instruments is the largest US-based analog IDM.

Industrial supernode *geographic*

A geographic concentration of industrial activity and capacity whose components interact and compound each other's value, producing system-level outcomes that no single node or facility could generate in isolation.

Inference *AI / compute*

The process of running a trained AI model on new input data to produce an output. Distinct from training, which builds the model. Most deployed AI systems run inference continuously.

ISRU (In-Situ Resource Utilization) *space*

The use of local materials at a destination to produce what would otherwise need to be transported from Earth. On Mars, ISRU produces methane and oxygen propellant from atmospheric CO_2 and subsurface water ice via the Sabatier reaction.

L

LCRA (Lower Colorado River Authority) *geographic / energy*

The Texas state agency managing water and energy resources along the Colorado River basin, including water allocations affecting the Austin metro corridor.

LEO (Low Earth Orbit) *space*

Orbital altitudes from approximately 200 to 2,000 km. Starlink operates in LEO. The radiation environment in LEO is partially shielded by Earth's magnetic field — more manageable than higher orbits or planetary surfaces without magnetic fields.

LLM (Large Language Model) *AI / compute*

An AI model trained on large text datasets to generate, classify, and reason about language. The foundation of current conversational AI systems including Claude, ChatGPT, and Gemini.

M

Mass driver *space*

An electromagnetic launcher that accelerates payloads to orbital or escape velocity using electrical power rather than chemical propellant. On the lunar surface, a mass driver could launch materials to orbit at orders-of-magnitude lower cost than chemical rockets.

Megablock *energy*

Tesla's integrated grid architecture product combining Megapack battery storage with Tesla-designed transformers and inverters as a unified grid infrastructure system on a skid.

Megapack *energy*

Tesla's commercial grid-scale battery energy storage product, deployed at utility and hyperscaler data center scale for frequency regulation, peak shaving, and renewable integration.

Megapod *compute*

Tesla's trademark for a modular, self-contained data center designed to handle AI workloads and provide edge computing. Co-located at Tesla Supercharger stations to take advantage of idle and excess stored energy. Forms a distributed data center topology.

Merchant foundry *semiconductor*

A semiconductor fabrication facility that manufactures chips for multiple external customers under a standard process design kit. TSMC and Samsung Foundry are the leading merchant foundries.

Moon Base *space*

The NASA term for the permanent continuous-operations human presence at the lunar South Pole. Realistically a 2035 to 2040 program given current Artemis timelines.

MSA (Metropolitan Statistical Area) *geographic*

A geographic region defined by the US Office of Management and Budget centered on an urbanized area with strong economic ties. The Austin-Round Rock-Georgetown MSA encompasses the core Austin metro geography.

N

Node (process node) *semiconductor*

A shorthand designation for a semiconductor manufacturing generation. Lower node numbers generally indicate more advanced process technology with smaller, more power-efficient transistors.

O

ODD (Operational Design Domain) *autonomy*

The specific conditions within which an autonomous system is designed to operate. Expanding the ODD requires additional training data and validation across the new conditions.

Optimus *autonomy*

Tesla's humanoid robot, being developed and trained inside the Giga Texas manufacturing environment — the first production-scale humanoid manufacturing operation on Earth.

OTA (Over-the-Air Update) *AI / compute*

Software updates delivered wirelessly to deployed devices without physical connection. Enables continuous improvement of autonomous systems — FSD, Optimus, Megapack — after deployment.

P

PDK (Process Design Kit) *semiconductor*

The set of design rules, device models, and process parameters that a foundry provides to chip designers. Chip-and-fab co-design modifies the PDK rather than accepting it as fixed.

Q

Quasi-sovereign state *geographic*

Used in this book to describe Texas's structural distinctiveness — independent power grid, retained public lands, no state income tax, international border — that gives it operating characteristics closer to a nation-state than a typical US state.

R

Rad-hard / Rad-tolerant *semiconductor / space*

Radiation-hardened chips are designed to survive high cumulative radiation doses in space or nuclear environments. Radiation-tolerant chips provide meaningful but less extreme resistance. D3 is being developed as a rad-tolerant chip for LEO orbital compute applications.

Raptor *space*

SpaceX's full-flow staged combustion methane-oxygen engine powering both the Super Heavy booster and the Starship upper stage. Produced and tested at McGregor, McLennan County. V3 at volume production and V4 in testing as of early 2026.

Robotaxi *autonomy*

An autonomous vehicle operating as a ride-hailing service without a human driver. Cybercab is Tesla's purpose-built robotaxi platform.

S

Sabatier reaction *space*

The chemical process — carbon dioxide plus hydrogen yields methane plus water — that enables Mars surface propellant production from atmospheric CO_2 and subsurface water ice. Makes Starship a reusable interplanetary transport rather than a one-way expenditure.

SAE autonomy levels *autonomy*

The Society of Automotive Engineers' classification of vehicle automation from Level 0 through Level 5. Most deployed commercial systems operate at Level 2 to Level 4.

SEE (Single-Event Effects) *space / semiconductor*

Radiation-induced malfunctions in semiconductor devices caused by individual high-energy particles — bit flips, latch-up, destructive burnout. Distinct from cumulative TID damage.

Singularity, the *coined*

Used in this book in the geographic sense: the convergence of multiple compounding systems in a specific place at a specific moment, producing a system whose density and self-sustaining momentum is qualitatively distinct from the sum of its components. Distinct from the Kurzweil technological singularity concept.

SMR (Small Modular Reactor) *energy*

A nuclear fission reactor design with generating capacity typically below 300 MW, designed for factory fabrication and modular deployment. Fermi America's behind-the-meter campus and lunar surface power applications both target SMR technology.

Solo-builder *coined*

An individual who deploys LLM and AI tooling as a cognitive amplifier to achieve research, content production, software development, and business operation outputs that previously required teams — operating from a home office as a one-person institution.

SpaceXAI *space / AI*

The division emerged within SpaceX in Feb 2026 following the integration of xAI. Encompasses the Grok frontier model series and Colossus AI training infrastructure under SpaceX's corporate structure alongside Starlink and Starship operations.

SSO (Sun-Synchronous Orbit) *space*

A polar orbit presenting a more demanding radiation environment than equatorial LEO due to weaker magnetic shielding at polar latitudes. Golden Dome's sensor layer likely requires SSO coverage.

Starbase *space*

SpaceX's integrated Starship production, testing, and launch facility at Boca Chica, Cameron County, Texas.

Starlink *space / AI*

SpaceX's global satellite internet constellation, exceeding 10,500 satellites as of mid-2026. Provides the connectivity layer. Operating under SpaceX alongside the SpaceXAI division.

Starmind *space / AI*

SpaceX's global data center satellite compute constellation, with 1 million satellites the goal as per FCC filing. Provides the orbital AI training and inference layer. Operating under SpaceX alongside the SpaceXAI division.

Starship *space*

SpaceX's fully reusable heavy-lift launch vehicle powered by Raptor engines. The load-bearing technology for orbital compute shell population, Artemis lunar landing, Mars colonization, and the broader cislunar economy.

Supplier ring *geographic*

The ecosystem of suppliers, contractors, and service providers whose operations are organized around an anchor facility's production requirements and may be co-located onsite.

T

Teleop / Teleoperation *autonomy*

Remote human operation of a robotic or autonomous system, typically used as a fallback when the system encounters a situation outside its operational design domain.

Terafab *semiconductor*

Tesla and SpaceX's captive semiconductor fabrication program comprising two facilities: a research and pilot fab within Giga Texas and a production fab at Gibbons Creek, Grimes County, operated under a SpaceX / Tesla joint venture.

Texas Triangle *geographic*

The roughly triangular geography defined by Dallas-Fort Worth, Houston, and San Antonio, enclosing an interior that contains Austin and the highway corridor infrastructure documented in this book.

TID (Total Ionizing Dose) *semiconductor / space*

The cumulative radiation damage absorbed by a semiconductor device over its operational lifetime. A primary failure mechanism for electronics in space environments.

TMR (Triple Modular Redundancy) *semiconductor*

A fault-tolerance technique in which three identical circuit blocks process the same input and a voting circuit selects the majority output. Used in radiation-hardened applications where a single-event upset cannot be tolerated.

Training *AI / compute*

The computational process of adjusting an AI model's parameters by exposing it to large datasets. Training runs are computationally intensive, consuming sustained high-power compute for days to weeks.

Trusted Foundry Program *semiconductor / defense*

A US Department of Defense radiation-level certification framework for semiconductor fabrication facilities approved as sources for defense and intelligence community silicon, ensuring supply chain integrity and process security.

TSIF (Texas Semiconductor Innovation Fund) *semiconductor*

A Texas state program providing targeted capital allocations to semiconductor fabrication, research, and supplier facilities in Texas, designed to complement the federal CHIPS Act program.

V

Von Neumann probe *space*

A theoretical self-replicating spacecraft that arrives at a destination, builds copies of itself from local materials, and dispatches the copies onward. The technology concept that makes interstellar civilization expansion geometrically rather than linearly scalable.

W

WFE (Wafer Fabrication Equipment) *semiconductor*

The specialized machinery used in semiconductor manufacturing. WFE supply chains are highly concentrated, with long lead times and allocation queues that new fab entrants must navigate.

Notes

These notes attach sources to specific load-bearing claims — recent developments with dated filings, contested figures, named institutional findings, and claims a knowledgeable reader may want to verify. Encyclopedic facts, common institutional knowledge, and general trend statements are not separately cited; readers seeking them can find them in one search. The notes are a credibility instrument for the analytical claims, not a fact-check trail for the encyclopedic ones.

Chapter 1: The Quasi-Sovereign State

1. The Six Seals (Capitol height and granite) — On the Texas State Capitol's 1888 completion, its standing as the largest state capitol in the United States by square footage, its 14.64-foot height advantage over the US Capitol, and the use of Sunset Red granite quarried from Granite Mountain in Burnet County and delivered to Austin on approximately 15,700 railroad carloads (via roughly 4,000 train trips) between 1886 and 1887: Texas State Preservation Board, "Capitol History, 1839–1888" and "Capitol Myths and Legends"; Texas Facilities Commission, "Building on Tradition: The Texas Capitol Complex Features Local Granite"; Texas Historical Commission Atlas marker for the Texas State Capitol (Atlas Number 5507014150).

2. Energy (ERCOT structural isolation) — On ERCOT's structural isolation from the Eastern and Western Interconnections and the resulting absence of FERC jurisdiction over Texas wholesale power markets: Federal Energy Regulatory Commission jurisdictional explainer; ERCOT corporate overview materials; Public Utility Commission of Texas regulatory framework documentation.

3. Land (federal ownership percentages) — On federal land ownership percentages by state — Nevada 80.1 percent, Utah 63.1 percent, California 45.9 percent, Texas approximately 1.8 percent: Congressional Research Service, Carol Hardy Vincent et al., *Federal Land Ownership: Overview and Data*, CRS Report R42346 (February 21, 2020 update).

4. Financial Posture (PUF and PSF) — On the Permanent University Fund and Permanent School Fund as Texas sovereign wealth instruments, the PUF's 2.1 million-acre land base derived from the 1839 Republic of Texas land grant and the 1876 Texas Constitution expansion, and the funds' combined assets exceeding $90 billion as of FY2024: Texas Permanent School Fund Corporation, FY2024 Annual Comprehensive Financial Report ($57.3 billion net position); Texas Legislative Budget Board, *Available University Fund Summary of Budget Recommendations*, 89th Legislature ($36.5 billion PUF corpus value as of August 31, 2024); University of Texas System, "Permanent University Fund Overview."

5. Geography (Port of Houston) — On the Port of Houston's #1 ranking among US ports in foreign waterborne tonnage (220.1 million short tons in 2024), a position held continuously since 2001: Port Houston, "Port Statistics"; US Army Corps of Engineers Navigation and Civil Works Decision Support Center; American Association of Port Authorities.

6. Regulations (county-line arbitrage) — On the structural pattern of operators siting facilities in adjacent counties after primary counties imposed restrictions, as documented across data center, semiconductor, and industrial buildouts: Community Impact regional reporting; Texas Comptroller economic development tracking; Texas county-by-county zoning and ordinance records.

7. The Accumulated Stage (brownfield substrate) — On the brownfield substrate inheritance pattern as a structural advantage of Texas industrial development — water rights, power infrastructure, and permitting frameworks inherited from prior-generation operators that compress new-facility timelines from years to months: This pattern is developed in detail in Chapter 2 (Gibbons Creek, Sandow Lakes) and Chapter 3 (the geological recursion); primary sources are cited in those chapters.

Chapter 2: The Energy Substrate

8. Opening (load profiles) — On the continuous baseline draw of hyperscale data center campuses (500 MW to 1 GW), leading-edge semiconductor fabs (200–600 MW), and Giga Texas (350–500 MW): Lawrence Berkeley National Laboratory, "United States Data Center Energy Usage Report" (December 2024); SEMI fabrication facility energy disclosures; Tesla load filings with Austin Water and ERCOT.

9. Opening (ERCOT 218 GW projection) — On ERCOT's April 2025 long-term load forecast projecting peak demand up to 218 GW by 2031, with an adjusted (planning-basis) forecast of 145 GW, both materially above the August 2023 demand record of 85.5 GW: Electric Reliability Council of Texas, Long-Term Load Forecast (April 2025); Data Center Dynamics, "ERCOT: Electricity demand in Texas set to soar to 218GW by 2031 fueled by data center growth"; Spectrum News, "Texas could see energy demand double by 2031."

10. The ERCOT Advantage (Senate Bill 6) — On Texas Senate Bill 6 of the 89th Legislature (signed June 2025) restructuring large-load interconnection standards, co-location rules, and grid-emergency curtailment authority for behind-the-meter loads above specified thresholds: Texas Legislature Online, S.B. 6 (89R) Bill Analysis and Enrolled Text; Baker Botts, "Texas Senate Bill 6: Understanding the Impacts to Large Loads and Co-located Generation" (July 2025); K&L Gates, "Senate Bill 6: A Texas Bill Impacting Large Load Development in ERCOT."

11. The ERCOT Advantage (1999 Senate Bill 7) — On Senate Bill 7 of the 76th Texas Legislature (1999) restructuring the retail electricity market and establishing the deregulated competitive structure under which ERCOT now operates: Texas Legislature Online, S.B. 7 (76R) Enrolled Text; Public Utility Commission of Texas regulatory history materials.

12. Energy Sovereignty (February 2021 Uri) — On Winter Storm Uri of February 2021 and the simultaneous loss of approximately one-third of ERCOT's installed generating capacity at the peak of the event, with the resulting argument that the structural advantage of grid isolation and the structural vulnerability are the same feature: ERCOT, "Preliminary Report on Causes of Generator Outages and Derates During the February 2021 Extreme Cold Weather Event"; FERC-NERC Inquiry into Bulk-Power System Operations During the February 2021 Cold Weather Event (November 2021).

13. The Electric Expressway — On the approximately $33 billion Strategic Transmission Expansion Plan, including three new 765-kV import paths into the Permian Basin (approved by PUCT on April 24, 2025) and the broader 765-kV core build-out of approximately 2,468 miles of new line: ERCOT, "2024 Regional Transmission Plan (RTP): 345-kV Plan and Texas 765-kV Strategic Transmission Expansion Plan"; Public Utility Commission of Texas Project No. 55718 (Permian Basin Reliability Plan); Texas Scorecard, "Permian Basin Power Fix Becomes $33 Billion Statewide Project."

14. The Brownfield Inheritance (Gibbons Creek / Terafab) — On Gibbons Creek Reservoir's coal-plant water rights flowing with the property to SpaceX Terafab, and on the broader pattern in which prior-generation industrial infrastructure compresses new-facility permitting timelines: Texas Commission on Environmental Quality water rights records for the former Gibbons Creek Steam Electric Station (Texas Municipal Power Agency); Texas Water Development Board reservoir capacity records.

15. The Brownfield Inheritance (Sandow Lakes / Alcoa) — On Sandow Lakes carrying 70 years of Alcoa water rights into the Xebec campus, and on the conversion of the former Rockdale aluminum smelter site into solar cell manufacturing capacity drawing on the same electrical and water infrastructure: Texas Commission on Environmental Quality water rights records for the Sandow Lakes Ranch; Xebec and T1 Energy corporate materials documenting the site's transition.

Chapter 3: The Critical Materials Substrate

16. The China Dependency — On China's approximate share of the global rare-earth supply chain — roughly 60 percent of mining, ~90 percent of separation and processing, and 90 percent or more of permanent magnet manufacturing — and on China's broader processing dominance in lithium, silicon metal, and cobalt: International Energy Agency, "With New Export Controls on Critical Minerals, Supply Concentration Risks Become Reality" (October 2025); Center for Strategic and International Studies, Gracelin Baskaran, "China's New Rare-earth and Magnet Restrictions Threaten US Defense Supply Chains" (2025); USGS Mineral Commodity Summaries 2025.

17. The China Dependency (export controls) — On China's 2010 rare-earth export controls and the global supply chain crisis that produced temporary price spikes of more than 1,000 percent for some elements, and on the 2025 export control announcements targeting lithium battery components, gallium, germanium, and rare-earth processing chemicals as the more recent expression of the same instrument: Council on Foreign Relations, "China's Rare-earth Stranglehold" (historical analysis); China Ministry of Commerce export control announcements (April 2025 and October 2025); Resources for the Future, "The Strategic Game of Rare-earths" (October 2025).

18. The Smackover Formation — On Smackover Formation brine lithium concentrations of 200 to 800 parts per million in the Texas and Arkansas portions, and on the USGS Arkansas resource estimate of 5.1 to 19 million tons of lithium (representing 35 to 136 percent of the current US lithium resource estimate, sufficient to exceed projected global demand by 2030): Knierim, Phelps, Brennwald, et al., "Evaluation of the lithium resource in the Smackover Formation brines of southern Arkansas using machine learning," *Science Advances* 10, eadp8149 (October 2024); US Geological Survey Fact Sheet 2024-3052, "Lithium Resource in the Smackover Formation Brines of Southern Arkansas."

19. EnergyX's Project Lonestar — On EnergyX's Direct Lithium Extraction demonstration plant at Hooks near Texarkana, its targeted 50,000 tons-per-year commercial capacity by 2030, and its approximately 47,500-acre Smackover position in Texas and Arkansas: EnergyX corporate disclosures and project documentation; Texas General Land Office leasing records; Arkansas Department of Energy and Environment, Office of the State Geologist, project filings.

20. The Permian's Water — On the Permian Basin's approximately 8 million barrels per day of formation water production, on dissolved lithium concentrations typically in the 15–20 parts-per-million range, and on the Texas-based extraction companies — Element3 in Fort Worth and Maverick Metals in San Antonio — pursuing recovery from this existing waste stream: Texas Railroad Commission produced water disposal records; Element3 (formerly Element3 Energy) corporate materials; Bureau of Economic Geology produced water analyses.

21. Round Top Mountain — On the Round Top deposit in Hudspeth County (Trans-Pecos Texas, approximately 80 miles southeast of El Paso) as the principal US heavy rare-earth resource — hosting 15 of 17 rare-earth elements together with lithium, uranium, thorium, beryllium, gallium, hafnium, and zirconium — and on USA Rare-earth's March 5, 2026 acquisition of Texas Mineral Resources Corp. (an all-stock transaction worth approximately $73 million at announcement) that consolidated 100 percent project ownership in advance of targeted commercial production in 2028: USA Rare-earth and Texas Mineral Resources joint announcement (March 5, 2026); Texas General Land Office press materials; Texas Mineral Resources Corp. and USA Rare-earth, "Preliminary Economic Assessment, Round Top Project" (August 2019).

22. MP Materials (Independence + DoD partnership + 10X Northlake) — On MP Materials' Fort Worth Independence facility (which began commercial metal production in 2024 and first finished magnet production in 2025), the July 2025 transformational public-private partnership with the US Department of Defense ($400 million DoD preferred-stock investment, $150 million DoD loan for heavy rare-earth separation at Mountain Pass, 10-year NdPr price floor, and 10-year offtake guarantee for 100 percent of magnets produced at the planned 10X Facility), and the February 26, 2026 announcement that 10X will be built on a 120-acre site in Northlake, Texas (less than 10 miles north of Independence) at a company investment exceeding $1.25 billion, scaling magnet capacity to 10,000 metric tons per year with commissioning targeted for 2028: MP Materials Corp. press releases (July 2025 DoD partnership; February 2026 Northlake site selection); Federation of American Scientists, "Unpacking the DoD and MP Materials Critical Minerals Partnership" (July 2025); *Fort Worth Report*, "MP Materials building $1.25B magnet plant just north of Fort Worth" (February 27, 2026).

23. Noveon Magnetics and Metallium — On Noveon Magnetics' San Marcos manufacturing center recovering neodymium, praseodymium, dysprosium, and terbium from end-of-life EV motors, wind turbine generators, hard drives, and defense systems for reuse in new sintered NdFeB magnets, and on Metallium's Houston operation applying flash-heating technology to recover critical minerals from discarded consumer electronics: Noveon Magnetics corporate materials and DOE Critical Materials Innovation Hub documentation; Metallium corporate materials and Rice University collaborative research disclosures.

24. The Geological Recursion (Gibbons Creek) — On REE concentrations in coal at the Gibbons Creek mine in Grimes County of up to approximately 2,800 parts per million (REY) on a dry-coal basis, with extractability of 63 to 93 percent using environmentally benign weak acid; on coal combustion concentrating REEs into ash at approximately four times higher concentrations than the original coal (with proportionally lower extractability from ash); and on the 470 MW Gibbons Creek Steam Electric Station operated by the Texas Municipal Power Agency from 1982 to 2020 whose reclaimed site and water rights are now flowing to the SpaceX Terafab production facility: Bryant, Scanlon, et al., "Rare-earth element potential in coal and coal ash in the US Gulf Coast,"*International Journal of Coal Science & Technology*" (2026); Bureau of Economic Geology at UT Austin, "Critical Minerals and Rare-earth Elements in Texas" (research summary and conference presentations); National Energy Technology Laboratory, "Assessment of Rare-earth Elements and Critical Minerals in Coal and Coal Ash in the US Gulf Coast."

25. Tesla Lithium Refinery (Robstown) — On Tesla's lithium refinery at Robstown near Corpus Christi — North America's first spodumene-to-lithium-hydroxide refinery and the first industrial deployment of an acid-free refining route — which broke ground in 2023, began commissioning in late 2024, and reached full operations as announced by Elon Musk on January 14, 2026, with a nameplate capacity sufficient for approximately one million electric vehicles annually (~50 GWh of battery capacity, ~50,000 tons LCE) at an investment of approximately $1 billion: Tesla blog, "Tesla Lithium Refinery Groundbreaking" (May 2023); *Electrek*, "Tesla releases rare look at its lithium refinery, confirms production start" (January 15, 2026); KXAN Austin, "Tesla lithium refinery now operating in Texas. Musk says it's 'largest lithium refinery in America'" (January 21, 2026).

CHAPTER 4: THE TEXAS TRIANGLE

26. Opening (Texas Triangle population and GDP) — On the Texas Triangle's combined population exceeding 22 million across the Dallas–Fort Worth, Houston, San Antonio, and Austin metropolitan regions; on its approximately $2.2 trillion in combined economic output (roughly 80 percent of Texas GDP); and on its standing as one of eleven US megaregions: Texas Department of Transportation, Texas Triangle Multimodal Strategic Plan (TTMSP) program documentation; Kinder Institute for Urban Research, "Triumph of the Triangle: How Texas can hold onto its urban economic powerhouse"; Texas 2036, "Texas isn't growing one city at a time" (March 2026); US Census Bureau Annual Population Estimates.

27. Dallas–Fort Worth (logistics and freight) — On Dallas–Fort Worth International Airport's standing among the top US cargo airports and on the BNSF and Union Pacific rail infrastructure converging on the metroplex as the bulk-freight backbone of the Triangle interior: DFW Airport annual statistics; Airports Council International North America cargo rankings; BNSF and Union Pacific corporate route maps.

28. Dallas–Fort Worth (Texas Instruments analog depth) — On Texas Instruments' anchor position in the metro's semiconductor identity — the Richardson headquarters, the 300mm RFAB analog facilities in Richardson, and the Sherman SM1 and SM2 300mm fabs under construction with CHIPS Act support — and on TI's broader $60 billion US manufacturing investment commitment: Texas Instruments corporate disclosures and capital expenditure plans; US Department of Commerce CHIPS Program Office award announcements; *Dallas Morning News* and *Sherman Herald Democrat* construction coverage.

29. Houston and the Gulf Coast (Port of Houston) — On the Port of Houston's #1 ranking among US ports in foreign waterborne tonnage and on the Houston Ship Channel petrochemical complex as the most concentrated industrial chemistry infrastructure in the Western Hemisphere (cross-reference Chapter 1 sourcing on the port; primary sources for petrochemical complex statistics include the Greater Houston Partnership *Global Houston Report* and the US Energy Information Administration Gulf Coast refining capacity disclosures).

30. Houston and the Gulf Coast (Gulf Coast corridor) — On the Gulf Coast industrial corridor extending south from the Houston Ship Channel through Robstown and Corpus Christi to Boca Chica near Brownsville, integrating petrochemical, lithium refining (Tesla Robstown, cross-reference Chapter 3), and launch infrastructure (SpaceX Starbase) along a single coastal axis: Texas Comptroller regional economic accounts; SpaceX corporate disclosures regarding the Starbase Boca Chica complex.

31. San Antonio (JBSA and cybersecurity) — On Joint Base San Antonio — comprising Randolph, Lackland, and Fort Sam Houston — as the Department of Defense's largest training installation, and on the cybersecurity employment concentration that has grown around the military's cyber operations presence to make San Antonio one of the top US cybersecurity markets: Joint Base San Antonio public affairs materials; Greater:SATX (San Antonio Economic Development Foundation) cybersecurity industry profile; 16th Air Force (Air Forces Cyber) command documentation.

32. San Antonio (San Marcos / New Braunfels corridor) — On CloudBurst Evolve's vertically integrated behind-the-meter AI campus near San Marcos powered by on-site natural gas, on Tract's Caldwell Technology Park as a multi-gigawatt master-planned data center foundation, and on New Braunfels' absorption of residential and data center growth between the San Antonio and Austin anchor metros: CloudBurst Evolve corporate materials; Tract company disclosures; *Community Impact* and *San Antonio Express-News* regional reporting on Caldwell County, Comal County, and Hays County development.

33. Austin (UT Austin and semiconductor cluster) — On UT Austin's Cockrell School of Engineering and the Texas Institute for Electronics as the research and workforce pipeline anchoring the Austin semiconductor ecosystem; on the cluster of approximately 30 major semiconductor companies in Austin proper (with broader Silicon Hills extension from San Marcos through The Domain) including AMD, Samsung Austin, NXP, Applied Materials, Silicon Labs, Cirrus Logic, and Infineon: Austin Chamber of Commerce semiconductor industry profile; University of Texas at Austin, Cockrell School of Engineering and Texas Institute for Electronics program materials.

34. Austin (The Waterline) — On the Waterline supertall at 98 Red River Street — a 74-story, 1,025-foot mixed-use tower that topped out on August 1, 2025 as the tallest building in Texas and the tallest in the Southern United States, opening in late 2026 — developed by Lincoln Property Company and Kairoi Residential with PSP Investments (the Public Sector Pension Investment Board, one of Canada's largest pension investment managers) as main equity partner and Blackstone Mortgage Trust providing construction financing: Lincoln Property Company press release, "Waterline Topping Out Brings Texas' New Tallest Tower to Full Height" (August 1, 2025); *REBusiness Online*, "Lincoln Property Co., Kairoi Top Out 74-Story Waterline Mixed-Use Tower in Downtown Austin"; PSP Investments News Hub, "Lincoln Property Company and Kairoi Residential Reveal Plans for Waterline" (September 2022).

Chapter 5: The Institutional Foundations

35. The Kozmetsky Foundation — On Kozmetsky's tenure as dean of the McCombs School of Business (1966–1982), his founding of the IC^2 Institute in 1977, his role in MCC's relocation to Austin in 1983, the recruitment of

Sematech in 1987, and his launch of the Austin Technology Incubator in 1989: Steve Brooks, "The Kozmetsky Effect," *Texas McCombs Magazine*, 2022 Centennial Edition; IC^2 Institute, "History of Innovation at UT Austin"; SEMATECH corporate history (incorporated August 1987; Austin operations began 1988).

36. The Italian Nun / Santa Rita No. 1 — On the May 28, 1923 oil strike at Santa Rita No. 1 on University of Texas land in Reagan County and the role of the well in capitalizing the Permanent University Fund: *Handbook of Texas Online* (Texas State Historical Association), "Santa Rita No. 1 Oil Well." On the rose-petal investor anecdote: also TSHA Handbook, "Santa Rita No. 1."

37. The Permanent University Fund — On the 1839 Republic of Texas land grant of approximately fifty leagues (~221,000 acres), the 1876 Texas Constitution expansion, and the present ~2.1 million-acre PUF land base: University of Texas System, "Permanent University Fund Overview"; UTIMCO annual reports.

38. The University Structure (TIE NGMM) — On the $840 million DARPA Next Generation Microelectronics Manufacturing award to the Texas Institute for Electronics at UT Austin (announced July 18, 2024), and the $552M Texas Legislature contribution that brings the total to $1.4 billion: University of Texas at Austin News, "UT's Texas Institute for Electronics Awarded $840M To Build a DoD Microelectronics Manufacturing Center" (July 18, 2024); DARPA NGMM program documentation.

39. The University Structure (T2COM) — On the activation of the US Army Transformation and Training Command at the LBJ Auditorium on the UT Austin campus on October 2, 2025, merging Army Futures Command and TRADOC under a single four-star command of approximately 350,000 personnel: US Army, "Transformation and Training Command activated," army.mil (November 14, 2025); Congressional Research Service R48606, "2025 Army Transformation Initiative (ATI) Force Structure and Organizational Proposals."

40. The Capital and Commercialization Layer — On the Austin Technology Incubator's three-decade history of deep-tech and university-spinout commercialization under IC^2: Austin Technology Incubator, ati.utexas.edu. On Capital Factory's 2009 founding and defense-technology accelerator presence: Capital Factory company materials.

41. Convergence at Startup Scale (Saronic) — On Saronic Technologies' $9.25 billion valuation following its $1.75B Series D led by Kleiner Perkins

(March 31, 2026), preceded by the $4B Series C of February 2025: "Saronic Closes $1.75B Series D at $9.25B Valuation," PRNewswire / Saronic Technologies (March 31, 2026).

42. Convergence at Startup Scale (Base Power) — On Base Power's $4 billion valuation following its $1 billion Series C led by Addition (October 8, 2025): Contrary Research, "Base Power Business Breakdown"; Sacra company profile.

43. Convergence at Startup Scale (Apptronik) — On Apptronik's $350 million Series A (February 2025, co-led by B Capital and Capital Factory with Google participation), and the company's 2016 spinout from the UT Austin Human Centered Robotics Laboratory at the Cockrell School. Co-founder and CEO Jeff Cardenas is a McCombs MSTC graduate: *TechCrunch,* "Apptronik, which makes humanoid robots, raises $350M as category heats up" (February 13, 2025); UT McCombs Magazine, "AI in Action" (Spring/Summer 2024).

44. Convergence at Startup Scale (Firefly Aerospace) — On Firefly Aerospace's Blue Ghost Mission 1 successful soft-landing in Mare Crisium on March 2, 2025, the first fully successful commercial lunar landing (Intuitive Machines' IM-1 landed in February 2024 but on its side): Firefly Aerospace press release, "Firefly Aerospace Becomes First Commercial Company to Successfully Land on the Moon" (March 2, 2025); KUT Austin coverage (March 3, 2025).

Chapter 6: The Metro Corridors

45. The I-35 Corridor (Fort Hood) — On Fort Hood's status as the largest US Army installation by population, its modernization program scale, and its role anchoring the I-35 corridor's defense identity: US Army Garrison Fort Hood public affairs; III Armored Corps and Fort Hood institutional documentation. The installation was redesignated Fort Cavazos in May 2023 and restored to Fort Hood in June 2025 under DoD direction.

46. The SH 130 Corridor (85 mph speed limit) — On the 85-mph posted speed limit on the SH 130 toll segment between Austin and Seguin, the highest in the United States: Texas Department of Transportation announcement (October 2012). On the SH 130 ownership and tolling structure: SH 130 Concession Company materials and TxDOT corridor documentation.

47. The US 79 Corridor (data center concentration) — On the 70-plus data center projects tracked between Temple and San Antonio with the densest concentration in the Williamson County section: Community Impact regional reporting; Texas Comptroller economic development tracking; Data Center Frontier project tracking.

48. The US 183 Corridor (Cedar Park aerospace cluster) — On Firefly Aerospace's manufacturing facility in Cedar Park, the Central Texas Spaceport and Defense Innovation Campus, and CesiumAstro's manufacturing presence: Firefly Aerospace corporate materials; City of Cedar Park economic development office; CesiumAstro company materials. (On Firefly's Blue Ghost Mission 1 specifically, see note to Chapter 5.)

49. The DFW Corridor System (AllianceTexas) — On Hillwood's AllianceTexas as the founding 18,000-acre Alliance development, the corridor's conversion from aviation/intermodal logistics into a data center concentration, and the proximity to Naval Air Station Fort Worth Joint Reserve Base: Hillwood Companies corporate materials and AllianceTexas annual economic impact reporting; Dallas Federal Reserve regional economic analysis.

50. The Houston Corridor System — On the I-10 Energy Corridor's concentration of supermajor and oilfield services engineering workforce, and on the Ship Channel industrial belt's status as the densest petrochemical and refining concentration in the United States: Greater Houston Partnership economic profile; Port Houston statistical reports; Texas Comptroller chemical-industry regional analysis.

Chapter 7: The Williamson County Buildout

51. Opening (Georgetown growth) — On Georgetown's growth from approximately 67,000 in the 2020 Census to more than 115,000 by 2026, ranking as the fastest-growing US city above 50,000 population in successive Census Bureau estimates: US Census Bureau, Annual Estimates of the Resident Population (2021, 2022, 2023, 2024); *World Population Review*, "Georgetown, Texas Population 2026"; Georgetown city demographics and statistics.

52. The Italian Nun (Liberty Hill growth) — On Liberty Hill's growth from approximately 1,400 residents in 2015 to more than 15,000 in 2026 — an order-of-magnitude increase in just over a decade: *World Population Review*, "Liberty Hill, Texas Population 2026" (15,377); US Census Bureau 2020 decennial count (3,646); *Handbook of Texas Online*, "Liberty Hill, Texas (Williamson County)."

53. The Italian Nun (Georgetown 100 percent renewable) — On Georgetown becoming one of the first US cities to reach 100 percent renewable power through its municipal utility in 2017, with long-term contracts including the 154 MW Buckthorn Solar project: City of Georgetown municipal utility materials; Texas Tribune coverage (2017); Buckthorn Solar Project documentation.

54. The Residential Phenomenon (Santa Rita Ranch) — On Santa Rita Ranch as the top-selling master-planned community in Greater Austin for six consecutive years: RCLCO Real Estate Consulting top-selling MPCs ranking; Northwest Williamson County developer materials.

55. The Residential Phenomenon (Wolf Ranch) — On Hillwood Communities' 2014 acquisition of the Wolf Ranch property from the Wolf family after 65 years of family stewardship, and the subsequent buildout of a 1,120-acre master-planned community with a projected full-buildout value of approximately $1.9 billion: Hillwood Communities, "Wolf Ranch master-planned community breaks ground" (December 10, 2015); Austin American-Statesman, "Final phase of 1,120-acre Wolf Ranch in Georgetown underway"; Wolf family legacy materials at wolfranchbyhillwood.com.

56. The Residential Phenomenon (Sun City Texas) — On Sun City Texas as the largest age-restricted master-planned community in Texas — approximately 4,750–5,400 acres depending on the phase counted, with an approved buildout of approximately 10,550 homes and a population reaching approximately 15,700 residents by 2021: Del Webb / PulteGroup community materials; 55places.com, "Sun City Texas Community of the Year" (2023); City of Georgetown planning records.

57. The Water Constraint — On the water consumption profiles of the three primary facility types (semiconductor fabs 5–10 million gallons per day; AI data centers 1–19 MGD; gigafactories ~1.5 MGD), the ultrapure water 40–60 percent input-to-output loss ratio, and Giga Texas's ~556 million gallon annual consumption that made it Austin Water's third-largest customer: Lawrence Berkeley National Laboratory, "United States Data Center Energy Usage Report"; SEMI water-use disclosures; Austin Water customer reports.

58. The Permanent Catch-Up (Williamson County road bond) — On the 2019 Williamson County Road Bond's $412 million authorization for multiple simultaneous corridor expansions in response to traffic outrunning the bond-funded projects: Williamson County, "2019 Road Bond Program" project records; Community Impact ongoing project tracking.

Chapter 8: The Singularity Center

59. Opening (Giga Texas address and scale) — On Tesla Giga Texas's corporate address at 1 Tesla Road, Austin, Texas 78725, its approximately 10 million square feet of floor area on a 2,100-acre site bordering the Colorado River, and its standing as one of the largest buildings by floor area in the United States: Tesla Corporate Contact (tesla.com/contact); Tesla Investor Relations; CoStar News, "Elon Musk's Tesla to expand sprawling Texas Gigafactory by 50 percent" (October 31, 2024). On the in-progress 5.2 million sq ft expansion permitted in 2024 for completion by year-end 2026: same CoStar reporting; Texas Department of Licensing and Regulation Tesla Gigafactory project records.

60. The Car Factory (manufacturing innovations) — On front and rear gigacastings replacing 70+ stamped and welded body parts each, the on-site 4680 cell line and structural battery pack, and the unboxed manufacturing approach introduced for Cybercab: Tesla 2024 and 2025 Investor Day presentations; Munro Live teardown analysis of Cybertruck and Model Y structural battery designs.

61. The AI Training Cluster (Cortex) — On Cortex as Tesla's hyperscale physical-AI training cluster inside the Giga Texas campus, with Cortex 1 operational and Cortex 2 partially operational by mid-2026, and on the closed captive loop in which Tesla fleet driving data trains models that propagate back to Tesla products: Tesla Investor Day and earnings call disclosures; SemiAnalysis coverage of Cortex compute scale.

62. The AI Training Cluster (FSD miles) — On Tesla's Full Self-Driving (Supervised) fleet crossing 11 billion cumulative miles in early May 2026, with the pace of accumulation having compressed from a 53-day interval between 7 and 8 billion (December 2025 to February 2026) to 31 days between 9 and 10 billion: Tesla Vehicle Safety Report; Teslarati, "Tesla FSD (Supervised) fleet passes 8.4 billion cumulative miles" (March 2026); eletric-vehicles.com, "Tesla FSD Hits 10 Billion Miles" (May 4, 2026).

63. The AI Training Cluster (Colossus and SpaceXAI) — On the post-merger structure under which Cortex and Colossus operate as captive training infrastructures under SpaceX's broader portfolio, with Cortex serving Tesla-side physical-AI workloads (Optimus, Cybercab, vehicle FSD) and Colossus in Memphis serving SpaceX-side workloads (Grok, SpaceXAI API): xAI–SpaceX merger announcements; The Information / Bloomberg coverage of the corporate structure.

64. The Optimus Factory (production trajectory) — On Optimus Gen 3 production scheduled to begin in late 2026 at an initial run of 5,000–10,000 units, scaling to approximately 250,000 in 2027, with Tesla's longer-horizon target of millions of units annually by the end of the decade: Tesla Q1 2026 and Q2 2026 earnings calls; Tesla Investor Day 2025; Reuters and Bloomberg coverage of Optimus production planning.

65. The Optimus Factory (humanoid robot market sizing) — On global humanoid robot market projections of approximately 40 billion by 2035 at the conservative end of analyst forecasts. Goldman Sachs, Morgan Stanley, and Citi have published 2035 TAM estimates ranging from approximately $38 billion to $150 billion: Goldman Sachs Equity Research, "Humanoid Robots: Sizing the Market"; Morgan Stanley Research, "The Humanoid 100"; Citi Global Insights humanoid robotics analysis.

66. The Terafab Research Fab (two-node architecture) — On Terafab as a two-node program — a Tesla-operated research and pilot facility within the Giga Texas campus footprint and a SpaceX-operated production facility at the former Gibbons Creek coal plant site in Grimes County, with the brownfield reservoir water rights providing the enabling water-infrastructure inheritance. On the AI5, AI6, and D3 (AI7) chip families and their respective deployment targets (vehicles, Optimus, SpaceX orbital systems): Tesla and SpaceX investor disclosures; Texas Commission on Environmental Quality permitting records for the Gibbons Creek site; SemiAnalysis coverage of the Terafab program.

Chapter 9: The Captive Fab

67. Opening (Samsung–Tesla agreement) — On the July 2025 Samsung–Tesla long-term wafer fabrication agreement at Taylor: Samsung Electronics regulatory filing (July 28, 2025); Mackenzie Hawkins and Sohee Kim, "Samsung to Make Tesla AI Chips in Multiyear Texas Deal," Bloomberg (July 28, 2025); the contract is valued at $16.5 billion and runs through December 31, 2033.

68. Opening (Taylor production start) — On Samsung's confirmation of H2 2027 mass production at Taylor: Han Jin-man, Samsung Foundry

President, shareholder meeting remarks (March 18, 2026), reported in Reuters and The Korea Economic Daily.

69. Samsung Taylor Campus (initial $17B commitment) — On Samsung's November 2021 announcement of the Taylor fab as an approximately $16.5 billion 4nm-process commitment, prior to the mid-construction 2nm pivot: Samsung Electronics press release, "Samsung Electronics Announces New Advanced Semiconductor Fab Site in Taylor, Texas" (November 23, 2021). On the headline $17B+ figure as the largest foreign direct investment in Texas history, see note to Chapter 7.

70. Samsung Taylor Campus 2 (2nm pivot and capital escalation) — On the mid-construction pivot from 4nm to first-generation 2nm and the resulting capital escalation to a range commonly cited at $37–$44 billion: US Department of Commerce CHIPS Direct Funding Agreement disclosure (December 20, 2024), citing $37 billion in committed Samsung Texas investment; Reuters reporting on the broader $44 billion total Taylor-complex investment.

71. The CHIPS Act Overlay (final $4.745B award) — On the December 2024 finalization of Samsung's CHIPS Act direct funding agreement: US Department of Commerce, "Biden-Harris Administration Announces CHIPS Incentives Award with Samsung Electronics" (December 20, 2024); the final award of $4.745 billion was revised down from the April 15, 2024 preliminary memorandum of terms of up to $6.4 billion.

72. The CHIPS Act Overlay (Texas Semiconductor Innovation Fund) — On the $250 million TSIF allocation to Samsung Taylor and the accompanying TSIF grants to surrounding suppliers: Office of the Governor of Texas, TSIF award announcements (2024); the TSIF is administered by the Texas Comptroller under SB 1759 (88th Legislature, 2023).

73. The Dual-Source Architecture (TSMC Arizona 2nm) — On TSMC's Arizona buildout to include a 2nm fab as the second-source counterpart to Samsung Taylor: TSMC press release on the US Department of Commerce CHIPS Act preliminary memorandum of terms (April 8, 2024); finalized November 15, 2024 with $6.6 billion in direct funding supporting three Arizona fabs, including a 2nm facility.

Chapter 10: The Co-Design Thesis

75. The Intel Partnership (ASML EUV constraints) — On EUV system production rates and pricing as a binding new-entrant constraint: ASML Holding N.V., 2024 Annual Report and Q4 2024 investor presentation; current-generation EUV systems (NXE:3800E and successors) ship in the 50–60 unit annual range, with list pricing above $200 million per scanner and High-NA EUV (EXE:5000) systems above $350 million.

76. The Intel Partnership (Intel 14A access) — On the Tesla–Intel partnership architecture and Intel 14A node access: Tesla quarterly earnings call commentary (Q4 2025 and Q1 2026); Intel Foundry roadmap disclosures, Intel Foundry Direct Connect (February 2024); SemiAnalysis coverage of the Intel 14A program and the Tesla foundry relationship.

77. The Trusted Foundry Program (DoD framework and limitations) — On the Defense Microelectronics Activity Trusted Foundry framework and its current limitations at leading-edge nodes: US Government Accountability Office, GAO-22-105117, Trusted Microelectronics: DoD Should Improve Risk Assessment and Oversight of an Emerging Approach (July 2022); DMEA program documentation.

78. The Two-Fab Architecture (Cyclotron Institute radiation qualification) — On Texas A&M's role in qualifying D3 against the LEO radiation environment: Texas A&M University Cyclotron Institute, single-event effects testing facility documentation (cyclotron.tamu.edu); the K500 superconducting cyclotron is one of the principal US facilities for heavy-ion radiation qualification of space-grade semiconductors. On the broader Terafab two-node architecture, see note to Chapter 8.

79. The Missile Defense Architecture (program origin) — On the program that drives the Trusted Foundry compute requirement: White House Executive Order 14186, "The Iron Dome for America" (January 27, 2025), subsequently rebranded Golden Dome; Missile Defense Agency program documentation and Congressional testimony (2025).

80. The Missile Defense Architecture (hypersonic glide vehicle velocities) — On the Mach-20 discrimination-window physics: DARPA Falcon HTV-2 flight-test data (3.6 miles per second sustained at Mach 17–22), reported in Mike Wall, "Superfast Military Aircraft Hit Mach 20 Before Ocean Crash, DARPA Says," Space.com (August 18, 2011). Mach 20 corresponds to approximately 15,000 mph (24,000 km/h), or roughly four miles per second at standard atmosphere.

81. The Silicon Chain (SiC boule physics) — On silicon carbide substrate manufacturing constraints and the industry 150mm-to-200mm wafer transition: SEMI International Standards; Wolfspeed Inc. investor materials and 10-K disclosures (2023–2025); US Department of Energy, Office of Energy Efficiency and Renewable Energy, Wide Bandgap Power Electronics Technology Assessment (2015, with subsequent updates).

82. The Silicon Chain (Texas Instruments Sherman complex) — On the world's largest concentration of analog and embedded processing semiconductor manufacturing: Texas Instruments Inc., 2023 Capital Management Investor Day presentation and Sherman SM1/SM2 facility disclosures; "Texas Instruments to Build Up to Four New 300mm Semiconductor Wafer Fabs in Sherman, Texas," TI press release (November 17, 2021).

Chapter 11: The AI Compute Layer

83. Opening (Texas data center census) — On Texas's 484 data centers (297 operational, 109 under construction, 124 planned) and the forecast capacity passing 70 GW by 2030: Baxtel data center database aggregation cited via University of Texas at Austin COMPASS (Computing Materials and Process Analysis for Sustainable Semiconductors) program (late 2025).

84. Opening (ERCOT installed capacity) — On ERCOT's installed generation capacity of roughly 140 GW: Electric Reliability Council of Texas, Capacity, Demand and Reserves Report (December 2025); ERCOT Capacity Changes by Fuel Type, monthly summary.

85. Opening (Vantage Frontier campus) — On the $25 billion, 1.4 GW, 1,200-acre Frontier campus in Shackelford County: Vantage Data Centers press release, "Vantage Data Centers Unveils Plans for 'Frontier,' a $25B Mega-campus in Texas to Meet Unprecedented AI Demand" (August 19, 2025); groundbreaking confirmed December 2025, with first building scheduled for second half of 2026.

86. The Power Constraint Solution (Fermi America Project Matador) — On the multi-phase Texas Panhandle private energy and AI data center campus: Fermi America Inc., corporate disclosures and TCEQ permit filings; the initial 6 GW gas-fired air permit was issued in November 2025, and a March 2026 filing announced expansion toward 17 GWe total, including four Westinghouse AP1000 reactors at 4.4 GWe (first unit targeted 2031–2032). See also "Fermi expands Texas energy campus," *Nuclear Engineering International* (March 17, 2026).

87. The Power Constraint Solution (Pacifico Energy GW Ranch) — On the 7.65 GW air permit issued for the Pecos County off-grid campus: Texas Commission on Environmental Quality air permit (issued January 2026); Pacifico Energy press release, "Pacifico Energy Secures 7.65 GW Power Generation Permit for GW Ranch Project" (January 26, 2026); first-power target Q1 2027 at 1 GW Phase 1, with 1.8 GW battery storage and 750 MWac solar integrated into full buildout.

88. The Power Constraint Solution (Nexus Hubbard) — On the 7.2 GW combined-cycle and simple-cycle gas project: Nexus Hubbard project filings, Hill County, Texas; project sited along State Highway 171 between Waco and the Dallas–Fort Worth metroplex.

89. The Power Constraint Solution (Sailfish Comanche Circle) — On the 5 GW Hood County proposal adjacent to Vistra Corp.'s Comanche Peak Nuclear Generating Station (~2,400 MW): Sailfish Investors filings with Hood County, Texas; first 600 MW via on-site power islands within 24 months, with subsequent phases targeted through 2028.

90. The Power Constraint Solution (Xebec Sandow Lakes) — On the 33,000-acre Milam County brownfield campus: Xebec Holdings master plan filings; the site incorporates water rights and infrastructure inherited from Alcoa's 1954 Sandow aluminum smelting operations and the retired Sandow Power Plant footprint.

91. The Power Constraint Solution (Stargate program) — On the $500 billion four-year OpenAI–SoftBank–Oracle program: joint announcement of The Stargate Project (January 21, 2025), with $100 billion committed to Abilene Phase 1; the subsequent October 2025 multi-site expansion added anchors in Shackelford County, Milam County (Freebird, with SB Energy), and Santa Teresa, New Mexico (Project Jupiter, $165 billion).

92. The Three Chokepoints (PJM interconnection queue) — On the roughly eight-year PJM large-load interconnection queue: PJM Interconnection generation interconnection queue reports (2024–2025); Lawrence Berkeley National Laboratory, Queued Up: 2024 Edition of Characteristics of Power Plants Seeking Transmission Interconnection (April 2024).

Chapter 12: The Autonomy Deployments

93. Tesla FSD The Data Moat (fleet mileage) — On Tesla's FSD (Supervised) fleet crossing 11 billion cumulative miles and the daily accumulation rate, see note to Chapter 8.

94. Tesla Cybercab Economics (Waymo comparison) — On Waymo's eleven-US-city operating footprint, approximately 4,000 vehicles, and 500,000-plus weekly paid rides: Waymo LLC operational metrics in Alphabet Inc. Q1 2026 earnings commentary and public disclosures.

95. Tesla Cybercab Economics (Zoox Austin expansion) — On Zoox's 2025 expansion of pre-commercial testing into Austin: Zoox Inc., "Zoox Expands Testing to Austin" announcement (2025); the Austin presence places Tesla's Robotaxi service, Waymo's commercial operation, and Zoox's pre-commercial testing in the same metropolitan deployment ground.

96. Megawatt Charging and Autonomous Frieght (CharIN MCS standard) — On the Megawatt Charging System specification targeting 1–3.75 MW: Charging Interface Initiative (CharIN) e.V., Megawatt Charging System (MCS) technical specification (2023, with subsequent revisions); the design target supports 1,250 V × 3,000 A continuous.

97. Megawatt Charging and Long-Haul Robotrucks (Texas Megacharger lead) — On Texas's 19 planned Megacharger sites leading the national network: Tesla Inc., Megacharger "Find Us" map update (February 2026), showing 66 sites across 15 states; the Pilot Travel Centers partnership announced January 2026 commits 4–8 stalls per Pilot site at up to 1.2 MW; Fred Lambert, "Tesla adds 64 Megacharger locations to map," *Electrek* (February 24, 2026).

98. Tesla Semi FSD (Aurora commercial launch) — On Aurora's May 2025 commercial driverless truck service launch: Aurora Innovation, Inc., "Aurora Begins Commercial Driverless Trucking in Texas, Ushering in a New Era of Freight" (press release, May 1, 2025); the initial route ran between Dallas and Houston, with El Paso expansion targeted by end of 2025.

99. Tesla Semi FSD (Nevada production launch) — On Tesla Semi volume production launching at the dedicated Nevada facility: Tesla Semi official communications (April 29, 2026), confirming volume production at the 1.7-million-square-foot facility adjacent to Gigafactory Nevada in Sparks; 50,000-unit annual capacity is the full-ramp target; "First Tesla Semi Rolls Off New High-Volume Production Line," *Tesla North* (April 30, 2026).

100. Texas Regulatory Posture (SB 2205) — On the Texas framework authorizing autonomous commercial vehicle operation without a human operator present: Texas Senate Bill 2205 (85th Legislature, 2017), codified in Texas Transportation Code, Chapter 545, Subchapter J; the framework places primary responsibility on the operator and shifts accountability from pre-deployment regulatory approval to post-deployment operator liability.

Chapter 13: The Space and Defense Layer

101. Opening (Falcon 9 booster recovery scale) — On Falcon 9 booster landing cadence and Starlink constellation size: SpaceX corporate disclosures and the Falcon 9 prospectus filed ahead of the company's June 2026 IPO; Falcon family successful landings exceeded 595 of 608 attempts as of mid-April 2026, with booster B1067 completing a record 35th flight on June 8, 2026; the Starlink active constellation passed 10,500 satellites by mid-2026 per Jonathan McDowell's tracking aggregation. See also "SpaceX surpasses 500 rocket landings," *Orlando Sentinel* (September 2025).

102. Starbase Complex (Starship and Super Heavy propellant load) — On the per-launch propellant requirement and the LOX/methane mass ratio: SpaceX Starship technical specifications and Federal Aviation Administration environmental review documentation for Starbase, Boca Chica; Super Heavy plus Starship combined propellant on the order of 4,600 metric tons at an approximately 78/22 LOX/methane mass split.

103. McGregor Facility (acquisition history and Raptor production scale) — On SpaceX's 2003 acquisition of the McGregor test site and current Raptor V3 production cadence at McGregor: SpaceX corporate history and FAA McGregor facility documentation; Super Heavy carries 33 Raptor engines per booster, with full-reusability operations requiring sustained high-cadence engine production at industrial scale. Raptor V4 testing is underway at McGregor as of early 2026.

104. Starship (V3 LEO capacity and full reusability) — On the 200-metric-ton LEO payload target for Starship V3 in fully reusable configuration: SpaceX Starship V3 specifications and Elon Musk public presentations; full reusability of both Super Heavy and Starship is the architectural feature that enables the launch economics on which the orbital infrastructure thesis depends.

105. Starlink Factory (Bastrop terminal manufacturing) — On the Bastrop user-terminal facility scale: SpaceX corporate disclosures and Steve

Jurvetson public commentary; the facility exceeds one million square feet and is the largest printed-circuit-board manufacturing operation in the United States, with annual production capacity above 4.7 million units.

106. Gigasat Complex (June 2026 announcement) — On SpaceX's June 8, 2026 unveiling of the Gigasat complex ahead of the company's mid-June IPO: Elon Musk on X (June 8, 2026); Anton Shilov, "SpaceX unveils 11-million-square-foot Gigasat factory," Tom's Hardware (June 2026); construction industry coverage at constructionreviewonline.com (June 2026); the campus spans more than 1,000 acres with over 11 million square feet of planned building capacity in Bastrop County, and is targeted to produce 1 GW of orbital AI compute capacity per year by the end of 2027.

107. Gigasat Complex (Project Echo Solar) — On the Bastrop solar cell facility, the first of five Gigasat buildings: Bastrop County permit filings under the name Project Echo Solar; the facility footprint is 1.1 million square feet at initial buildout, expandable to roughly 1.6 million square feet, with annual capacity of 10 GW across two 5-GW production floors. Bloomberg reporting (May 2026); SpaceX Director of Solar Production Noah Cowles, public statements.

108. Bastrop Compound (Snailbrook) — On Snailbrook, Musk's planned company town adjacent to the SpaceX Bastrop compound: Bastrop County property records and Snailbrook Inc. corporate filings; the master plan covers approximately 3,500 acres, with restaurant, pub, salon, and recreational facilities currently operational. The *Wall Street Journal*, "Elon Musk's Latest Project: A Texas Town Called Snailbrook" (March 9, 2023, with subsequent reporting through 2025–2026).

109. Blue Origin — On Blue Origin's reported interest in a Hutto manufacturing and R&D campus: Hutto Economic Development Corp. public hearing on Project Blue Hub (May 11, 2026); reporting in the Austin Business Journal, KXAN Austin, Bisnow Austin–San Antonio, and Taylor Press identifying Blue Origin as the unnamed proposer; project specifications described as 1.3 million square feet, more than 2,000 jobs at an average salary of $88,000, and capital investment exceeding $650 million (reported up to $1 billion).

110. Blue Origin (Launch Site One) — On Blue Origin's West Texas launch facility: Corn Ranch / Launch Site One in Culberson County, Texas, approximately 30 miles north of Van Horn; the 165,000-acre site was purchased by Jeff Bezos beginning in 2004, and New Shepard has flown from the facility since April 2015. Blue Origin corporate materials; Wikipedia, "Corn Ranch" (composite primary-source aggregation).

111. Blue Origin (Project Sunrise and TeraWave) — On Blue Origin's two announced orbital constellations: Project Sunrise (50,000-satellite orbital data center constellation, the direct competitive response to SpaceX's million-satellite filing) and TeraWave (5,000-satellite connectivity constellation delivering six terabytes per second to data centers, governments, and businesses, the direct competitive response to Starlink). Blue Origin corporate announcements (2025–2026).

112. Starshield Program — On the Golden Dome missile defense architecture and its driving role in the Trusted Foundry compute requirement, see notes to Chapter 10.

Chapter 14: The Energy Manufacturing Layer

113. Brookshire Campus (siting and Houston-corridor advantages) — On the Tesla Energy Brookshire campus at the Empire West Business Park in Waller County, approximately 35 miles west of Houston: Tesla corporate disclosures; Houston Chronicle reporting on the Empire West Business Park anchor tenant additions; the campus co-locates two Tesla Energy manufacturing operations (Megapack Megafactory and Brookshire Solar Manufacturing Complex) and benefits from Port of Houston equipment-import access and the petrochemical labor pool.

114. The Megapack Factory (50 GWh annual capacity) — On the Brookshire Megapack Megafactory: Tesla Q1 2026 update letter, which confirms construction progress and notes that Megapack 3 start of production at the new Houston-area Megafactory is on track for later in the year; planned ramp toward 50 GWh of annual capacity beginning in the second half of 2026.

115. The Megapack Factory (Megablock architecture) — On the Megablock as the factory-built grid node combining four Megapack 3 units (approximately 20 MWh), a Tesla-built transformer, the Tesla inverter stack, medium-voltage switchgear, integrated thermal management, on-skid controls, and an energy management system: Tesla 2025 Investor Day presentations and Tesla Energy product disclosures; the architecture targets a 1 GWh commissioning timeline of 20 business days.

116. Tesla Solar Factory (100 GW target and ingot-to-panel vertical integration) — On Tesla's announced 100 GW annual domestic solar manufacturing target with full vertical integration from silicon ingot growth through finished panels: *Electrek* reporting (May 19, 2026) on Tesla's Brookshire solar confirmation; Tesla 2025–2026 investor commentary; the target represents a manufacturing program more than two orders of magnitude larger than Tesla's prior Buffalo, NY operation acquired in the 2016 SolarCity transaction.

117. Tesla Solar Factory (Chinese equipment procurement) — On Tesla's reported $2.9 billion solar manufacturing equipment procurement program from Chinese suppliers: Reuters reporting (2026); solar manufacturing equipment is one of the few categories in the solar supply chain not subject to US tariffs, which makes the Chinese-equipment route the fastest available path to domestic production capability. Initial capital expenditure across the simultaneous Brookshire facility builds reported above $250 million, separate from the equipment procurement program.

118. The T1 Energy Factory (G2_Austin TOPCon cell plant) — On T1 Energy Inc. and the G2_Austin solar cell facility under construction at Rockdale, Milam County: T1 Energy press release, "T1 Energy Starts Construction on Texas Solar Cell Fab" (December 17, 2025); Phase 1 capacity of 2.1 GW of high-efficiency TOPCon cells targeted for end-of-2026 production, on a $400–$425 million investment supporting up to 1,800 advanced manufacturing jobs. Phase 1 alone exceeds the current US capacity for silicon-based solar cell manufacturing.

119. T1 Energy Solar Factory (G1_Dallas module plant and domestic supply chain closure) — On the G1_Dallas module-manufacturing facility (acquired from Trina Solar in late 2024) and T1's upstream polysilicon and wafer supply arrangements: T1 Energy investor disclosures; August 2025 supply agreement with Corning Inc. and Hemlock Semiconductor for polysilicon and wafers produced in Michigan, completing the end-to-end domestic supply chain from polysilicon through finished modules without leaving the country.

Chapter 15: The Capital Stack

120. The Federal Layer (CHIPS and Science Act of 2022) — On the CHIPS and Science Act of 2022 as the federal anchor of the domestic semiconductor-and-AI capital stack: Public Law 117-167, August 9, 2022;

Congressional Research Service R47523, "The CHIPS and Science Act: Implementation and Issues for Congress." Total program $52 billion in financial assistance plus tax credits.

121. The Private Institutional Layer (Digital Realty hyperscale data center fund) — On Digital Realty Trust's March 30, 2026 final close of its inaugural US hyperscale data center fund: Digital Realty Trust, "Digital Realty Announces Final Close of $3.25 billion US Hyperscale Data Center Fund," GlobeNewswire (March 30, 2026); $3.25 billion in total equity commitments from a global investor base including sovereign wealth funds, public pensions, endowments, foundations, corporate pensions, insurance and asset managers, and family offices; Digital Realty retains a 20 percent ownership interest as fund manager.

122. The State Layer (TSIF and PUF) — On the Texas Semiconductor Innovation Fund as the state-level instrument paralleling the federal CHIPS Act, see notes to Chapter 9. On the Permanent University Fund as the sovereign-wealth research-funding anchor for UT Austin, see notes to Chapter 5.

123. The State Layer (Texas Enterprise Fund) — On the Texas Enterprise Fund and the governor's discretionary economic development programs as the deal-closing margin in major facility location decisions: Office of the Texas Governor, Texas Enterprise Fund grants database and annual legislative reports.

124. The County and Municipal Layer (Williamson County Road Bond) — On Williamson County's 2019 Road Bond as the canonical illustration of county-level capital committed on general trajectory confidence two years before the catalyst it inadvertently preceded (Samsung Taylor), see notes to Chapter 7.

Chapter 16: The Workforce

125. The University Anchor (UT Austin Cockrell School) — On the Cockrell School of Engineering's sustained top-tier ranking and the discipline mix most directly tied to the buildout's workforce requirements: US News & World Report Best Graduate Engineering Schools rankings (2024 and 2025 editions); Cockrell School of Engineering annual research and placement reports.

126. The University Anchor (Texas Institute for Electronics) — On the Texas Institute for Electronics at UT Austin as the DARPA- and state-funded prototyping infrastructure that doubles as a workforce development instrument, see notes to Chapter 5.

127. The University Anchor (Texas A&M Cyclotron Institute) — On Texas A&M's Cyclotron Institute as the radiation-qualification facility and specialized nuclear and radiation physics workforce environment, see notes to Chapter 10.

128. The Engineering Labor Market (Austin technology in-migration) — On the pandemic-era acceleration of engineering in-migration to Austin from the Bay Area, Seattle, Boston, and New York: US Census Bureau Current Population Survey migration data; Austin Chamber of Commerce regional economic profile; Federal Reserve Bank of Dallas regional analysis on Texas labor market dynamics (2023–2025).

129. The Demographic Base (Texas population and median age) — On Texas as the second most populous state with the youngest median age among large states: US Census Bureau, American Community Survey 5-Year Estimates; Texas Demographic Center annual population projections.

130. The Technician Pipeline (TSIF semiconductor workforce cluster) — On the coordinated TSIF-funded semiconductor workforce development cluster within roughly 40 miles of Samsung Taylor: Office of the Texas Governor, TSIF award announcements (2024–2026); awards include Austin Community College District ($3.6 million for the Round Rock campus), Temple College ($9.8 million for the Central Texas Chips Hub), UT Austin's Texas Quantum Institute and Prairie View A&M University's Roy G. Perry College of Engineering (research and engineering pipeline grants), and Texas State Technical College's Hutto campus ($3.5 million, January 8, 2026).

131. The Technician Pipeline (TSTC Hutto Accelerated Semiconductor Technician Training Program) — On the January 8, 2026 TSIF award to TSTC's Williamson County campus in Hutto: Office of the Governor of Texas, "Governor Abbott Announces Texas Semiconductor Innovation Fund Grant To Texas State Technical College" (January 8, 2026); the $3.5 million grant funds the Accelerated Semiconductor Technician Training Program, a ten-week modular curriculum producing equipment technicians, process technicians, and gas control system specialists.

132. The Construction Trades (CHIPS Act workforce conditions) — On the CHIPS Act workforce development conditions attached to Samsung

Taylor and other major facility awards as the federal mechanism shaping the technician training response, see notes to Chapter 9.

Chapter 17: The Trickle-Down Engine

133. Winter Storm Uri (February 2021 grid failure and West Lake Hills context) — On the cascading grid and gas-system failures during the February 2021 winter storm and their distribution across the western Travis County luxury enclaves: Federal Energy Regulatory Commission and North American Electric Reliability Corporation, The February 2021 Cold Weather Outages in Texas and the South Central United States: FERC, NERC and Regional Entity Staff Report, November 2021. See also note to Chapter 1 on ERCOT's structural isolation and the Uri grid event.

134. The Fleet Opportunity (Tesla Cybercab $30,000 target unit price) — On Tesla's publicly stated Cybercab pricing target: Tesla, Inc., "We, Robot" event, October 10, 2024, Warner Bros. Studio, Burbank, California, at which CEO Elon Musk cited a target unit price of approximately $30,000. Production timing and the robotaxi fleet program were discussed further across Tesla's subsequent earnings cycles and Q1 2026 update letter.

135.The Fleet Opportunity (Tesla Optimus $30,000 target unit price) — On Tesla's stated Optimus unit-cost target: Tesla Investor Day presentation, March 1, 2023; subsequent earnings remarks across 2024–2025; the company has cited an eventual unit price in the $20,000–$30,000 range at scale, below the annual cost of a full-time household employee in most US labor markets.

Chapter 18: The Solo-Builders

136. Austin Educational Attainment (top-tier MSA ranking) — On the educational attainment of the Austin–Round Rock–Georgetown metropolitan statistical area: US Census Bureau, American Community Survey 5-Year Estimates; bachelor's-degree-or-higher attainment ranks among the top large US metros, reflecting both the University of Texas at Austin pipeline and a decade of in-migration by credentialed technology professionals from the Bay Area, Seattle, Boston, and New York.

137. South by Southwest (Austin technology cultural anchor) — On the SXSW conference and festival ecosystem as a recurring annual aggregator of technology, music, and film communities in Austin: SXSW LLC annual disclosures; Greater Austin Chamber of Commerce economic-impact reporting. SXSW has operated continuously since 1987 and is one of several institutional inputs to Austin's entrepreneurial culture.

138. Williamson County Master-Planned Communities (Liberty Hill, Leander, Georgetown, Hutto growth) — On the residential growth corridor extending across northwest Travis and Williamson counties: Texas Demographic Center population estimates; US Census Bureau population data; Community Impact reporting on master-planned community development in Liberty Hill, Leander, Georgetown, and Hutto, 2020–2025. Liberty Hill and Hutto have ranked among the fastest-growing US cities during this period.

Chapter 19: The Austin Supercluster

139. Frontera (2019 academic supercomputer ranking) — On the Frontera supercomputer at the Texas Advanced Computing Center: TACC, "Frontera Named 5th Fastest Supercomputer in the World," June 17, 2019; Top500.org, June and November 2019 lists. Frontera achieved 23.5 petaflops on the High Performance LINPACK benchmark and held the #1 academic supercomputer ranking globally on both 2019 Top500 lists. The system was funded by a $60 million National Science Foundation award.

140. Horizon at Sabey Round Rock (NSF Leadership-Class Computing Facility flagship) — On the Horizon supercomputer entering production in Spring 2026: TACC, "NSF LCCF Horizon Supercomputer To Power Breakthroughs for the Nation's Leading Scientists," November 17, 2025; "Horizon Takes Shape at TACC as NSF Leadership-Class Facility Moves Forward," HPCwire, November 19, 2025. Horizon is installed at Sabey Data Centers' Round Rock campus (SDC Austin) — a 430,000-square-foot, 85 MW facility built for high-density deployments — and uses the Nvidia Grace Blackwell platform to deliver approximately 300 petaflops of simulation performance and roughly 100 times Frontera's AI capability.

141. Austin Semiconductor Design Cluster (cluster composition and scale) — On the approximately 40 semiconductor design and EDA companies operating in the Austin core: Greater Austin Chamber of Commerce semiconductor industry profile; Austin Regional Manufacturers Association member directory. Estimates vary by inclusion criteria; the cluster includes AMD, NXP, Silicon Labs, Cirrus Logic, Arm, Apple, Intel design operations, and a long tail of fabless and design-services firms.

142. Texas Institute for Electronics (NGMM Center designation) — On the Texas Institute for Electronics and its role under the National Advanced Packaging Manufacturing Program: TIE program announcements; University of Texas at Austin disclosures, 2024–2025. TIE coordinates state, federal, and industry funding into a consortium spanning UT Austin, Texas A&M, multiple semiconductor operators, and DARPA, with the NGMM Center under construction in Austin for advanced packaging research.

143. Firefly Aerospace (Cedar Park product portfolio) — On Firefly Aerospace, Inc. (Nasdaq: FLY) and its operational and in-development products: Firefly Aerospace Form 10-K (fiscal year 2024); Firefly Q3 2025 earnings release, November 12, 2025; "Firefly Aerospace Debuts Elytra Orbital Vehicles," Firefly press release, March 2024. The product portfolio comprises Alpha (small-lift launch vehicle), Eclipse (medium-lift launch vehicle in development with Northrop Grumman, formerly designated MLV), Blue Ghost (lunar lander, with Mission 1 achieving the first fully successful commercial lunar landing on March 2, 2025), and Elytra (orbital transfer vehicle / space tug).

144. Central Texas Spaceport Development Corporation (Cedar Park campus) — On the CTSDC and the planned Central Texas Spaceport and Defense Innovation Campus: Williamson County, "WilCo Partners with Cedar Park to Create Central Texas Spaceport Development Corporation," August 20, 2025; "Space Industry Campus Planned Cedar Park," CRE Daily, April 16, 2026. The CTSDC was formed under Chapter 507 of the Texas Local Government Code as the sixth Texas spaceport development corporation. The planned 115,000-square-foot, four-building campus (estimated at $78–$108 million across three phases) would be developed in partnership with the UT Austin Cockrell School of Engineering and private commercial-aerospace and defense partners.

145. TSMC Fab 21 (Phoenix advanced-node production) — On Taiwan Semiconductor Manufacturing Company's Arizona production complex: TSMC Arizona project disclosures and investor materials. Fab 21 in Phoenix entered Phase 1 volume production at the 4nm node in 2024–2025, with subsequent phases planned for 3nm and 2nm; the Phoenix complex is the largest US-based foreign-direct-investment advanced-node manufacturing project under the CHIPS Act.

146. Intel Ocotillo (Fab 52 and Fab 62) — On Intel Corporation's Arizona campus capacity: Intel Arizona disclosures and CHIPS Act program filings.

Fab 52 and Fab 62 at the Ocotillo campus in Chandler, Arizona, host Intel's advanced-node manufacturing including the Intel 18A process generation.

147. Hsinchu Science Park (TSMC supplier ring) — On Hsinchu's role as the longest-established advanced-node semiconductor fabrication and supplier concentration in the world: Chris Miller, Chip War: The Fight for the World's Most Critical Technology (Scribner, 2022); Hsinchu Science Park Bureau, Taiwan.

Chapter 20: The Step Function

148. Pittsburgh Steel Concentration (late-19th-century industrial supernode) — On the consolidation of Pittsburgh's steel, railroad, coal, and financial concentration in the 1880s as a comparator industrial supernode: Kenneth Warren, Triumphant Capitalism: Henry Clay Frick and the Industrial Transformation of America (University of Pittsburgh Press, 1996); see also Thomas Misa, A Nation of Steel: The Making of Modern America, 1865–1925 (Johns Hopkins University Press, 1995).

149. Ford Highland Park (moving assembly line, October 1913) — On Ford Motor Company's first operation of the moving assembly line at its Highland Park, Michigan plant on October 7, 1913: Ford Motor Company corporate history, "Highland Park: Birthplace of Moving Assembly Line." The introduction reduced Model T chassis assembly time from approximately 12.5 hours to 93 minutes and is the canonical inflection point for the Detroit automotive supernode the chapter references.

150. Samsung Taylor (2021 announcement and prior Austin presence) — On Samsung Electronics Co., Ltd.'s November 23, 2021 selection of Taylor, Texas for its advanced semiconductor fab: Samsung Electronics, "Samsung Electronics Announces New Advanced Semiconductor Fab Site in Taylor, Texas," press release, November 23, 2021. Samsung's Austin Semiconductor operations have run continuously since 1997, providing the workforce and supplier substrate that informed the Taylor site decision.

151. Tesla FSD (11 billion cumulative-miles milestone) — On Tesla's Full Self-Driving (Supervised) fleet surpassing 11 billion cumulative miles on May 4, 2026: Tesla, Inc., Vehicle Safety Report, May 2026; coverage in Teslarati and *Electrek*, May 3–4, 2026. Approximately 3.76 billion of the cumulative miles were logged on city streets. The fleet was adding approximately 29 million miles per day at the milestone, with the prior billion miles added in the preceding 31 days.

152. Amazon Data Services Cedar Creek (1,300-acre Bastrop County acquisition) — On Amazon Data Services Inc.'s May 2026 acquisition of multiple parcels totaling at least 1,300 acres in the Cedar Creek area of Bastrop County: Olivia Lueckemeyer, "Amazon scoops up 1,300 acres in Bastrop County," Austin Business Journal, May 6, 2026; "Amazon Data Services buys 1,300 acres of land outside Austin, Texas," Data Center Dynamics, May 9, 2026. The land, located off State Highway 21 West approximately 30 miles southeast of Austin, was previously owned by CTX Capital Partners and slated for the Creekside master-planned community.

153. Blue Origin Project Blue Hub (Hutto site evaluation, May 2026) — On Blue Origin's evaluation of a proposed 1.3 million-square-foot manufacturing, research-and-development, warehouse, and logistics campus in Hutto: Olivia Lueckemeyer, "Blue Origin eyes Williamson County for massive project," Austin Business Journal, May 13, 2026; KXAN and Bisnow reporting, May 2026. The Hutto Economic Development Corp. held a May 11, 2026 public hearing on incentives for "Project Blue Hub" with no vote taken; total project parameters were reported as more than 2,000 jobs at an average salary of $88,000 and capital investment exceeding $650 million. Blue Origin had not publicly confirmed Hutto as the selected site as of mid-June 2026.

154. EdgeConneX Cedar Creek (first Bastrop County building completion, June 2026) — On EdgeConneX's first Bastrop County data center building: Matthew Brooks, "EdgeConneX plans second data center campus in Bastrop County," Community Impact, March 31, 2026; subsequent Community Impact and Connect CRE reporting on the three planned EdgeConneX campuses spanning approximately 1,500 acres. The first building at FM 535 and Wolf Lane — approximately 578,000 square feet representing approximately $440 million in capital improvements — was scheduled for June 2026 completion as the opening node of a planned 2.8 million-square-foot Cedar Creek complex.

Chapter 21: Orbital Compute

155. The Kármán Line (atmosphere–space boundary) — On the 100-km altitude line separating atmospheric flight from orbital mechanics: Fédération Aéronautique Internationale convention; National Oceanic and Atmospheric Administration atmospheric profile data. The line marks the practical altitude above which aerodynamic lift cannot sustain flight.

156. SpaceX Orbital Data Center Application (January 2026 FCC filing) — On Space Exploration Holdings, LLC, "Application for Authority to Launch and Operate the SpaceX Orbital Data Center System," ICFS File No. SAT-LOA-20260108-00016, filed January 30, 2026: Federal Communications Commission Space Bureau Public Notice DA 26-113, released February 4, 2026; Jeff Foust, "SpaceX files plans for million-satellite orbital data center constellation," SpaceNews, January 31, 2026. The application proposes up to one million satellites in 50-km-thick orbital shells at altitudes from 500 to 2,000 km, split between sun-synchronous and 30-degree inclinations.

157. AI1 Starmind Satellite Specification (compute, power, and thermal architecture) — On the AI1 satellite payload disclosures: SpaceX FCC filing (SAT-LOA-20260108-00016) and Introl industry analysis, "SpaceX Files for Million-Satellite Orbital Data Center," February 3, 2026. Peak compute payload of approximately 150 kW, average compute payload of approximately 120 kW, deployable solar array of approximately 70-meter wingspan, and a deployable liquid radiator of approximately 110-square meters serving as the thermal rejection surface.

158. Launch Complex 39A (Starship dual-site launch architecture) — On the modification of NASA's Launch Complex 39A at Kennedy Space Center for Starship operations: SpaceX program disclosures; NASA Kennedy Space Center site documentation. LC-39A was the launch pad for Apollo 11 in July 1969 and the final Space Shuttle missions; SpaceX has operated Falcon 9 and Falcon Heavy from the pad since 2017 and has modified it for Starship, providing a Florida launch site complementary to Boca Chica with different inclination access.

159. Starlink Inter-Satellite Optical Links (laser mesh network) — On the laser-based optical inter-satellite links carried by Starlink Generation 2 satellites: SpaceX Starlink technical documentation and FCC filings. The optical mesh enables continuous global routing of constellation traffic without dependence on individual ground-station passes, which at LEO altitudes are typically only a fraction of each orbit.

160. Project Suncatcher (Google 81-satellite TPU cluster demonstration) — On Google's Project Suncatcher: Travis Beals et al., "Towards a future space-based, highly scalable AI infrastructure system design," Google Research preprint, November 4, 2025; Sundar Pichai blog post, November 4, 2025; SpaceNews and Data Center Dynamics coverage, November–December 2025. The proposed architecture envisions 81

Trillium-TPU–equipped satellites in a 1-km-radius cluster at approximately 650-km dawn-dusk sun-synchronous orbit; two prototype satellites are planned for launch in partnership with Planet Labs in early 2027. Google's published analysis identifies approximately $200 per kilogram as the launch-cost equilibrium at which orbital compute becomes economically comparable to terrestrial data center energy costs, with that threshold projected for the mid-2030s.

161. Project Sunrise (Blue Origin orbital compute filing, March 2026) — On Blue Origin's Project Sunrise FCC application filed March 19, 2026: Jeff Foust, "Blue Origin joins the orbital data center race," SpaceNews, March 19, 2026. The filing proposes up to 51,600 satellites operating in sun-synchronous orbits at altitudes between 500 and 1,800 km, communicating via optical inter-satellite links with Blue Origin's separately proposed TeraWave broadband constellation.

162. Colossus Data Center Leases (Anthropic and Google compute agreements with SpaceX) — On the Anthropic and Google compute lease arrangements with SpaceX: SpaceX S-1 IPO prospectus disclosures, May–June 2026; Lora Kelley and Hayden Field, "Google will pay SpaceX $920M per month for compute," *TechCrunch,* June 5, 2026; Yahoo Finance and CNBC coverage, June 5, 2026. Anthropic's agreement leases full available Colossus 1 compute capacity in Memphis, Tennessee at $1.25 billion per month through 2029; Google's separate 32-month agreement covers roughly half that capacity at $920 million per month and provides bridge capacity for the Gemini Enterprise platform.

Chapter 22: The Moon

163. Apollo Lunar Surface Operations (six missions, 1969–1972) — On the six crewed Apollo lunar landings between July 1969 (Apollo 11) and December 1972 (Apollo 17): NASA Apollo program archives, Lyndon B. Johnson Space Center. Twelve Apollo astronauts walked on the lunar surface across the six missions, with surface stays ranging from approximately 21 hours (Apollo 11) to approximately 75 hours (Apollo 17).

164. Artemis III (current crewed lunar return timeline) — On NASA's Artemis III mission as the program of record for the first crewed lunar landing since Apollo 17: NASA Artemis program disclosures, 2024–2026. The mission architecture uses the Space Launch System for crew launch and either SpaceX's Starship Human Landing System or Blue Origin's HLS variant for surface descent and ascent. Current planning targets the 2027–2028 window, subject to Starship qualification milestones.

165. Commercial Lunar Payload Services (NASA CLPS program) — On NASA's Commercial Lunar Payload Services initiative as the procurement vehicle for commercial lunar surface delivery: NASA CLPS contractor roster and task order awards, 2018–2026. The program contracts a roster of commercial providers — including Firefly Aerospace, Intuitive Machines, Astrobotic, and others — to deliver NASA science instruments, technology demonstrations, and precursor hardware to the lunar surface.

166. Firefly Blue Ghost Mission 1 (first fully successful commercial lunar landing) — On Firefly Aerospace's Blue Ghost Mission 1, named Ghost Riders in the Sky: Firefly Aerospace, "Firefly Aerospace Becomes First Commercial Company to Successfully Land on the Moon," March 2, 2025; NASA, "Blue Ghost Lands on Moon," March 2, 2025. The lander touched down at 3:34 a.m. EST on March 2, 2025 near Mons Latreille in Mare Crisium, delivered ten NASA payloads, and operated for 346 hours of daylight plus approximately five hours into the lunar night before its solar-powered batteries depleted. The mission was awarded the 2025 Robert J. Collier Trophy in March 2026 as the first fully successful commercial lunar landing in history.

167. Lunar South Pole Water Ice (orbital confirmation and "peaks of eternal light") — On the confirmation of water ice in permanently shadowed lunar polar craters and the near-continuous solar illumination at adjacent crater rims: NASA LCROSS impact mission, 2009; Lunar Reconnaissance Orbiter mission results, 2009–present; ShadowCam imagery from the Korea Pathfinder Lunar Orbiter (Danuri), 2022–present. The combination of stable cryogenic shadowed regions with adjacent solar-rich rim sites is the primary scientific and operational basis for South Pole base architecture across both NASA Artemis and commercial planning frameworks.

168. Lunar Escape Velocity (mass driver economics) — On lunar escape velocity of approximately 2.38 km/s versus Earth escape velocity of approximately 11.2 km/s: NASA Goddard Space Flight Center planetary fact sheets. The order-of-magnitude difference is the physical basis for electromagnetic mass-driver economics: surface acceleration to escape velocity using electrical energy from local power infrastructure, with no chemical propellant required.

169. Kilopower / Fission Surface Power (NASA–DOE lunar reactor program) — On NASA and Department of Energy fission surface power development for lunar applications: NASA Kilopower technology demonstration program; subsequent Fission Surface Power request-for-proposal awards, 2022–2025. Current systems target the 10 to 40-kW class for initial base operations, with successor systems required for the industrial-scale power demands of mass driver operation and large-scale in-situ resource utilization.

CHAPTER 23: MARS

170. Musk 2018 Tesla Compensation (grant valuation, Delaware litigation history) — On the 2018 Tesla CEO Performance Award: Tesla, Inc., Schedule 14A proxy statement, 2018; Tornetta v. Musk, Delaware Court of Chancery Memorandum Opinion, January 30, 2024 (Chancellor Kathaleen McCormick), rescinding the award. The package granted Elon Musk options on approximately 304 million Tesla shares (representing approximately nine percent of outstanding stock) vesting against a combination of market capitalization milestones and revenue/EBITDA operational milestones, with the headline projected value at approximately $56 billion at the original grant. Mars colonization was not among the contractual vesting milestones; the vesting structure was financial and operational.

171. Tesla Shareholder Ratification (June 13, 2024 vote) — On the June 13, 2024 shareholder vote ratifying the 2018 award following its Delaware rescission and Tesla's Texas reincorporation: Tesla 2024 Annual Meeting proxy materials and Form 8-K disclosure, June 14, 2024. The ratification was paired with Tesla's redomiciliation from Delaware to Texas, approved at the same meeting.

172. Delaware Supreme Court Restoration (December 19, 2025 reinstatement of 2018 package) — On the Delaware Supreme Court's restoration of the 2018 Tesla CEO Performance Award: Tornetta v. Musk, Delaware Supreme Court opinion, December 19, 2025. The 49-page ruling reversed the Court of Chancery's rescission remedy as "improper and inequitable," awarding nominal damages of $1 to the plaintiff and reinstating the package in full.

173. November 2025 Tesla Performance Award (twelve-tranche $1 trillion-scale comp plan) — On Tesla's 2025 CEO Performance Award approved by shareholders at the November 2025 annual meeting: Tesla Form

8-K and Form S-8 filings, November 2025; Definitive Proxy Statement, October 2025. The award consists of twelve tranches vesting against twelve market capitalization milestones (rising from $2 trillion to $8.5 trillion) and six operational milestones (20 million vehicles delivered, 10 million active FSD subscriptions, 1 million bots delivered, 1 million robotaxis in commercial operation, and adjusted EBITDA tiers from $50 billion to $400 billion). Mars colonization is not among the operational milestones; the vesting structure remains financial and operational. The award is contextualized in board communications and Bloomberg interviews with Chair Robyn Denholm as the compensation architecture for Tesla's Master Plan IV phase.

174. Mars Physical Parameters (24.6-hour day, 38 percent gravity, 43 percent solar irradiance, 95 percent CO2 atmosphere) — On Mars's surface and orbital parameters: NASA Goddard Space Flight Center, Mars Fact Sheet. Mars's sidereal day is 24.62 hours, surface gravity is 0.38 Earth gravity, mean solar irradiance is approximately 43 percent of Earth's value at one astronomical unit, and atmospheric composition is approximately 95 percent carbon dioxide at a mean surface pressure of approximately 0.6 percent of Earth's.

175. Sabatier Reaction (Mars ISRU propellant production) — On the Sabatier reaction as the in-situ resource utilization pathway for Martian methane-oxygen propellant production: Paul Sabatier, original synthesis work, late 19th century; subsequent NASA technical reports on Mars ISRU architectures. The reaction combines atmospheric carbon dioxide with hydrogen (produced by electrolysis of water ice) to yield methane and water, providing the chemical basis for producing Starship-compatible propellant from Martian resources using electrical power.

176. Mars Synodic Period (26-month Earth–Mars launch alignment cycle) — On the Mars–Earth opposition cycle and its implications for high-cadence transit planning: NASA Jet Propulsion Laboratory orbital mechanics documentation. Mars and Earth reach favorable launch alignment approximately every 26 months; the duration and energy requirements of each transit window vary with specific orbital geometry, with most efficient transits using Hohmann or near-Hohmann transfer trajectories.

177. Earth–Mars Communication Delay (3-to-22-minute one-way light travel time) — On the one-way light-time delay between Earth and Mars: NASA Deep Space Network operational documentation. Earth–Mars distance varies between approximately 55 million km at closest approach and approximately 400 million km at superior conjunction, yielding one-way signal travel times ranging from approximately three minutes at closest approach to approximately 22 minutes at maximum separation.

CHAPTER 24: ALPHA CENTAURI

178. Alpha Centauri System (4.37 light-year distance, triple-star structure, Proxima b) — On the Alpha Centauri triple-star system: European Southern Observatory and Gaia mission stellar parallax data. Alpha Centauri A (G-type) and B (K-type) form a binary pair with an orbital period of approximately 79.91 years, separated by roughly the Saturn–Sun distance at closest approach. Proxima Centauri, a red dwarf gravitationally bound to the binary at approximately 0.21 light-years, is the closest of the three to Earth at approximately 4.24 light-years and hosts Proxima Centauri b, a potentially rocky planet within the star's habitable zone discovered in 2016 (Anglada-Escudé et al., *Nature*, August 2016).

179. Parker Solar Probe (fastest human-made object record) — On the Parker Solar Probe's peak velocity at solar perihelion: NASA Johns Hopkins Applied Physics Laboratory mission disclosures. The probe achieved a peak heliocentric velocity of approximately 692,000 km/h during its December 2024 closest approach to the Sun — the fastest velocity ever achieved by a human-made object.

180. Breakthrough Starshot (2016 lightsail initiative) — On the Breakthrough Starshot initiative announced April 12, 2016: Breakthrough Initiatives, founding announcement by Stephen Hawking, Yuri Milner, and Mark Zuckerberg, Manhattan, New York. The $100 million research program proposes accelerating gram-scale reflective sails to approximately 20 percent of the speed of light using a ground-based phased-array laser system in the 100-GW class, enabling Alpha Centauri flyby transit times of approximately 20 years plus the 4.37-year signal-return time.

181. Orion Nuclear Pulse Propulsion (1963 Partial Nuclear Test Ban Treaty halt) — On Project Orion: Freeman Dyson and Theodore Taylor, General Atomics, 1957–1965; Project Orion archival materials. The nuclear-pulse propulsion concept proposed sequential nuclear detonations behind a pusher plate to achieve specific impulses thousands of times higher than chemical propulsion. US government development effort was effectively halted by the Limited Test Ban Treaty (also known as the Partial Nuclear Test

Ban Treaty) of August 5, 1963, which prohibited nuclear explosions in the atmosphere, in outer space, and underwater.

182. Fusion Propulsion Progress (CFS HTS magnets and NIF ignition) — On recent terrestrial fusion progress relevant to long-term interstellar propulsion concepts: Commonwealth Fusion Systems, high-temperature superconducting magnet test results, 2021–2024; Lawrence Livermore National Laboratory, National Ignition Facility scientific breakeven announcement, December 13, 2022, and subsequent ignition shots. These results constitute the strongest evidence in roughly five decades that terrestrial fusion power is approaching commercial demonstration; interstellar fusion propulsion is a post-terrestrial-demonstration program whose timeline depends on the demonstration phase being completed.

183. Von Neumann Probe Concept (Tipler analysis, galactic colonization timescales) — On the Von Neumann self-replicating probe and its interstellar implications: Frank J. Tipler, "Extraterrestrial Intelligent Beings Do Not Exist," Quarterly Journal of the Royal Astronomical Society, vol. 21, 1980; subsequent treatments in the astrobiology and SETI literature. The mathematical framework demonstrates that a self-replicating probe traveling at modest fractions of light speed could, in principle, propagate technological presence to every star in the Milky Way within geologically brief timescales of order ten million years.

INDEX

D

NOTE ON THE AUTHOR

Daniel M. Sisson is an electrical engineer and technical writer whose forty-year career has traversed the industries whose convergence this book documents. He spent a decade at Black & Veatch as a systems engineer, specifying controls for GE nuclear reactors, wind tunnels for NASA, shipyard SCADA systems for the US Navy, and high-voltage substations. He then spent nearly three decades at Microsoft as a technical writer documenting enterprise server and hyperconverged cloud platforms.

He operates as a solo-builder — an architecture decision, not a constraint. In an era of AI-amplified knowledge work, controlling the taxonomy, the editorial thesis, the analytical framework, and the development stack end to end produces a coherence that distributed teams rarely achieve. He lives in Liberty Hill, Texas.

www.ingramcontent.com/pod-product-compliance
Lightning Source LLC
LaVergne TN
LVHW100723061126
851857LV00011B/374

* 9 7 9 8 9 9 6 6 6 0 5 1 3 *